Springer Proceedings in Mathematics & Statistics

Volume 423

This book series features volumes composed of selected contributions from workshops and conferences in all areas of current research in mathematics and statistics, including data science, operations research and optimization. In addition to an overall evaluation of the interest, scientific quality, and timeliness of each proposal at the hands of the publisher, individual contributions are all refereed to the high quality standards of leading journals in the field. Thus, this series provides the research community with well-edited, authoritative reports on developments in the most exciting areas of mathematical and statistical research today.

Vladimir Vasilyev
Editor

Differential Equations, Mathematical Modeling and Computational Algorithms

DEMMCA 2021, Belgorod, Russia, October 25–29

Springer

Editor
Vladimir Vasilyev
Applied Mathematics and Computer Modeling
Belgorod State National Research University
Belgorod, Russia

ISSN 2194-1009 ISSN 2194-1017 (electronic)
Springer Proceedings in Mathematics & Statistics
ISBN 978-3-031-28504-2 ISBN 978-3-031-28505-9 (eBook)
https://doi.org/10.1007/978-3-031-28505-9

Mathematics Subject Classification: 34A34, 35A22, 47G20, 65M22, 74H55, 92F05

This Springer imprint is published by the registered company Springer Nature Switzerland AG
The registered company address is: Gewerbestrasse 11, 6330 Cham, Switzerland

Preface

This volume contains some of the reports which were presented at the Conference "Differential Equations, Mathematical Modeling and Computational Algorithms". This conference was held in Belgorod, Russia, October 25–29, 2021, at Belgorod State National Research University. A lot of people from different cities and countries participated in the conference. Unfortunately, not all could submit the paper for the Proceeding, and we have collected some of them only.

There were certain plenary talks which were presented by leading mathematicians, for example

- Michael Ruzhansky (Belgium) "Nonharmonic Pseudo-Differential Analysis"
- Josef Diblik (Czech Republic) "Global Solutions to Functional Differential Equations of Mixed Type"
- Sandra Pinelas (Portugal) "Difference and Differential Equations: Oscillatory Behavior"
- Vladimir Rabinovich (Mexico) "Dirac Operators on $\mathbb{R}^n$ with Singular Potentials"
- Eugene Tyrtyshnikov (Russia) "On Correct Statements of Ill-Posed Problems"
- Alexander Soldatov (Russia) "Generalized Cauchy—Riemann Equations with Power-Law Singularities in Coefficients of Lower Order"
- Tynysbek Kalmenov (Kazakhstan) "Minimality Criteria for the Laplacian"
- Armen Sergeev (Russia) "On Ginzburg–Landau Equation"
- Victor Nistor (France) "On Some Results of Kondratiev and Extensions and Applications to Singular Spaces and Numerical Methods"
- Dumitru Baleanu (Turkey) "New Trends in Fractional Differential Equations"

and others. The papers of these authors were published in separate issues of mathematical journals.

Conference topics have included the following sections:

1. Linear and nonlinear operators in function spaces
2. Differential, integral and operator equations
3. Initial and boundary value problems for differential equations
4. Numerical methods in theory of differential equations and their applications

5. Mathematical and computer modeling
6. Mathematical physics and modeling in physics
7. Questions of differential equations and mathematical modeling in pedagogical research.

All topics (excluding the 7th) are reflected in the Proceedings.
A reader can look at Table of Contents to be sure that it is true.
I would not like to describe briefly the annotations of papers; you can find them in the text.

Finally, I can say that the Conference was a success, and this my opinion is shared by many participants.

Belgorod, Russia
November 2022

Vladimir Vasilyev

Contents

Some Classes of Quasilinear Equations with Gerasimov—Caputo Derivatives

Vladimir E. Fedorov and Kseniya V. Boyko

Abstract The Cauchy problem for resolved with respect to the oldest derivative quasilinear multi-term equations in Banach spaces with the fractional Gerasimov—Caputo derivatives, with bounded linear operators at them and with locally Lipschitzian nonlinear operator is studied. Theorem on the local existence and uniqueness of a solution to the Cauchy problem is proved. This result applied to study of the so-called degenerate (i.e. with a degenerate linear operator at the oldest derivative) equations of the similar form. Using the reduction of a special initial value problem for a degenerate equation to the Cauchy problem for two non-degenerate equations on two subspaces under four types of additional conditions on nonlinear locally Lipschitzian operator four theorems on local unique solvability are proved. Abstract results are illustrated by initial-boundary value problems for partial differential systems of equations with Gerasimov—Caputo derivatives in time.

Keywords Multi-term fractional differential equation · Quasilinear equation · Gerasimov—Caputo fractional derivative · Fixed point theorem · Initial boundary value problem

1 Introduction

Over the past few decades, there has been a sharp increase in the interest of researchers in fractional differential equations, primarily due to their increasing importance in modeling various phenomena that arise in physics, chemistry, mathematical biology, engineering [1, 2]. For more details on fractional differential equations and closely related Volterra integro-differential equations see the monographs [3–9]. The unique solvability issues for initial value problems to some types of equations in Banach spaces with Gerasimov—Caputo fractional derivatives were researched in works

V. E. Fedorov (✉) · K. V. Boyko
Mathematical Analysis Department, Chelyabinsk State University, 129, Kashirin Brothers St., Chelyabinsk 454001, Russia
e-mail: kar@csu.ru

V. Vasilyev (ed.), *Differential Equations, Mathematical Modeling and Computational Algorithms*, Springer Proceedings in Mathematics & Statistics 423,
https://doi.org/10.1007/978-3-031-28505-9_1

[10–13], equations with fractional Riemann—Liouville derivatives were also studied in [14–16]. Analogous investigations for linear multi-term equations in Banach spaces were carried out in [17] for Gerasimov—Caputo derivatives and in [18–20] for Riemann—Liouville derivatives.

Here we consider a multi-term equation with a nonlinearity, which depends on fractional derivatives of lower orders

$$D^{\alpha} Lz(t) = \sum_{k=1}^{n} D^{\alpha_k} M_k z(t) + N(t, D^{\gamma_1} z(t), D^{\gamma_2} z(t), \dots, D^{\gamma_r} z(t)), \qquad (1)$$

where D^{γ} is the Gerasimov—Caputo derivative, if $\gamma > 0$, or the Riemann—Liouville integral in the case $\gamma < 0$, $m - 1 < \alpha \leq m \in \mathbf{N}$, $\alpha_1 < \alpha_2 < \cdots < \alpha_n < \alpha$, $\gamma_1 < \gamma_2 < \cdots < \gamma_r < \alpha$, $\mathcal{X}$ and $\mathcal{Y}$ are Banach spaces, $L, M_k : \mathcal{X} \to \mathcal{Y}, k = 1, 2, \dots, n-1$, are linear and continuous operators, a linear closed densely defined in $\mathcal{X}$ operator $M_n : \mathcal{X} \to \mathcal{Y}$ is $(L, 0)$-bounded, a nonlinear map N is continuous, locally Lipschitzian with respect to phase variables and satisfies some additional conditions of four types. Equation (1) under condition $\ker L \neq \{0\}$ will be called degenerate.

In the second section, preliminary results are given that will be used below. In the third one, we obtain a theorem on the local unique solvability of the Cauchy problem for a non-degenerate Eq. (1), i.e. for Eq. (1) with $\mathcal{X} = \mathcal{Y}$, $L = I$. For this aim we use the fixed point theorem in a specially constructed metric space. The fourth section begins with known results on the pairs of the invariant subspaces under condition of $(L, 0)$-boundedness of the operator M_n, after which theorems on the local unique solvability of the special initial value problem for degenerate Eq. (1) was proved for every of four types of additional conditions on the operator N. In the fifth section, examples of initial boundary value problems for systems of partial differential equations illustrate the abstract results.

2 Non-degenerate Linear Equation

Let $\mathcal{Z}$ be a Banach space, $h : (t_0, T) \to \mathcal{Z}$, for $\beta > 0$, $t > t_0$,

$$J^{\beta} h(t) := \int_{t_0}^{t} \frac{(t-s)^{\beta-1}}{\Gamma(\beta)} h(s) ds$$

be the Riemann—Liouville integral of the order $\beta > 0$, J^0 be the identity operator by definition. Let $m - 1 < \alpha \leq m \in \mathbf{N}$, D^m be the derivative of the order $m \in \mathbf{N}$, D^{α} be the Gerasimov—Caputo derivative [6, 21, 22], i.e.

$$D^\alpha h(t) := D^m J^{m-\alpha}\left(h(t) - \sum_{k=0}^{m-1} h^{(k)}(t_0)\frac{(t-t_0)^k}{k!}\right), \quad t > t_0.$$

For $\beta \leq 0$ denote $D^\beta h(t) := J^{-\beta} h(t)$.

Denote the Laplace transform of a function $h : \mathbf{R}_+ \to \mathcal{Z}$ by $\widehat{h}$. For the fractional integral and the fractional derivative of Gerasimov—Caputo we have the equalities [5, 6].

$$\widehat{J^\alpha h}(\lambda) = \lambda^{-\alpha}\widehat{h}(\lambda), \quad \widehat{D^\alpha h}(\lambda) = \lambda^\alpha \widehat{h}(\lambda) - \sum_{k=0}^{m-1} h^{(k)}(0)\lambda^{\alpha-1-k}.$$

Let $m-1 < \alpha \leq m \in \mathbf{N}$, $n \in \mathbf{N}$, $\alpha_1 < \alpha_2 < \cdots < \alpha_n < \alpha$, $m_k - 1 < \alpha_k \leq m_k \in \mathbf{Z}$, $k = 1, 2, \ldots, n$, $A_1, A_2, \ldots, A_n \in \mathcal{L}(\mathcal{Z})$, where $\mathcal{L}(\mathcal{Z})$ is the Banach space of bounded linear operators on $\mathcal{Z}$. Denote $n_l := \min\{k \in \{1, 2, \ldots, n\} : l \leq m_k - 1\}$ for $l = 0, 1, \ldots, m-1$. If set $\{k \in \{1, 2, \ldots, n\} : l \leq m_k - 1\}$ is empty for some $l \in \{0, 1, \ldots, m-1\}$ (this is done exactly when $\alpha_n \leq m-1$), then $n_l := n+1$.

Lemma 1 ([23]) *Let $l - 1 < \beta \leq l \in \mathbf{N}$. Then*

$$\exists C > 0 \quad \forall h \in C^l([t_0, t_1]; \mathcal{Z}) \quad \|D^\beta h\|_{C([t_0,t_1];\mathcal{Z})} \leq C\|h\|_{C^l([t_0,t_1];\mathcal{Z})}.$$

In this paper, we consider the Cauchy problem

$$z^{(l)}(t_0) = z_l, \quad l = 0, 1, \ldots, m-1, \tag{2}$$

to the inhomogeneous linear multi-term fractional differential equation

$$D^\alpha z(t) = \sum_{k=1}^{n} D^{\alpha_k} A_k z(t) + f(t), \quad t \in [t_0, T], \tag{3}$$

where $f \in C([t_0, T]; \mathcal{Z})$. A function $z \in C^{m-1}([t_0, t_1]; \mathcal{Z})$ is called a solution of Cauchy problem (2), (3), if $D^\alpha z, D^{\alpha_k} z \in C([t_0, t_1]; \mathcal{Z})$, $k = 1, 2, \ldots, n$, equality (3) is valid for all $t \in [t_0, T]$ and conditions (2) are satisfied.

Denote $\Gamma := \{Re^{i\varphi} : \varphi \in (-\pi, \pi)\} \cup \{\rho e^{i\pi} : \rho \in [R, \infty)\} \cup \{\rho e^{-i\pi} : \rho \in [R, \infty)\}$ with large enough $R > 0$,

$$R_\lambda := \left(I - \sum_{k=1}^{n} \lambda^{\alpha_k - \alpha} A_k\right)^{-1} : \mathcal{Z} \to \mathcal{Z},$$

for $l = 0, 1, \ldots, m-1$

$$Z_l(t) = \frac{1}{2\pi i}\int_\Gamma R_\lambda\left(\lambda^{-l-1}I - \sum_{k=n_l}^{n}\lambda^{\alpha_k-\alpha-l-1}A_k\right)e^{\lambda t}d\lambda, \quad t > t_0,$$

$$Z(t) := \frac{1}{2\pi i}\int_\Gamma \lambda^{-\alpha}R_\lambda e^{\lambda t}d\lambda, \quad t > t_0.$$

Theorem 1 ([17, Theorem 2]) *Let $m-1 < \alpha \le m \in \mathbf{N}$, $\alpha_1 < \alpha_2 < \dots < \alpha_n < \alpha$, $A_1, A_2, \dots, A_n \in \mathcal{L}(\mathcal{Z})$, $z_l \in \mathcal{Z}$, $l = 0, 1, \dots, m-1$, $f \in C([t_0, T]; \mathcal{Z})$. Then there exists an unique solution to (2), (3). It has the form*

$$z(t) = \sum_{l=0}^{m-1} Z_l(t-t_0)z_l + \int_{t_0}^{t} Z(t-s)f(s)ds.$$

3 Quasilinear Equation

Let $r \in \mathbf{N}$, $\gamma_1 < \gamma_2 < \dots < \gamma_r < \alpha$, $n_i - 1 < \gamma_i \le n_i \in \mathbf{Z}$, $i = 1, 2, \dots, r$. Let U be an open set in $\mathbf{R} \times \mathcal{Z}^r$, $B : U \to \mathcal{Z}$, $z_l \in \mathcal{Z}$, $l = 0, 1, \dots, m-1$, $t_0 \in \mathbf{R}$.

A function $z \in C^{m-1}([t_0, t_1]; \mathcal{Z})$ is called a solution of Cauchy problem (2) to quasilinear fractional differential equation

$$D^\alpha z(t) = \sum_{k=1}^{n} D^{\alpha_k}A_k z(t) + B(t, D^{\gamma_1}z(t), D^{\gamma_2}z(t), \dots, D^{\gamma_r}z(t)) \tag{4}$$

on $[t_0, t_1]$, if $D^\alpha z, D^{\alpha_k}z, D^{\gamma_i}z \in C([t_0, t_1]; \mathcal{Z})$, $k = 1, 2, \dots, n$, $i = 1, 2, \dots, r$, the inclusion $(t, D^{\gamma_1}z(t), \dots, D^{\gamma_r}z(t)) \in U$ and equality (4) are valid for all $t \in [t_0, t_1]$ and conditions (2) are fulfilled.

Denote $\bar{x} := (x_1, x_2, \dots, x_r) \in \mathcal{Z}^r$, $S_\delta(\bar{x}) := \{\bar{y} \in \mathcal{Z}^r : \|y_i - x_i\|_{\mathcal{Z}} \le \delta, i = 1, 2, \dots, r\}$. A mapping $B : U \to \mathcal{Z}$ is called locally Lipschitzian in $\bar{x}$, if for every $(t, \bar{x}) \in U$ there exist $\delta > 0, q > 0$, such that $[t-\delta, t+\delta] \times S_\delta(\bar{x}) \subset U$, and for all $(s, \bar{y}), (s, \bar{v}) \in [t-\delta, t+\delta] \times S_\delta(\bar{x})$ the inequality

$$\|B(s, \bar{y}) - B(s, \bar{v})\|_{\mathcal{Z}} \le q\sum_{i=1}^{r}\|y_i - v_i\|_{\mathcal{Z}}$$

is satisfied.

Using initial data $z_0, z_1, \dots, z_{m-1}$, define a polynomial

$$\tilde{z}(t) := z_0 + (t-t_0)z_1 + \frac{(t-t_0)^2}{2!}z_2 + \dots + \frac{(t-t_0)^{m-1}}{(m-1)!}z_{m-1}$$

and vectors $\tilde{z}_i := D^{\gamma_i}|_{t=t_0}\tilde{z}(t)$, $i = 1, 2, \dots, r$. Note that $\tilde{z}_i = 0$, if we have $\gamma_i \notin \{0, 1, \dots, m-1\}$. In cases $\gamma_i \in \{0, 1, \dots, m-1\}$ $\tilde{z}_i = z_{\gamma_i}$. So the starting point of the solution trajectory will be considered as $(t_0, \tilde{z}_1, \tilde{z}_2, \dots, \tilde{z}_r)$.

Lemma 2 *Let $A_1, A_2, \dots, A_n \in \mathcal{L}(\mathcal{Z})$, U be an open set in $\mathbf{R} \times \mathcal{Z}^r$, $B \in C(U, \mathcal{Z})$, $(t_0, \tilde{z}_1, \tilde{z}_2, \dots, \tilde{z}_r) \in U$. Then a function $z \in C([t_0, t_1]; \mathcal{Z})$ is a solution to problem (2), (4) on $[t_0, t_1]$, if and only if $D^{\gamma_i} z \in C([t_0, t_1]; \mathcal{Z})$, $i = 1, 2, \dots, r$, for all $t \in [t_0, t_1]$ it satisfies the inclusion $(t, D^{\gamma_1} z(t), D^{\gamma_2} z(t), \dots, D^{\gamma_r} z(t)) \in U$ and the equality*

$$z(t) = \sum_{l=0}^{m-1} Z_l(t - t_0) z_l + \int_{t_0}^{t} Z(t - s) B(s, D^{\gamma_1} z(s), D^{\gamma_2} z(s), \dots, D^{\gamma_r} z(s)) ds. \tag{5}$$

Proof If z is a solution to problem (2), (4) on $[t_0, t_1]$, then for all $t \in [t_0, t_1]$ $(t, D^{\gamma_1} z(t), D^{\gamma_2} z(t), \dots, D^{\gamma_r} z(t)) \in U$ and the mapping

$$t \to B(t, D^{\gamma_1} z(t), D^{\gamma_2} z(t), \dots, D^{\gamma_r} z(t)) \tag{6}$$

acts continuously from $[t_0, t_1]$ into $\mathcal{Z}$. As in the proof of Theorem 1 it can be shown that a solution satisfies Eq. (5).

Let $D^{\gamma_i} z \in C^{m-1}([t_0, t_1]; \mathcal{Z})$, $i = 1, 2, \dots, r$, for all $t \in [t_0, t_1]$ the inclusion $(t, D^{\gamma_1} z(t), D^{\gamma_2} z(t), \dots, D^{\gamma_r} z(t)) \in U$ holds and equality (5) is valid. Then (6) belongs to the class $C([t_0, t_1]; \mathcal{Z})$. By repeating word to word the proof of Theorem 1 and Lemma 2 in [17] we will obtain the required statement.

Theorem 2 *Let $m - 1 < \alpha \le m \in \mathbf{N}$, $n, r \in \mathbf{N}$, $\alpha_1 < \alpha_2 < \dots < \alpha_n < \alpha$, $\gamma_1 < \gamma_2 < \dots < \gamma_r < \alpha$, $A_1, A_2, \dots, A_n \in \mathcal{L}(\mathcal{Z})$, U be an open set in $\mathbf{R} \times \mathcal{Z}^r$, an operator $B \in C(U, \mathcal{Z})$ be locally Lipschitzian in $\bar{x}$, $(t_0, \tilde{z}_1, \tilde{z}_2, \dots, \tilde{z}_r) \in U$. Then for some $t_1 > t_0$ problem (2), (4) has a unique solution on $[t_0, t_1]$.*

Proof Denote $k_* := \min\{k \in \{1, 2, \dots, n\} : \alpha_k > m - 1\}$, if the set $\{k \in \{1, 2, \dots, n\} : \alpha_k > m - 1\}$ is not empty, and $k_* := n + 1$ otherwise, $i_* := \min\{i \in \{1, 2, \dots, r\} : \gamma_i > m - 1\}$, if the set $\{i \in \{1, 2, \dots, r\} : \gamma_i > m - 1\}$ is not empty, and $i_* := r + 1$ otherwise. Take for some $t_1 > t_0$ $C^{m-1,\{\alpha_k\},\{\gamma_i\}}([t_0, t_1]; \mathcal{Z}) := \{z \in C^{m-1}([t_0, t_1]; \mathcal{Z}) : D^{\alpha_k} z, D^{\gamma_i} z \in C([t_0, t_1]; \mathcal{Z}), k = k_*, \dots, n, i = i_*, \dots, r\}$ and endow this space by the norm

$$\|z\|_{C^{m-1,\{\alpha_k\},\{\gamma_i\}}([t_0,t_1];\mathcal{Z})} = \|z\|_{C^{m-1}([t_0,t_1];\mathcal{Z})} +$$

$$+ \sum_{k=k_*}^{n} \|D^{\alpha_k} z\|_{C([t_0,t_1];\mathcal{Z})} + \sum_{i=i_*}^{r} \|D^{\gamma_i} z\|_{C([t_0,t_1];\mathcal{Z})}.$$

Let $\{z_l\}$ be a fundamental sequence in $C^{m-1,\{\alpha_k\},\{\gamma_i\}}([t_0, t_1]; \mathcal{Z})$, then there exist a limit $z \in C^{m-1}([t_0, t_1]; \mathcal{Z})$ of $\{z_l\}$ in $C^{m-1}([t_0, t_1]; \mathcal{Z})$, limits x_k of $\{D^{\alpha_k} z_l\}$ in

$C([t_0, t_1]; \mathcal{Z})$, $k = k_*, k_* + 1, \dots, n$, y_i of $\{D^{\gamma_i} z_l\}$ in $C([t_0, t_1]; \mathcal{Z})$, $i = i_*, i_* + 1, \dots, r$. Hence,

$$J^{\alpha_k} x_k(t) = \lim_{l\to\infty} J^{\alpha_k} D^{\alpha_k} z_l(t) = \lim_{l\to\infty} \left(z_l(t) - \sum_{j=0}^{m-1} z_l^{(j)}(t_0) \frac{(t-t_0)^j}{j!} \right) =$$

$$= z(t) - \sum_{j=0}^{m-1} z^{(j)}(t_0) \frac{(t-t_0)^j}{j!}, \quad x_k = D^{\alpha_k} z \in C([t_0, t_1]; \mathcal{Z});$$

$$J^{\gamma_i} y_i(t) = \lim_{l\to\infty} J^{\gamma_i} D^{\gamma_i} z_l(t) = \lim_{l\to\infty} \left(z_l(t) - \sum_{j=0}^{m-1} z_l^{(j)}(t_0) \frac{(t-t_0)^j}{j!} \right) =$$

$$= z(t) - \sum_{j=0}^{m-1} z^{(j)}(t_0) \frac{(t-t_0)^j}{j!}, \quad y_i = D^{\gamma_i} z \in C([t_0, t_1]; \mathcal{Z}).$$

Thus, $C^{m-1,\{\alpha_k\},\{\gamma_i\}}([t_0, t_1]; \mathcal{Z})$ is a Banach space.

Take $\tau > 0$ and $\delta > 0$, such that $[t_0, t_0 + \tau] \times S_\delta(\bar{\tilde{z}}) \subset U$, where $\bar{\tilde{z}} = (\tilde{z}_1, \tilde{z}_2, \dots, \tilde{z}_n)$. Denote by $\mathcal{S}$ the set of $z \in C^{m-1,\{\alpha_k\},\{\gamma_i\}}([t_0, t_0 + \tau]; \mathcal{Z})$, such that $\|D^{\gamma_i} z(t) - \tilde{z}_i\|_{\mathcal{Z}} \le \delta$ for $t_0 \le t \le t_0 + \tau$. Note here that due to Lemma 1 for a function $z \in C^{m-1,\{\alpha_k\},\{\gamma_i\}}([t_0, t_0 + \tau]; \mathcal{Z})$ we have $D^{\alpha_k} z, D^{\gamma_i} z \in C([t_0, t_1]; \mathcal{Z})$ for all $k = 1, 2, \dots, n$, $i = 1, 2, \dots, r$. Define on the set $\mathcal{S}$ a metrics $d(x, y) := \|x - y\|_{C^{m-1,\{\alpha_k\},\{\gamma_i\}}([t_0, t_0+\tau]; \mathcal{Z})}$, then $\mathcal{S}$ is a complete metric space. Note that $\tilde{z} \in \mathcal{S}$ for small $\tau > 0$.

For $z \in \mathcal{S}$ define a mapping

$$G(z)(t) := \sum_{l=0}^{m-1} Z_l(t - t_0) z_l + \int_{t_0}^{t} Z(t - s) B(s, D^{\gamma_1} z(s), D^{\gamma_2} z(s), \dots, D^{\gamma_r} z(s)) ds,$$

$t \in [t_0, t_0 + \tau]$. Reasoning as in the proof of Theorem 1 and Lemma 2 in [17], we obtain that $G(z) \in C^{m-1}([t_0, t_0 + \tau]; \mathcal{Z})$, $D^{\alpha_k} G(z) \in C([t_0, t_0 + \tau]; \mathcal{Z})$, $k = 1, 2, \dots, n$, $[G(z)]^{(k)}(t_0) = z_k$ for $k = 0, 1, \dots, m - 1$.

Further we have for $\gamma_i < l$

$$D^{\gamma_i} Z_l(0) := \lim_{t\to 0+} \frac{1}{2\pi i} \int_{\Gamma} \lambda^{-l-1+\gamma_i} R_\lambda \left(I - \sum_{k=n_l}^{n} \lambda^{\alpha_k - \alpha} A_k \right) e^{\lambda t} d\lambda = 0,$$

since

$$-l - 1 + \gamma_i < -1, \quad \left\| R_\lambda \left(I - \sum_{k=n_l}^{n} \lambda^{\alpha_k - \alpha} A_k \right) \right\|_{\mathcal{L}(\mathcal{Z})} \le C.$$

For $\gamma_i > l$

$$\widehat{D^{\gamma_i} Z_l}(\lambda) = \lambda^{-l-1+\gamma_i} R_\lambda \left(I - \sum_{k=n_l}^{n} \lambda^{\alpha_k - \alpha} A_k \right) - \lambda^{\gamma_i - 1 - l} =$$

$$= \lambda^{-l-1+\gamma_i} R_\lambda \sum_{k=1}^{n_l - 1} \lambda^{\alpha_k - \alpha} A_k,$$

$$D^{\gamma_i} Z_l(0) = \lim_{t \to 0+} \frac{1}{2\pi i} \int_\Gamma \lambda^{-l-1+\gamma_i} R_\lambda \sum_{k=1}^{n_l - 1} \lambda^{\alpha_k - \alpha} A_k e^{\lambda t} d\lambda = 0,$$

since for $k \le n_l - 1$ we have $l > m_k - 1$, hence, $l \ge \alpha_k$, $-l - 1 + \gamma_i + \alpha_k - \alpha < -1$.

For $\gamma_i > 0$ we have

$$J^{n_i - \gamma_i} Z(t) = \frac{1}{2\pi i} \int_\Gamma \lambda^{-\alpha + \gamma_i - n_i} R_\lambda e^{\lambda t} d\lambda,$$

for $l = 0, 1, \ldots, n_i - 1$

$$D^l J^{n_i - \gamma_i} Z(0) = \lim_{t \to 0+} \frac{1}{2\pi i} \int_\Gamma \lambda^{-\alpha + \gamma_i - n_i + l} R_\lambda e^{\lambda t} d\lambda = 0, \tag{7}$$

due to $-\alpha + \gamma_i - n_i + l \le -\alpha + \gamma_i - 1 < -1$. From here it follows also that $\|D^{n_i} J^{n_i - \gamma_i} Z(t)\|_{\mathcal{L}(\mathcal{Z})} \le C t^{\alpha - \gamma_i - 1}$ for some $C > 0$ and every $t > 0$. Denote $B^z(s) := B(s, D^{\gamma_1} z(s), D^{\gamma_2} z(s), \ldots, D^{\gamma_r} z(s))$, then due to (7) and equalities

$$D^l|_{t=t_0} \int_{t_0}^{t} Z(t-s) B^z(s) ds = 0, \;\; l = 0, 1, \ldots, m-1,$$

which are proved in Lemma 2 [17], we have

$$D^{\gamma_i} \int_{t_0}^{t} Z(t-s) B^z(s) ds = D^{n_i} J^{n_i - \gamma_i} \int_{t_0}^{t} Z(t-s) B^z(s) ds =$$

$$= D^{n_i} \int_{t_0}^{t} J^{n_i - \gamma_i} Z(t-s) B^z(s) ds = \int_{t_0}^{t} D^{n_i} J^{n_i - \gamma_i} Z(t-s) B^z(s) ds,$$

$$\left\| \lim_{t\to t_0+} D^{\gamma_i} \int_{t_0}^{t} Z(t-s)B^z(s)ds \right\|_{\mathcal{Z}} \leq \lim_{t\to t_0+} C_1(t-t_0)^{\alpha-\gamma_i} \max_{s\in[t_0,t_0+\tau]} \|B^z(s)\|_{\mathcal{Z}} = 0.$$

Thus, $G(z) \in C^{m-1,\{\alpha_k\},\{\gamma_i\}}([t_0, t_1]; \mathcal{Z})$ and $D^{\gamma_i} G(z)(t_0) = \tilde{z}_i, i = 0, 1, \dots, r$. Therefore, $G(z) \in \mathcal{S}$ for a small enough $\tau > 0$.

For $x, y \in \mathcal{S}, l = 0, 1, \dots, m-1$

$$D^l Z(t) = \frac{1}{2\pi i} \int_{\Gamma} \lambda^{-\alpha+l} R_\lambda e^{\lambda t} d\lambda,$$

hence, $D^l Z(0) = 0$ for $l = 0, 1, \dots, m-2$, $\|D^l Z(t)\|_{\mathcal{L}(\mathcal{Z})} \leq C t^{\alpha-l-1}$. Then for a small enough $\tau > 0$

$$\|[G(x)]^{(l)}(t) - [G(y)]^{(l)}(t)\|_{\mathcal{Z}} = \left\| \int_{t_0}^{t} D^l Z(t-s)[B^x(s) - B^y(s)]ds \right\|_{\mathcal{Z}} \leq$$

$$\leq C_l(t-t_0)^{\alpha-l} \sum_{i=1}^{r} \sup_{t\in[t_0,t_0+\tau]} \|D^{\gamma_i}(x(t) - y(t))\|_{\mathcal{Z}} \leq \frac{d(x,y)}{2(n-k_*+r-i_*+2+m)},$$

$$\|D^{\gamma_i} G(x)(t) - D^{\gamma_i} G(y)(t)\|_{\mathcal{Z}} = \left\| \int_{t_0}^{t} D^{\gamma_i} Z(t-s)[B^x(s) - B^y(s)]ds \right\|_{\mathcal{Z}} \leq$$

$$\leq C_{\gamma_i}(t-t_0)^{\alpha-\gamma_i} \sum_{j=1}^{r} \sup_{t\in[t_0,t_0+\tau]} \|D^{\gamma_j}(x(t) - y(t))\|_{\mathcal{Z}} \leq \frac{d(x,y)}{2(n-k_*+r-i_*+2+m)},$$

where $n - k_* + r - i_* + 2$ is the quantity of α_k and γ_i, which are greater than $m-1$. Here we used Lemma 1. Thus, $d(G(y), G(v)) \leq d(y, v)/2$ and the mapping G has a unique fixed point z in the metric space $\mathcal{S}$. It is a unique solution of (5) in $C^{m-1,\{\alpha_k\},\{\gamma_i\}}([t_0, t_0+\tau]; \mathcal{Z})$, therefore, due to Lemma 2 it is a unique solution of problem (2), (4) on the chosen segment $[t_0, t_0+\tau]$.

4 Degenerate Quasilinear Equation

Let $\mathcal{X}, \mathcal{Y}$ be a Banach space, $\mathcal{L}(\mathcal{X}; \mathcal{Y})$ be the space of bounded linear operators from $\mathcal{X}$ into $\mathcal{Y}$, and let $Cl(X; Y)$ be the set of all linear closed operators densely defined in the space $\mathcal{X}$, acting into $\mathcal{Y}$. Let $n \in \mathbb{N}$, $L, M_1, \dots, M_{n-1} \in \mathcal{L}(\mathcal{X}; \mathcal{Y})$, $\ker L \neq \{0\}$, $M_n \in Cl(\mathcal{X}; \mathcal{Y})$, D_{M_n} be the domain of the operator M_n, on which the norm of the graph $\|\cdot\|_{D_{M_n}} := \|\cdot\|_{\mathcal{X}} + \|M_n \cdot\|_{\mathcal{Y}}$ be set. Denote

$$\rho^L(M_n) := \left\{\mu \in \mathbf{C} : (\mu L - M_n)^{-1} \in \mathcal{L}(\mathcal{Y};\mathcal{X})\right\}, \ \sigma^L(M_n) := \mathbf{C} \setminus \rho^L(M_n),$$

$$R^L_\mu(M_n) := (\mu L - M_n)^{-1} L, \quad L^L_\mu(M_n) := L(\mu L - M_n)^{-1}.$$

An operator M_n is called (L,σ)-bounded if $\sigma^L(M_n) \subset \{\mu \in \mathbf{C} : |\mu| \le a\}$ for some $a > 0$. Under the condition of (L,σ)-boundedness of the operator M_n there exist projections

$$P = \frac{1}{2\pi i}\int_\gamma R^L_\mu(M_n)d\mu \in \mathcal{L}(\mathcal{X}), \quad Q = \frac{1}{2\pi i}\int_\gamma L^L_\mu(M_n)d\mu \in \mathcal{L}(\mathcal{Y}), \tag{8}$$

where $\gamma := \{\mu \in \mathbf{C} : |\mu| = r > a\}$ [24]. Let $\mathcal{X}^0 = \ker P$, $\mathcal{X}^1 = \operatorname{Im} P$, $\mathcal{Y}^0 = \ker Q$, $\mathcal{Y}^1 = \operatorname{Im} Q$. Denote for short $P_0 := I - P$, $Q_0 := I - Q$ and L_k $(M_{l,k})$ restriction of the operator L (M_l) on $\mathcal{X}^k$ $(D_{M_{n,k}} := D_{M_n} \cap \mathcal{X}^k$ for $l = n)$, $k = 0, 1$, $l = 1, 2, \dots, n$. It is known (see [24]) that $LP = QL$, $M_n Px = QM_n x$ for $x \in D_{M_n}$, therefore, $M_{n,1} \in \mathcal{L}(\mathcal{X}^1;\mathcal{Y}^1)$, $M_{n,0} \in \mathcal{C}l(\mathcal{X}^0;\mathcal{Y}^0)$, $L_k \in \mathcal{L}(\mathcal{X}^k;\mathcal{Y}^k)$, $k = 0, 1$. Moreover, in this case there are operators $M_{n,0}^{-1} \in \mathcal{L}(\mathcal{Y}^0;\mathcal{X}^0)$, $L_1^{-1} \in \mathcal{L}(\mathcal{Y}^1;\mathcal{X}^1)$. An operator M_n will be called $(L, 0)$-bounded if L_0 is zero operator.

Consider the initial problem

$$x^{(l)}(t_0) = x_l, l = 0, 1, \dots, m_n - 1, (Px)^{(l)}(t_0) = x_l, l = m_n, m_n + 1, \dots, m - 1, \tag{9}$$

for a quasilinear fractional order equation

$$D^\alpha Lx(t) = \sum_{k=1}^{n} D^{\alpha_k} M_k x(t) + N(t, D^{\gamma_1}x(t), D^{\gamma_2}x(t), \dots, D^{\gamma_r}x(t)), \tag{10}$$

which is called degenerate in the case $\ker L \ne \{0\}$. It is assumed that, as before, $n, r \in \mathbf{N}$, $m - 1 < \alpha \le m \in \mathbf{N}$, $\alpha_1 < \alpha_2 < \cdots < \alpha_n < \alpha$, $m_l - 1 < \alpha_l \le m_l \in \mathbf{Z}$, $l = 1, 2, \dots, n$, $\gamma_1 < \gamma_2 < \cdots < \gamma_r < \alpha$, U be an open set in $\mathbf{R} \times \mathcal{X}^r$, $N \in C(U, \mathcal{Z})$.

A solution to problem (9), (10) on a segment $[t_0, t_1]$ is a function $x \in C^{m_n-1}([t_0, t_1];\mathcal{X}) \cap C([t_0, t_1]; D_{M_n})$, such that $D^\alpha Lx, D^{\alpha_k} M_k x \in C([t_0, t_1];\mathcal{Y})$, $k = 1, 2, \dots, n$, $D^{\gamma_i} x \in C([t_0, t_1];\mathcal{X})$, $i = 1, 2, \dots, r$, the inclusion

$$(t, D^{\gamma_1}x(t), D^{\gamma_2}x(t), \dots, D^{\gamma_r}x(t)) \in U$$

and equality (10) are valid for all $t \in [t_0, t_1]$ and conditions (9) are satisfied.

In the definition of a solution we used the fact that for $(L, 0)$-bounded operator M_n the smoothness of Px is the same as for Lx, since $Px = L_1^{-1}Lx$, $Lx = LPx$. And the condition $D^\alpha Lx \in C([t_0, t_1];\mathcal{Y})$ due to the definition of the Gerasimov—Caputo derivative means, in partial, that $Lx \in C^{m-1}([t_0, t_1]; \mathcal{Y})$, hence, $Px \in C^{m-1}([t_0, t_1]; \mathcal{X})$.

Denote $V = U \cap (\mathbf{R} \times (\mathcal{X}^1)^r)$, $\tilde{v}_i := D^{\gamma_i}|_{t=t_0}\tilde{v}(t)$, $i = 1, 2, \dots, r$, where

$$\tilde{v}(t) := Px_0 + (t - t_0)Px_1 + \frac{(t-t_0)^2}{2!}Px_2 + \dots + \frac{(t-t_0)^{m-1}}{(m-1)!}Px_{m-1}.$$

Theorem 3 *Let $L, M_k \in \mathcal{L}(\mathcal{X}; \mathcal{Y})$, $k = 1, 2, \dots, n-1$, $M_n \in \mathcal{C}l(\mathcal{X}; \mathcal{Y})$ be $(L, 0)$-bounded, $M_kP = QM_k$, $k = 1, 2, \dots, n-1$, $N : U \to \mathcal{Y}$, for every $(t, z_1, z_2, \dots, z_r) \in U$, such that $(t, Pz_1, Pz_2, \dots, Pz_r) \in U$, $N(t, z_1, z_2, \dots, z_r) = N_1(t, Pz_1, Pz_2, \dots, Pz_r)$ for some operator $N_1 \in C(V; \mathcal{Y})$, which be locally Lipschitzian in $v_1, v_2, \dots, v_r$, $(t_0, \tilde{v}_1, \tilde{v}_2, \dots, \tilde{v}_r) \in V$, $x_l \in \mathcal{X}^1$ for $l = m_n, m_n + 1, \dots, m-1$. Then there exists a unique solution of problem (9), (10).*

Proof Conditions $M_lP = QM_l$ for $l = 1, 2, \dots, n-1$ immediately imply that $M_{l,k} \in \mathcal{L}(\mathcal{X}^k; \mathcal{Y}^k)$, $k = 0, 1$, $l = 1, 2, \dots, n-1$. Act on Eq. (10) by $L_1^{-1}Q \in \mathcal{L}(\mathcal{Y}^1; \mathcal{X}^1)$ and obtain the equation

$$D^{\alpha}v(t) = \sum_{k=1}^{n} D^{\alpha_k}L_1^{-1}M_kv(t) + L_1^{-1}QN_1(t, D^{\gamma_1}v(t), \dots, D^{\gamma_r}v(t)), \tag{11}$$

where $v(t) = Px(t)$. After acting by $M_{n,0}^{-1}Q_0 \in \mathcal{L}(\mathcal{Y}^0; \mathcal{X}^0)$ we obtain

$$D^{\alpha_n}w(t) = -\sum_{k=1}^{n-1} D^{\alpha_k}M_{n,0}^{-1}M_kw(t) - M_{n,0}^{-1}Q_0N_1(t, D^{\gamma_1}v(t), \dots, D^{\gamma_r}v(t)) \tag{12}$$

with $w(t) = P_0x(t)$. Here we used the equalities $M_kP_0 = M_k - QM_k = Q_0M_k$, $k = 1, 2, \dots, n-1$.

Due to (9) Eqs. (11) and (12) are endowed by the initial conditions

$$v^{(l)}(t_0) = Px_l, \; l = 0, 1, \dots, m-1, \tag{13}$$

$$w^{(l)}(t_0) = P_0x_l, \; l = 0, 1, \dots, m_n - 1, \tag{14}$$

respectively. By Theorem 2 Cauchy problem (11), (13) has a unique solution on a segment $[t_0, t_1]$. Then Eq. (12) is inhomogeneous linear, and due to Theorem 1 Cauchy problem (14) for it has a unique solution also.

Denote $W = U \cap (\mathbf{R} \times (\mathcal{X}^0)^r)$, $\tilde{w}_i := D^{\gamma_i}|_{t=t_0}\tilde{w}(t)$, $i = 1, 2, \dots, r$, where

$$\tilde{w}(t) := P_0x_0 + (t - t_0)P_0x_1 + \frac{(t-t_0)^2}{2!}P_0x_2 + \dots + \frac{(t-t_0)^{m_n-1}}{(m_n-1)!}P_0x_{m_n-1}.$$

Theorem 4 *Let $\gamma_r < \alpha_n$, $L, M_k \in \mathcal{L}(\mathcal{X}; \mathcal{Y})$, $k = 1, 2, \dots, n-1$, $M_n \in \mathcal{C}l(\mathcal{X}; \mathcal{Y})$ be $(L, 0)$-bounded, $M_kP = QM_k$, $k = 1, 2, \dots, n-1$, $N : U \to \mathcal{Y}$, for every*

$(t, z_1, z_2, \ldots, z_r) \in U$, such that $(t, P_0 z_1, P_0 z_2, \ldots, P_0 z_r) \in U$, $N(t, z_1, z_2, \ldots, z_r) = N_0(t, P_0 z_1, P_0 z_2, \ldots, P_0 z_r)$ for some operator $N_0 \in C(W; \mathcal{Y})$, which be locally Lipschitzian in $w_1, w_2, \ldots, w_r$, $(t_0, \tilde{w}_1, \tilde{w}_2, \ldots, \tilde{w}_r) \in W$, $x_l \in \mathcal{X}^1$ for $l = m_n, m_n + 1, \ldots, m-1$. Then there exists a unique solution of problem (9), (10).

Proof As in the proof of the previous theorem reduce problem (9), (10) to problem (13), (14) for the system of equations

$$D^\alpha v(t) = \sum_{k=1}^{n} D^{\alpha_k} L_1^{-1} M_k v(t) + L_1^{-1} Q N_0(t, D^{\gamma_1} w(t), \ldots, D^{\gamma_r} w(t)), \tag{15}$$

$$D^{\alpha_n} w(t) = -\sum_{k=1}^{n-1} D^{\alpha_k} M_{n,0}^{-1} M_k w(t) - M_{n,0}^{-1} Q_0 N_0(t, D^{\gamma_1} w(t), \ldots, D^{\gamma_r} w(t)) \tag{16}$$

with $v(t) = Px(t)$, $w(t) = P_0 x(t)$. By Theorem 2 Cauchy problem (14), (16) has a unique solution on a segment $[t_0, t_1]$, by Theorem 1 there exists a unique solution of problem (13) for linear equation (15) on $[t_0, t_1]$.

Denote $\tilde{x}_i := D^{\gamma_i}|_{t=t_0} \tilde{x}(t)$, $i = 1, 2, \ldots, r$, where

$$\tilde{x}(t) := x_0 + (t - t_0)x_1 + \frac{(t - t_0)^2}{2!} x_2 + \cdots + \frac{(t - t_0)^{m_n - 1}}{(m_n - 1)!} x_{m_n - 1}.$$

Theorem 5 *Let $L, M_k \in \mathcal{L}(\mathcal{X}; \mathcal{Y})$, $k = 1, 2, \ldots, n-1$, $M_n \in \mathcal{C}l(\mathcal{X}; \mathcal{Y})$ be $(L, 0)$-bounded, $M_k P = Q M_k$, $k = 1, 2, \ldots, n-1$, $N \in C(U; \mathcal{Y}^1)$ be locally Lipschitzian in $x_1, x_2, \ldots, x_r$, $(t_0, \tilde{x}_1, \tilde{x}_2, \ldots, \tilde{x}_r) \in U$, $x_l \in \mathcal{X}^1$ for $l = m_n, m_n + 1, \ldots, m-1$. Then there exists a unique solution of problem (9), (10).*

Proof Anologously to the proof of the previous two theorems we get

$$D^\alpha v(t) = \sum_{k=1}^{n} D^{\alpha_k} L_1^{-1} M_k v(t) + L_1^{-1} Q N(t, D^{\gamma_1}(v(t) + w(t)), \ldots, D^{\gamma_r}(v(t) + w(t))), \tag{17}$$

$$D^{\alpha_n} w(t) = -\sum_{k=1}^{n-1} D^{\alpha_k} M_{n,0}^{-1} M_k w(t). \tag{18}$$

Then due to Theorem 1 there exists a unique solution of (14), (18) on $[t_0, \infty)$, and by Theorem 2 problem (13), (17) has a unique solution on some segment $[t_0, t_1]$. We used that the nonlinear mapping

$$(v_1, v_2, \ldots, v_r) \to L_1^{-1} Q N(t, v_1 + w_1, v_2 + w_2, \ldots, v_r + w_r)$$

with given $w_1, w_2, \ldots, w_r$ satisfies the conditions of Theorem 2.

Theorem 6 *Let $\gamma_r < \alpha_n$, $L, M_k \in \mathcal{L}(\mathcal{X};\mathcal{Y})$, $k = 1, 2, \dots, n-1$, $M_n \in \mathcal{C}l(\mathcal{X};\mathcal{Y})$ be $(L, 0)$-bounded, $M_k P = QM_k$, $k = 1, 2, \dots, n-1$, $N \in C(U; \mathcal{Y}^0)$ be locally Lipschitzian in $x_1, x_2, \dots, x_r$, $(t_0, \tilde{x}_1, \tilde{x}_2, \dots, \tilde{x}_r) \in U$, $x_l \in \mathcal{X}^1$ for $l = m_n, m_n + 1, \dots, m-1$. Then there exists a unique solution of problem (9), (10).*

Proof Here we have a system of equations

$$D^{\alpha} v(t) = \sum_{k=1}^{n} D^{\alpha_k} L_1^{-1} M_k v(t), \tag{19}$$

$$D^{\alpha_n} w(t) = -\sum_{k=1}^{n-1} D^{\alpha_k} M_{n,0}^{-1} M_k w(t) +$$

$$+ M_0^{-1} Q_0 N(t, D^{\gamma_1}(v(t) + w(t)), \dots, D^{\gamma_r}(v(t) + w(t))). \tag{20}$$

Theorem 1 implies the unique solvability of problem (13), (19) and due to Theorem 2 there exists a unique solution of (14), (20).

Remark 1 From the proof of theorems it follows that the Cauchy problem

$$x^{(l)}(0) = x_l, \ l = 0, 1, \dots, m-1,$$

for (10) is solvable only if the conditions

$$P_0 x_{m_n+l} = w^{(m_n+l)}(0), \quad l = 0, 1, \dots, m - m_n - 1,$$

is fulfilled.

Remark 2 It is easy to show that, if M_n is $(L, 0)$-bounded, then the conditions $(Px)^{(l)}(0) = x_l \in \mathcal{X}^1$ for $l = 1, 2, \dots, m-1$ are equivalent to the conditions $(Lx)^{(l)}(0) = y_l := Lx_l \in \mathcal{Y}^1$.

5 Examples

Here we consider simple examples of degenerate systems of partial differential equations, which illustrate four considered in the previous section cases.

Let $\Omega \subset \mathbf{R}^d$ is a bounded region with a smooth boundary $\partial\Omega$. Consider the initial boundary value problem

$$\frac{\partial^k x_1}{\partial t^k}(\xi, t_0) = x_{1k}(\xi), \ k = 0, 1, \dots, m-1, \ \xi \in \Omega, \tag{21}$$

$$\frac{\partial^k x_2}{\partial t^k}(\xi, t_0) = x_{2k}(\xi),\ \ k = 0, 1, \dots, m_n - 1,\ \ \xi \in \Omega, \tag{22}$$

$$x_i(\xi, t) = 0,\ \ (\xi, t) \in \partial\Omega \times [t_0, t_1],\ \ i = 1, 2, \tag{23}$$

$$\begin{aligned} D_t^{\alpha}\Delta x_1 &= \sum_{k=1}^{n} a_k D_t^{\alpha_k} x_1 + h_1\left(\xi, t, D_t^{\gamma_1}x_1, D_t^{\gamma_1}x_2, \dots, D_t^{\gamma_r}x_1, D_t^{\gamma_r}x_2\right), \\ 0 &= \sum_{k=1}^{n} b_k D_t^{\alpha_k} x_2 + h_2\left(\xi, t, D_t^{\gamma_1}x_1, D_t^{\gamma_1}x_2, \dots, D_t^{\gamma_r}x_2\right),\ (\xi, t) \in \Omega \times [t_0, t_1], \end{aligned} \tag{24}$$

where $n, r \in \mathbf{N}$, $m - 1 < \alpha \le m \in \mathbf{N}$, $\alpha_1 < \alpha_2 < \dots < \alpha_n < \alpha$, $m_n - 1 < \alpha_n \le m_n$, $\gamma_1 < \gamma_2 < \dots < \gamma_r$, D_t^{β} is the Gerasimov—Caputo derivative with respect to t.

Let $A := \sum\limits_{j=1}^{d} \frac{\partial^2}{\partial \xi_j^2}$ be the Laplace operator with a domain $H_0^2(\Omega) = \{z \in H^2(\Omega) : z(s) = 0,\ s \in \partial\Omega\} \subset L_2(\Omega)$, $\{\varphi_k\}$ be an orthonormal in $L_2(\Omega)$ system of its eigenfunctions, corresponding to eigenvalues $\{\lambda_k\}$ of A, numbered in descending order, taking into account their multiplicities.

Reduce problem (21)–(24) to (9), (10), taking the spaces

$$\mathcal{X} = H_0^2(\Omega) \times L_2(\Omega),\quad \mathcal{Y} = (L_2(\Omega))^2, \tag{25}$$

and the operators

$$L = \begin{pmatrix} \Delta & 0 \\ 0 & 0 \end{pmatrix} \in \mathcal{L}(\mathcal{X}; \mathcal{Y}),\quad M_k = \begin{pmatrix} a_k I & 0 \\ 0 & b_k I \end{pmatrix} \in \mathcal{L}(\mathcal{X}; \mathcal{Y}),\ k = 1, 2, \dots, n. \tag{26}$$

Lemma 3 *Let spaces (25) and operators (26) are given, $b_n \neq 0$. Then the operator M_n is $(L, 0)$-bounded and projections has the form*

$$P = \begin{pmatrix} 1 & 0 \\ 0 & 0 \end{pmatrix},\quad Q = \begin{pmatrix} 1 & 0 \\ 0 & 0 \end{pmatrix}. \tag{27}$$

Proof If $\mu \neq a_n\lambda_k^{-1}$ for all $k \in \mathbf{N}$, then

$$(\mu L - M_n)^{-1} = \sum_{k=1}^{\infty} \langle \cdot, \varphi_k \rangle \varphi_k \begin{pmatrix} (\mu\lambda_k - a_n)^{-1} & 0 \\ 0 & -b_n^{-1} \end{pmatrix},$$

hence, for $|\mu| > |a_n||\lambda_1|^{-1}$ the operator $(\mu L - M_n)^{-1} : \mathcal{Y} \to \mathcal{X}$ is bounded,

$$R_{\mu}^{L}(M_n) = \sum_{k=1}^{\infty} \langle \cdot, \varphi_k \rangle \varphi_k \begin{pmatrix} \lambda_k(\mu\lambda_k - a_n)^{-1} & 0 \\ 0 & 0 \end{pmatrix},$$

$$L^L_\mu(M_n) = \sum_{k=1}^{\infty} \langle \cdot, \varphi_k \rangle \varphi_k \begin{pmatrix} \lambda_k(\mu\lambda_k - a_n)^{-1} & 0 \\ 0 & 0 \end{pmatrix}.$$

These equalities due to formulas (8) and the residue theorem imply form (27) of projections, since for large enough $|\mu|$

$$\lambda_k(\mu\lambda_k - a_n)^{-1} = \mu^{-1}(1 - a_n\mu^{-1}\lambda_k^{-1})^{-1} = \sum_{j=0}^{\infty} a_n^j \mu^{-j-1}\lambda_k^{-j}.$$

Therefore, $\mathcal{X}^1 = H_0^2(\Omega) \times \{0\}$, $\mathcal{X}^0 = \{0\} \times L_2(\Omega)$, $\mathcal{Y}^1 = L_2(\Omega) \times \{0\}$, $\mathcal{Y}^0 = \{0\} \times L_2(\Omega)$, $L_0 = 0$ and the operator M_n is $(L, 0)$-bounded.

The first formula in (27) implies that initial conditions (21), (22) has form (9). The system of equation

$$\begin{aligned} D_t^\alpha \Delta x_1 &= \sum_{k=1}^{n} a_k D_t^{\alpha_k} x_1 + h_1\left(\xi, t, D_t^{\gamma_1} x_1, D_t^{\gamma_2} x_1, \dots, D_t^{\gamma_r} x_1\right), \\ 0 &= \sum_{k=1}^{n} b_k D_t^{\alpha_k} x_2 + h_2\left(\xi, t, D_t^{\gamma_1} x_1, D_t^{\gamma_2} x_1, \dots, D_t^{\gamma_r} x_1\right) \end{aligned}$$

satisfies the conditions of Theorem 3; the system

$$\begin{aligned} D_t^\alpha \Delta x_1 &= \sum_{k=1}^{n} a_k D_t^{\alpha_k} x_1 + h_1\left(\xi, t, D_t^{\gamma_1} x_2, D_t^{\gamma_2} x_2, \dots, D_t^{\gamma_r} x_2\right), \\ 0 &= \sum_{k=1}^{n} b_k D_t^{\alpha_k} x_2 + h_2\left(\xi, t, D_t^{\gamma_1} x_2, D_t^{\gamma_2} x_2, \dots, D_t^{\gamma_r} x_2\right), \ \gamma_r < \alpha_n, \end{aligned}$$

corresponds to Theorem 4; conditions of Theorem 5 are valid for the system

$$\begin{aligned} D_t^\alpha \Delta x_1 &= \sum_{k=1}^{n} a_k D_t^{\alpha_k} x_1 + h_1\left(\xi, t, D_t^{\gamma_1} x_1, D_t^{\gamma_1} x_2, D_t^{\gamma_2} x_1, D_t^{\gamma_2} x_2, \dots, D_t^{\gamma_r} x_2\right), \\ 0 &= \sum_{k=1}^{n} b_k D_t^{\alpha_k} x_2; \end{aligned}$$

and Theorem 6 can be applied to the system of equations

$$\begin{aligned} D_t^\alpha \Delta x_1 &= \sum_{k=1}^{n} a_k D_t^{\alpha_k} x_1, \\ 0 &= \sum_{k=1}^{n} b_k D_t^{\alpha_k} x_2 + h_2\left(\xi, t, D_t^{\gamma_1} x_1, D_t^{\gamma_1} x_2, \dots, D_t^{\gamma_r} x_1, D_t^{\gamma_r} x_2\right), \ \gamma_r < \alpha_n. \end{aligned}$$

Notes and Comments. The work is funded by the Russian Science Foundation and Chelyabinsk Region, project no. 22-21-20095.

References

1. Tarasov VE (2011) Fractional dynamics: applications of fractional calculus to dynamics of particles, fields and media. Springer, New York
2. Uchaykin VV (2013) Fractional derivatives for physicists and engineers, vol I. Background and theory. Higher Education Press, Beijing
3. Samko SG, Kilbas AA, Marichev OI (1993) Fractional integrals and derivatives. Theory and applications. Gordon and Breach Science Publishers, Philadelphia
4. Prüss J (1993) Evolutionary integral equations and applications. Birkhauser-Verlag, Basel
5. Podlubny I (1999) Fractional differential equations. Academic, Boston
6. Bajlekova EG (2001) Fractional evolution equations in banach spaces: PhD thesis. Eindhoven University of Technology, University Press Facilities, Eindhoven
7. Pskhu AV (2005) Partial differential equations of fractional order. Nauka Publ, Moscow [In Russian]
8. Kilbas AA, Srivastava HM, Trujillo JJ (2006) Theory and applications of fractional differential equations. Elsevier Science Publishing, Amsterdam
9. Kostić M (2015) Abstract volterra integro-differential equations. CRC Press, Boca Raton
10. Romanova EA, Fedorov VE (2016) Resolving operators of a linear degenerate evolution equation with Caputo derivative. The sectorial case. Math Notes NEFU 23(4):58–72
11. Plekhanova MV (2016) Nonlinear equations with degenerate operator at fractional Caputo derivative. Math Methods Appl Sci 40:41–44. https://doi.org/10.1002/mma.3830
12. Plekhanova MV, Baybulatova GD (2019) Semilinear equations in Banach spaces with lower fractional derivatives. Springer Proc Math Stat 292:81–93. https://doi.org/10.1007/978-3-030-26987-6_6
13. Fedorov VE, Phuong TD, Kien BT, Boyko KV, Izhberdeeva EM (2020) A class of semilinear distributed order equations in Banach spaces. Chelyabinsk Phys Math J 5(3):343–351. https://doi.org/10.47475/2500-0101-2020-15308
14. Fedorov VE, Plekhanova MV, Nazhimov RR (2018) Degenerate linear evolution equations with the Riemann – Liouville fractional derivative. Sib Math J 59(1):136–146. https://doi.org/10.17377/smzh.2018.59.115
15. Fedorov VE, Avilovich AS (2019) A Cauchy type problem for a degenerate equation with the Riemann – Liouville derivative in the sectorial case. Sib Math J 60(2):359–372. https://doi.org/10.33048/smzh.2019.60.216
16. Fedorov VE, Avilovich AS (2021) Semilinear fractional-order evolution equations of Sobolev type in the sectorial case. Complex Var Elliptic Equ 66(6–7):1108–1121. https://doi.org/10.1080/17476933.2020.1833870
17. Fedorov VE, Boyko KV, Phuong TD (2021) Initial value problems for some classes of linear evolution equations with several fractional derivatives. Mat Notes NEFU 28(3):85–104. https://doi.org/10.25587/SVFU.2021.75.46.006
18. Fedorov VE, Turov MM (2021) The defect of a Cauchy type problem for linear equations with several Riemann – Liouville derivatives. Sib Math J 62(5):925–942. https://doi.org/10.33048/smzh.2021.62.514
19. Fedorov VE, Du W-S, Turov MM (2022) On the unique solvability of incomplete Cauchy type problems for a class of multi-term equations with the Riemann — Liouville derivatives. Symmetry 14(75). https://doi.org/10.3390/sym14010075
20. Fedorov VE, Kien BT, Turov MM (2022) A class of quasilinear equations with Riemann – Liouville derivatives and bounded operators. Axioms. 11(3):96. https://doi.org/10.3390/axioms11030096
21. Gerasimov AN (1948) Generalization of linear laws of deformation and their application to problems of internal friction. Appl Math Mech 12(3):251–260 [In Russian]
22. Caputo M (1967) Linear model of dissipation whose Q is almost frequency independent. II. Geophys J Roy Astron Soc 13:529–539

23. Plekhanova MV, Baybulatova GD (2019) Semilinear equations in Banach spaces with lower fractional derivatives. Springer Proc Math Stat 292:81–93. https://doi.org/10.1007/978-3-030-26987-6_6
24. Sviridyuk GA, Fedorov VE (2003) Linear sobolev type equations and degenerate semigroups of operators. Utrecht; Boston, VSP

On the Solvability of Initial Problems for Abstract Singular Equations Containing Fractional Derivatives

Alexander Glushak

Abstract With the help of integral representations of the Poisson type, it is established that the Cauchy problem for a number of abstract singular equations with fractional derivatives reduces to a simpler problem for a non-singular equation.

Keywords Abstract singular equations · Fractional derivatives · Transformation operator · Cauchy problem

1 Introduction

One of the methods for studying differential equations is the method of transformation operators. Using conversion operators, many important results are established for various classes differential equations, including those for singular differential equations containing the Bessel differential expression

$$\frac{d^2}{dt^2} + \frac{k}{t}\frac{d}{dt}, \quad k \in R.$$

So in the monograph [1] the singular equation of Euler–Poisson–Darboux in partial derivatives

$$\frac{\partial^2 u(t,x)}{\partial t^2} + \frac{k}{t}\frac{\partial u(t,x)}{\partial t} = \Delta u(t,x), \quad k > 0, \ x \in R^n,$$

where Δ is the Laplace operator in space variables, investigated by reduction with the help of a suitable transformation operator to a simpler wave equation when $k = 0$. In this case, the formulas for the solution are written using spherical averages over spatial variables.

A. Glushak (✉)
Belgorod State National Research University, 85, Pobedy St., Belgorod 308015, Russia
e-mail: aleglu@mail.ru

V. Vasilyev (ed.), *Differential Equations, Mathematical Modeling and Computational Algorithms*, Springer Proceedings in Mathematics & Statistics 423,
https://doi.org/10.1007/978-3-031-28505-9_2

The review paper [2] presents the results of studies in which transformation operators are used in more general situation, when in the Euler–Poisson–Darboux equation the Laplace operator in space variables is replaced by some abstract operator A acting in a Banach space, as well as for some other singular equations of integer order. In these studies, a class of operators A is described for which the corresponding initial value problem is well-posed and an explicit representation is established for the enabling operator.

In this paper, the method of transformation operators is applied to abstract singular differential equations, containing fractional derivatives (see [3, Sect. 5], [4, Chap. 2]).

2 Generalized Euler–Poisson–Darboux Differential Equation

Let A be a closed operator in a Banach space E with dense in E domain $D(A)$. For $k \geq 0,\ 0 < \alpha < 1$, consider abstract singular equation with fractional derivatives

$$B_{k,\alpha}u(t) \equiv \frac{d}{dt}\partial^{\alpha}_{0,t}u(t) + \frac{k}{t}\partial^{\alpha}_{0,t}u(t) = Au(t), \quad t > 0, \tag{1}$$

where $\partial^{\alpha}_{0,t}u(t)$ is the fractional Caputo derivative defined by the equality

$$\partial^{\alpha}_{0,t}u(t) = D^{\alpha}_{0,t}\left(u(t) - u(0)\right), \quad \partial^{\alpha}_{0,t}u(0) = \lim_{t\to 0} \partial^{\alpha}_{0,t}u(t),$$

wherein

$$D^{\alpha}_{0,t}\left(u(t) - u(0)\right) = \frac{d}{dt}I^{1-\alpha}_{0,t}\left(u(t) - u(0)\right), \; I^{1-\alpha}_{0,t}u(t) = \frac{1}{\Gamma(1-\alpha)}\int\limits_0^t \frac{u(\tau)}{(t-\tau)^{\alpha}}d\tau$$

respectively, the left-hand fractional derivative and the fractional Riemann–Liouville integral, $\Gamma(\cdot)$ is the gamma function.

If $\alpha = 1$, then the Eq. (1) becomes the Euler–Poisson–Darboux equation

$$u''(t) + \frac{k}{t}u'(t) = Au(t), \quad t > 0, \tag{2}$$

for which the abstract Cauchy problem with conditions

$$u(0) = u_0, \quad u'(0) = 0 \tag{3}$$

previously explored in detail in [5–7] (see also [2]). In these papers there is a review of the studies of the Euler–Poisson–Darboux equation, the class G_k of operators A is described, with which the problem (2), (3) is uniformly well-posed, the construction of the resolving operator of the problem (2), (3), which is called the operator Bessel function and which we denote by $Y_{k,1}(t)$.

In this paper, we present the setting of initial conditions for an equation with fractional derivatives (1), let us describe the class of operators A with which the corresponding initial problems are solvable and establish a number of properties of the solutions.

We will look for a solution to the Eq. (1) that satisfies the initial conditions

$$u(0) = u_0, \quad \partial^{\alpha}_{0,t} u(0) = 0. \tag{4}$$

Definition 1 A solution of the problem (1), (4) is a function continuous for $t > 0$ $u(t)$ such that for $t > 0$ the functions $I^{1-\alpha}u(t)$ are twice continuously differentiable, the function $u(t)$ takes values from the domain $D(A)$ of the operator A and satisfies the equalities (1), (4).

We begin the study of the solvability of the problem (1), (4) from the case when the parameter $k = 0$ in the Eq. (1) and describe the class considered operators A.

Condition 1 If $\operatorname{Re}\lambda > \omega \geq 0$ and $0 < \alpha \leq 1$, then $\lambda^{\alpha+1}$ belongs to the resolvent set $\rho(A)$ of the operator A and for all integers $n \geq 0$ the resolution $R(\lambda) = (\lambda I - A)^{-1}$ satisfies the inequalities

$$\left\| \frac{d^n}{d\lambda^n} \left(\lambda^{\alpha} R\left(\lambda^{\alpha+1}\right)\right) \right\| \leq \frac{Mn!}{(\operatorname{Re}\lambda - \omega)^{n+1}}. \tag{5}$$

Theorem 1 *Let $k = 0$, $0 < \alpha \leq 1$, $u_0 \in D(A)$ and the operator A satisfies Condition 1. Then the problem (1), (4) uniquely resolvable.*

Proof After applying to the Eq. (1) the integration operator $I^1_{0,t}$ and fractional differentiation $D^{1-\alpha}_{0,t}$ the problem (1), (4) reduces to the next initial problem

$$u'(t) = \frac{1}{\Gamma(\alpha)} \int_0^t (t-s)^{\alpha-1} Au(s)\, ds, \quad t \geq 0, \tag{6}$$

$$u(0) = u_0. \tag{7}$$

Problem (6), (7) is a special case of the problem studied in [8]. In Theorem 3 of [8], it is established that Condition 1 is necessary and sufficient condition on the operator A, which, under the assumptions made in the theorem being proved, ensures the unique solvability problem (6), (7), and thus the equivalent problem (1), (4). The resolving operator of the problem (6), (7) will be denoted by $Y_{0,\alpha}(t)$, while $u(t) = Y_{0,\alpha}(t)u_0$. For $Y_{0,\alpha}(t)$ in [8] the representation and estimate are set respectively

$$Y_{0,\alpha}(t)u_0 = \frac{1}{2\pi i}\int\limits_{\sigma-i\infty}^{\sigma+i\infty} e^{\lambda t}\,\lambda^{\alpha}R\left(\lambda^{\alpha+1}\right)u_0\,d\lambda,\quad u_0\in D\left(A^2\right), \tag{8}$$

$$\left\|Y_{0,\alpha}(t)\right\| \le M\,e^{\sigma t},\quad \sigma>\omega.$$

□

Let us proceed to consider the case $k > 0$ and introduce the Poisson-type transformation operator

$$P_{k,\alpha}u(t) = c_{k,\alpha}\int\limits_0^1 \left(1-s^{\alpha+1}\right)^{k/(\alpha+1)-1} u(ts)\,ds, \tag{9}$$

where $B(\cdot,\cdot)$ is the beta function,

$$c_{k,\alpha} = \frac{\alpha+1}{B(k/(\alpha+1),\,1/(\alpha+1))}.$$

The Poisson-type transformation operator is expressed in terms of the Erdelyi–Kober fractional integral $I^{\gamma}_{0+,\sigma,\eta}$ (see [3, Sect. 18]) as follows

$$P_{k,\alpha}u(t) = \frac{\Gamma\left((k+1)/(\alpha+1)\right)}{\Gamma\left(1/(\alpha+1)\right)}\, I^{k/(\alpha+1)}_{0+,\alpha+1,-\alpha/(\alpha+1)}u(t),$$

and the constant $c_{k,\alpha}$ is chosen so that

$$\lim_{t\to 0} P_{k,\alpha}u(t) = u(0).$$

Theorem 2 *Let $k > 0$, $0 < \alpha \le 1$ and the function $u(t)$ be that there is a fractional derivative of the form $\left(\partial^{\alpha}_{0,t}u(t)\right)'$. Then the equality*

$$B_{k,\alpha}\,P_{k,\alpha}u(t) = P_{k,\alpha}\left(\partial^{\alpha}_{0,t}u(t)\right)' + \frac{c_{k,\alpha}}{t}\,\partial^{\alpha}_{0,t}u(0). \tag{10}$$

Proof Applying the operator $B_{k,\alpha}$ to (9), after integrating by parts we get

$$B_{k,\alpha}\,P_{k,\alpha}u(t) = c_{k,\alpha}\int\limits_0^1 \left(1-s^{\alpha+1}\right)^{k/(\alpha+1)-1} s^{\alpha+1}\,\frac{d}{d(ts)}\,\partial^{\alpha}_{0,ts}u(t)\,ds+$$

$$+\frac{k\,c_{k,\alpha}}{t}\int\limits_0^1 \left(1-s^{\alpha+1}\right)^{k/(\alpha+1)-1} s^{\alpha}\,\partial^{\alpha}_{0,ts}u(t)\,ds =$$

$$= c_{k,\alpha} \int_0^1 \left(1 - s^{\alpha+1}\right)^{k/(\alpha+1)-1} s^{\alpha+1} \frac{d}{d(ts)} \partial^{\alpha}_{0,ts} u(t)\, ds +$$

$$+ \frac{c_{k,\alpha}}{t} \partial^{\alpha}_{0,t} u(0) + c_{k,\alpha} \int_0^1 \left(1 - s^{\alpha+1}\right)^{k/(\alpha+1)} \frac{d}{d(ts)} \partial^{\alpha}_{0,ts} u(t)\, ds =$$

$$= c_{k,\alpha} \int_0^1 \left(1 - s^{\alpha+1}\right)^{k/(\alpha+1)-1} \left(s^{\alpha+1} + 1 - s^{\alpha+1}\right) \frac{d}{d(ts)} \partial^{\alpha}_{0,ts} u(t)\, ds +$$

$$+ \frac{c_{k,\alpha}}{t} \partial^{\alpha}_{0,t} u(0) = P_{k,\alpha} \left(\partial^{\alpha}_{0,t} u(t)\right)' + \frac{c_{k,\alpha}}{t} \partial^{\alpha}_{0,t} u(0).$$

□

An immediate consequence of Theorem 2 is a theorem that establishes the solvability of the problem (1), (4) for $k > 0$.

Theorem 3 *Let $k > 0$, $0 < \alpha \le 1$, $u_0 \in D(A)$ and operator A satisfy Condition 1. Then the function*

$$u(t) = P_{k,\alpha} Y_{0,\alpha}(t) u_0 = c_{k,\alpha} \int_0^1 \left(1 - s^{\alpha+1}\right)^{k/(\alpha+1)-1} Y_{0,\alpha}(ts) u_0\, ds \tag{11}$$

is a solution to the problem (1), (4).

In what follows, for $k > 0$, $0 < \alpha \le 1$ we will use the notation

$$Y_{k,\alpha}(t) = P_{k,\alpha} Y_{0,\alpha}(t).$$

Example 1 If the operator A is bounded and $0 < \alpha \le 1$, then it is easy to verify directly that the function

$$Y_{0,\alpha}(t) u_0 = E_{\alpha+1,1}(t^{\alpha+1} A) u_0 = \sum_{j=0}^{\infty} \frac{t^{(\alpha+1)j} A^j u_0}{\Gamma((\alpha+1)j + 1)},$$

where $E_{\alpha,\beta}(\cdot)$ is the Mittag–Leffler function, is the solution to the problem

$$\frac{d}{dt} \partial^{\alpha} u(t) = A u(t), \quad u(0) = u_0 \in E, \;\; \partial^{\alpha} u(0) = 0.$$

By virtue of Theorem 3, the function

$$u(t) = Y_{k,\alpha}(t)u_0 = P_{k,\alpha}Y_{0,\alpha}(t)u_0 =$$

$$= \frac{\Gamma((k+1)/(\alpha+1))}{\Gamma(1/(\alpha+1))} \sum_{j=0}^{\infty} \frac{\Gamma(j+1/(\alpha+1))\, t^{(\alpha+1)j} A^j w_0}{\Gamma((\alpha+1)j+1)\, \Gamma(j+(k+1)/(\alpha+1))} =$$

$$= \frac{\Gamma((k+1)/(\alpha+1))}{\Gamma(1/(\alpha+1))}\, {}_2\Psi_2 \left[\begin{matrix} (1/(\alpha+1), 1)\,,\ (1, 1) \\ (1, \alpha+1)\,,\ ((k+1)/(\alpha+1), 1) \end{matrix} \middle| t^{\alpha} A \right] w_0, \quad (12)$$

where ${}_p\Psi_q(\cdot)$ is the Fox–Wright function (see [9, 10]) is the solution to the problem (1), (4).

Note that for $\alpha = 1$ the series in the formula (12) turns into the operator Bessel function (see [2, 5–7])

$$Y_{k,1}(t) = \Gamma(k/2+1/2) \sum_{j=0}^{\infty} \frac{\left(t\sqrt{A}/2\right)^{2j}}{j!\, \Gamma(j+k/2+1/2)} =$$

$$= \Gamma(k/2+1/2) \left(t\sqrt{A}/2\right)^{1/2-k/2} I_{k/2-1/2}\left(t\sqrt{A}\right),$$

where $I_\nu(\cdot)$ is the modified Bessel function.

Example 2 The operator function $Y_{0,\alpha}(t)$ satisfies the principle of subordination, which for the Eq. (1) with $k = 0$ was actually established in Chap. 3 of [11]. Let $0 \le \beta < \alpha \le 1$, then the following shift formula with respect to the second parameter is valid

$$Y_{0,\beta}(t) = \frac{1}{t^{(1+\beta)/(1+\alpha)}} \int_0^{\infty} \phi\left(-\frac{1+\beta}{1+\alpha}, \frac{\alpha-\beta}{1+\alpha}; -\frac{\tau}{t^{(1+\beta)/(1+\alpha)}}\right) Y_{0,\alpha}(\tau)\, d\tau,$$

in which the Wright function is used

$$\phi(\mu, \nu; z) = \sum_{n=0}^{\infty} \frac{z^n}{n!\, \Gamma(\mu n + \nu)}.$$

In particular, if the operator A is the generator of the operator cosine function $C(t; A)$, then for $\alpha = 1$ we obtain

$$Y_{0,\beta}(t) = \frac{1}{t^{(1+\beta)/2}} \int_0^{\infty} \phi\left(-\frac{1+\beta}{2}, \frac{1-\beta}{2}; -\frac{\tau}{t^{(1+\beta)/2}}\right) C(\tau; A)\, d\tau, \quad (13)$$

$$Y_{k,\beta}(t) = P_{k,\beta} Y_{0,\beta}(t).$$

In the limiting case, when $\beta = 0$, $\alpha = 1$, the equality (13) becomes the well-known semigroup connection formula $T(t; A)$ and cosine of the operator-function $C(t; A)$ generated by the operator A, which has the form

$$T(t; A) = \frac{1}{\sqrt{\pi t}} \int_0^\infty \exp\left(-\frac{\tau^2}{4t}\right) C(\tau; A)\, d\tau. \tag{14}$$

The operator function $Y_{k,\alpha}(t)$ also satisfies the shift formula with respect to the first parameter.

Theorem 4 *Let $m > k \geq 0$, $0 < \alpha \leq 1$ and operator A satisfy Condition 1. Then there is an equality*

$$Y_{m,\alpha}(t) = \frac{\alpha + 1}{B((m-k)/(\alpha+1), (k+1)/(\alpha+1))} \times$$

$$\times \int_0^1 s^k \,(1 - s^{\alpha+1})^{(m-k)/(\alpha+1)-1}\, Y_{k,\alpha}(ts)\, ds. \tag{15}$$

Proof After a series of obvious transformations, using the integral 2.2.5.1 [12], we obtain

$$\int_0^1 s^k \,(1 - s^{\alpha+1})^{(m-k)/(\alpha+1)-1}\, Y_{k,\alpha}(ts)\, ds =$$

$$= \int_0^t \tau^k \,(t^{\alpha+1} - \tau^{\alpha+1})^{(m-k)/(\alpha+1)-1}\, Y_{k,\alpha}(\tau)\, d\tau =$$

$$= c_{k,\alpha} \int_0^t \tau^\alpha \,(t^{\alpha+1} - \tau^{\alpha+1})^{(m-k)/(\alpha+1)-1} \times$$

$$\times \int_0^\tau \left(\tau^{\alpha+1} - \xi^{\alpha+1}\right)^{k/(\alpha+1)-1} Y_{0,\alpha}(\xi)\, d\xi d\tau = c_{k,\alpha} \times$$

$$\times \int_0^t Y_{0,\alpha}(\xi) \int_\xi^t \tau^\alpha \,(t^{\alpha+1} - \tau^{\alpha+1})^{(m-k)/(\alpha+1)-1} \left(\tau^{\alpha+1} - \xi^{\alpha+1}\right)^{k/(\alpha+1)-1} d\tau d\xi =$$

$$= \frac{c_{k,\alpha}}{\alpha+1} \int\limits_0^t Y_{0,\alpha}(\xi) \int\limits_{\xi^{\alpha+1}}^{t^{\alpha+1}} (t^{\alpha+1} - \eta)^{(m-k)/(\alpha+1)-1} \left(\eta - \xi^{\alpha+1}\right)^{k/(\alpha+1)-1} \, d\eta d\xi =$$

$$= \frac{c_{k,\alpha}\, B((m-k)/(\alpha+1), k/(\alpha+1))}{\alpha+1} \int\limits_0^t (t^{\alpha+1} - \xi)^{m/(\alpha+1)-1} Y_{0,\alpha}(\xi) \, d\xi =$$

$$= \frac{\Gamma\left((m-k)/(\alpha+1)\right)\, \Gamma\left((k+1)/(\alpha+1)\right)}{(\alpha+1)\, \Gamma\left((m+1)/(\alpha+1)\right)} \, t^{m-\alpha} Y_{m,\alpha}(t).$$

Consequently,

$$Y_{m,\alpha}(t) = \frac{(\alpha+1)\Gamma\left((m+1)/(\alpha+1)\right) t^{\alpha-m}}{\Gamma\left((k+1)/(\alpha+1)\right) \Gamma\left((m-k)/(\alpha+1)\right)} \times$$

$$\times \int\limits_0^t \tau^k \left(t^{\alpha+1} - \tau^{\alpha+1}\right)^{(m-k)/(\alpha+1)-1} Y_{k,\alpha}(\tau) \, d\tau =$$

$$= \frac{\alpha+1}{B((m-k)/(\alpha+1), (k+1)/(\alpha+1))} \int\limits_0^1 s^k \left(1 - s^{\alpha+1}\right)^{(m-k)/(\alpha+1)-1} Y_{k,\alpha}(ts) \, ds,$$

and the required equality (15) is established. □

3 Generalized Functional-Differential Bessel–Struve Equation

Let us proceed to the study of the case of a nonzero second initial condition $\partial^{\alpha}_{0,t} u(0) \neq 0$ and we will study the following initial problem for the functional differential equation

$$\frac{d}{dt} \partial^{\alpha}_{0,t} u(t) + \frac{k}{t} \left(\partial^{\alpha}_{0,t} u(t) - \partial^{\alpha}_{0,t} u(0)\right) = Au(t), \quad t > 0, \tag{16}$$

$$u(0) = 0, \quad \partial^{\alpha}_{0,t} u(0) = u_1. \tag{17}$$

For $\alpha = 1$ the problem (16), (17) becomes the initial problem for the Bessel–Struve equation, which was previously investigated by the author in [13].

Let us first consider the case when the parameter $k = 0$ in the Eq. (16).

Theorem 5 *Let $k = 0$, $0 < \alpha \le 1$, $u_1 \in D(A)$ and the operator A satisfies Condition 1. Then the problem (16), (17) is uniquely solvable.*

Proof After applying to the Eq. (1) the integration operator $I^1_{0,t}$ and fractional differentiation $D^{1-\alpha}_{0,t}$ problem (16), (17) reduces to the following initial problem for the inhomogeneous equation

$$u'(t) = \frac{1}{\Gamma(\alpha)} \int_0^t (t-s)^{\alpha-1} Au(s)\, ds + \frac{t^{\alpha-1}}{\Gamma(\alpha)} u_1, \quad t \ge 0, \tag{18}$$

$$u(0) = 0. \tag{19}$$

Just like task (6), (7), task (18), (19) is is a special case of the problem investigated in [8] and is uniquely solvable. The resolving operator of the problem (18), (19) will be denoted by $L_{0,\alpha}(t)$, and $u(t) = L_{0,\alpha}(t)u_1$, and $L_{0,\alpha}(t)$ in [8] is set to the representation

$$L_{0,\alpha}(t) = I^{\alpha}_{0,t} Y_{0,\alpha}(t) = \frac{1}{\Gamma(\alpha)} \int_0^t (t-s)^{\alpha-1} Y_{0,\alpha}(s)\, ds. \tag{20}$$

□

An immediate consequence of Theorems 5 and 2 is the solvability of the problem (16), (17) for $k > 0$. For $0 < \alpha \le 1$ we introduce the following notation:

$$d_{k,\alpha} = \frac{k}{\alpha+1} B\left(\frac{k}{\alpha+1}, \frac{1}{\alpha+1}\right), \quad L_{k,\alpha}(t) = d_{k,\alpha} P_{k,\alpha} L_{0,\alpha}(t).$$

Theorem 6 *Let $k > 0$, $0 < \alpha \le 1$, $u_1 \in D(A)$ and operator A satisfy Condition 1. Then the function*

$$u(t) = L_{k,\alpha}(t)u_1 = d_{k,\alpha} P_{k,\alpha} L_{0,\alpha}(t) u_1 \tag{21}$$

is a solution to the problem (16), (17).

Example 3 If $0 < \alpha \le 1$ and A is a bounded operator, then

$$L_{k,\alpha}(t) = \Gamma\left(k/(\alpha+1)+1\right) \sum_{j=0}^{\infty} \frac{\Gamma(j+1)\, t^{(\alpha+1)j+\alpha} A^j}{\Gamma\left((\alpha+1)j+\alpha+1\right)\, \Gamma\left(j+k/(\alpha+1)+1\right)}. \tag{22}$$

For $\alpha = 1$, the series on the right-hand side (22) is expressed in terms of the Struve function

$$L_k(t) = \frac{\sqrt{\pi}}{2}\, \Gamma\,(k/2+1) \sum_{j=0}^{\infty} \frac{\left(t\sqrt{A}/2\right)^{2j}}{\Gamma\,(j+3/2)\,\Gamma\,(j+k/2+1)} =$$

$$= \frac{2^{k/2-1/2}\sqrt{\pi}\;\Gamma\,(k/2+1)}{A^{k/4+1/4}\, t^{k/2-1/2}}\, \mathbf{L}_{k/2-1/2}\left(t\sqrt{A}\right),$$

where $\mathbf{L}_\nu(\cdot)$ is the modified Struve function ([14], p. 655).

Example 4 If $0 < \beta < 1$ and the operator A is the generator of the operator cosine function $C(t; A)$, then

$$L_{k,\beta}(t) = d_{k,\beta} P_{k,\beta} I^{\beta}_{0,t} Y_{0,\beta}(t),$$

where the operator function $Y_{0,\beta}(t)$ is defined by the equality (13).

The operator function $L_{k,\alpha}(t)$ satisfies the shift formula with respect to the first parameter, whose proof is carried out in the same way as in Theorem 4.

Theorem 7 *Let $m > k \geq 0$, $0 < \alpha \leq 1$ and operator A satisfy Condition 1. Then*

$$L_{m,\alpha}(t) = \frac{\alpha+1}{B((m-k)/(\alpha+1), k/(\alpha+1)+1)} \times$$

$$\times \int_0^1 s^k \left(1 - s^{\alpha+1}\right)^{(mk)/(\alpha+1)-1} L_{k,\alpha}(ts)\, ds.$$

The constructed operator functions $Y_{k,\alpha}(t)$, $L_{k,\alpha}(t)$, as well as Theorems 3 and 6 allow us to establish the following statement about the solvability of the general initial problem for the Eq. (16).

Theorem 8 *Let $k \geq 0$, $0 < \alpha \leq 1$, $u_0, u_1 \in D(A)$ and the operator A satisfies Condition 1. Then the function $u(t) = Y_{k,\alpha}(t)u_0 + L_{k,\alpha}(t)u_1$ is a solution to the Eq. (16) satisfying the conditions*

$$u(0) = u_0, \quad \partial^{\alpha}_{0,t} u(0) = u_1. \tag{23}$$

Theorems 3, 6, 8 do not contain a statement about the uniqueness of the solution. To prove the uniqueness of the solution of these problems, we make an additional assumption. We assume that $A \in G_k$, i.e., with the operator A, the Cauchy problem (2), (3) is uniformly well-posed for the Euler–Poisson–Darboux equation, and the resolving operator of this problem, as indicated earlier, is denoted by $Y_{k,1}(t)$.

Theorem 9 *Let $k \geq 0$, $0 < \alpha \leq 1$ and operator $A \in G_k$. Then the solutions of problems (1), (4) and (16), (23) are unique.*

Proof Proof of the uniqueness of the solution to the problem (16), (23) we will lead from the contrary. If $u_1(t)$ and $u_2(t)$ are two solutions to the problem (16), (23), then consider a function of two variables $w(t,s) = f\,(Y_k(s)\,(u_1(t) - u_2(t)))$, where $f \in E^*$ (E^* is the dual space), $t, s \geq 0$. She, obviously satisfies the equation

$$B_{k,\alpha} w(t,s) = \frac{\partial^2 w(t,s)}{\partial s^2} + \frac{k}{s}\,\frac{\partial w(t,s)}{\partial s}, \quad t, s > 0 \tag{24}$$

and conditions

$$\lim_{t\to 0} w(t,s) = \lim_{t\to 0} \partial^{\alpha}_{0,t} w(t,s) = \lim_{s\to 0} \frac{\partial w(t,s)}{\partial s} = 0. \tag{25}$$

As was done in [15], we interpret $w(t,s)$ as a generalized function of moderate growth and on the variable s we apply the Fourier–Bessel transformation

$$\hat{w}(t,\lambda) = \int_0^\infty s^{2p+1}\, j_p(\lambda s)\, w(t,s)\, ds, \quad w(t,s) = \gamma_p \int_0^\infty \lambda^{2p+1}\, j_p(\lambda s)\, \hat{w}(t,\lambda)\, d\lambda,$$

$$p = \frac{1-k}{2}, \quad \gamma_p = \frac{1}{2^{2p}\, \Gamma^2(p+1)}, \quad j_p(s) = \frac{2^p\, \Gamma(p+1)}{s^p}\, J_p(s),$$

where $J_p(\cdot)$ is the Bessel function.

From (24), (25) for the image $\hat{w}(t,\lambda)$ we get the following problem

$$B_{k,\alpha} \hat{w}(t,\lambda) = -\lambda^2\, \hat{w}(t,\lambda), \quad t > 0, \tag{26}$$

$$\lim_{t\to 0} \hat{w}(t,\lambda) = \lim_{t\to 0} \partial^{\alpha}_{0,t} \hat{w}(t,\lambda) = 0. \tag{27}$$

By virtue of Examples 1 and 3, the general solution of the Eq. (26) has the form

$$\hat{w}(t,\lambda) = \frac{d_1(\lambda)\Gamma\,((k+1)/(\alpha+1))}{\Gamma\,(1/(\alpha+1))} \sum_{j=0}^{\infty} \frac{\Gamma\,(j+1/(\alpha+1))\, t^{(\alpha+1)j} (-\lambda^2)^j}{\Gamma\,((\alpha+1)j+1)\, \Gamma\,(j+(k+1)/(\alpha+1))} +$$

$$+ d_2(\lambda)\, \Gamma\,(k/(\alpha+1)+1) \sum_{j=0}^{\infty} \frac{\Gamma(j+1)\, t^{(\alpha+1)j+\alpha} (-\lambda^2)^j}{\Gamma\,((\alpha+1)j+\alpha+1)\ \Gamma\,(j+k/(\alpha+1)+1)},$$

and the initial conditions (27) imply the equalities $d_1(\lambda) = d_2(\lambda) = 0$. Hence $\hat{w}(t,\lambda) = w(t,s) = 0$ for any $s \geq 0$. Since the functional $f \in E^*$ is arbitrary, for $s = 0$ we obtain the equality $u_1(t) \equiv u_2(t)$, and the uniqueness of the solution of the considered problems is established. □

As an application of Theorem 8 consider the problem (16), (23) with the operator which is a fractional power of the operator A. Let A be the generator of a uniformly bounded cosine-operator function. Then one can define a positive fractional power of the operator $-A$ (see, for example, [16, p. 358])

$$-\left(-A\right)^{\gamma}x=\frac{\sin\gamma\pi}{\pi}\int\limits_{0}^{\infty}\lambda^{\gamma-1}\left(\lambda I-A\right)^{-1}Ax\,d\lambda, \tag{28}$$

where $\gamma\in(0,1)$, $x\in D(A)$.

Moreover, if $y\in E$, $\mu>0$, then the resolvent of the operator $A_{\gamma}=-(-A)^{\gamma}$ satisfies the representation

$$\left(\mu I-A_{\gamma}\right)^{-1}y=\frac{\sin\gamma\pi}{\pi}\int\limits_{0}^{\infty}\frac{\lambda^{\gamma}\left(\lambda I-A\right)^{-1}y\,d\lambda}{\mu^{2}-2\mu\lambda^{\gamma}\cos\gamma\pi+\lambda^{2\gamma}}. \tag{29}$$

Next, we establish the solvability of the initial problem (16), (23) with the operator A_{γ}, where the exponent is $\gamma=(\alpha+1)/2$.

Theorem 10 *Let $\gamma=(\alpha+1)/2$, $0<\alpha<1$, $u_0,u_1\in D(A)$, the operator A — generator of uniformly bounded cosine-operator function $C(t;A)$ and operator A_{γ} defined by (28). Then the solution of the initial problem*

$$\frac{d}{dt}\partial^{\alpha}_{0,t}u(t)+\frac{k}{t}\left(\partial^{\alpha}_{0,t}u(t)-\partial^{\alpha}_{0,t}u(0)\right)=A_{\gamma}u(t),\quad t>0, \tag{30}$$

$$u(0)=u_0,\quad \partial^{\alpha}_{0,t}u(0)=u_1. \tag{31}$$

is the function $u(t)=Y_{k,\alpha}(t;A_{\gamma})u_0+L_{k,\alpha}(t;A_{\gamma})u_1$, where

$$Y_{0,\alpha}(t;A_{\gamma})=\frac{\sin\gamma\pi}{\gamma\pi}\int\limits_{0}^{\infty}\frac{C\left(ts^{-1/(2\gamma)};A\right)\,ds}{s^{2}-2s\cos\gamma\pi+1}, \tag{32}$$

while the operator functions $Y_{k,\alpha}(t;A_{\gamma})$, $L_{0,\alpha}(t;A_{\gamma})$, $L_{k,\alpha}(t;A_{\gamma})$ are defined respectively by the formulas (11), (20), (21).

Proof The operator A is the generator of a uniformly bounded cosine operator function, and in order to to use Theorem 8, one should check the fulfillment of Condition 1 for the operator A_{γ}. In the In our case, this condition is that for Re $\mu>0$ the resolvent $\left(\mu^{\alpha+1}I-A_{\gamma}\right)^{-1}$ satisfied the inequality

$$\left\|\frac{d^{n}\left(\mu^{\alpha}\left(\mu^{\alpha+1}I-A_{\gamma}\right)^{-1}\right)}{d\mu^{n}}\right\|\le\frac{Mn!}{(\operatorname{Re}\mu)^{n+1}}. \tag{33}$$

Given the representation (29), after the change of variables, we get

$$\mu^\alpha \left(\mu^{\alpha+1} I - A_\gamma\right)^{-1} y = \frac{\mu \sin \gamma\pi}{\gamma\pi} \int\limits_0^\infty \frac{s^{1/\gamma} \left(\mu^2 s^{1/\gamma} I - A\right)^{-1} y \, ds}{s^2 - 2s \, \cos \gamma\pi + 1} =$$

$$= \frac{\sin \gamma\pi}{\gamma\pi} \int\limits_0^\infty \frac{s^{1/(2\gamma)} \xi \left(\xi^2 I - A\right)^{-1} y \, ds}{s^2 - 2s \, \cos \gamma\pi + 1},$$

where $\xi = \mu s^{1/(2\gamma)}$ and hence

$$\frac{d^n \left(\mu^\alpha \left(\mu^{\alpha+1} I - A_\gamma\right)^{-1} y\right)}{d\mu^n} =$$

$$= \frac{\sin \gamma\pi}{\gamma\pi} \int\limits_0^\infty \frac{s^{(1+n)/(2\gamma)}}{s^2 - 2s \, \cos \gamma\pi + 1} \frac{d^n}{d\xi^n} \left(\xi \left(\xi^2 I - A\right)^{-1} y\right) \, ds. \tag{34}$$

Since for the resolvent of the generator of a uniformly bounded cosine-operator function for $\operatorname{Re} \xi > 0$ there is an estimate

$$\left\| \frac{d^n}{d\xi^n} \left(\xi \left(\xi^2 I - A\right)^{-1} y\right) \right\| \le \frac{M_1 \, n!}{(\operatorname{Re} \xi)^{n+1}}, \tag{35}$$

then (34), (35) implies the validity of the inequality

$$\left\| \frac{d^n}{d\mu^n} \left(\mu^\alpha \left(\mu^{\alpha+1} I - A_\gamma\right)^{-1}\right) \right\| \le \frac{M_1 \, n!}{(\operatorname{Re} \mu)^{n+1}} \int\limits_0^\infty \frac{ds}{s^2 - 2s \, \cos \gamma\pi + 1} \le \frac{M n!}{(\operatorname{Re} \mu)^{n+1}},$$

and thus the inequality (33) is proved, and with it the solvability of the problem (30), (31).

It remains for us to obtain the representation (32) for the operator function $Y_{0,\alpha}(t; A_\gamma)$. Using (8), (29), for $u_0 \in D(A^2)$ we get

$$Y_{0,\alpha}(t; A_\gamma) u_0 = \frac{1}{2\pi i} \int\limits_{\sigma - i\infty}^{\sigma + i\infty} e^{\lambda t} \, \lambda^\alpha \left(\lambda^{\alpha+1} I - A_\gamma\right)^{-1} u_0 \, d\lambda =$$

$$= \frac{\sin \gamma\pi}{\gamma\pi} \int\limits_0^\infty \frac{s^{1/\gamma}}{s^2 - 2s \, \cos \gamma\pi + 1} \frac{1}{2\pi i} \int\limits_{\sigma - i\infty}^{\sigma + i\infty} \lambda e^{\lambda t} \left(\lambda^2 s^{1/\gamma} I - A\right)^{-1} u_0 \, d\lambda ds =$$

$$= \frac{\sin \gamma\pi}{\gamma\pi} \int\limits_0^\infty \frac{C\left(t s^{-1/(2\gamma)}; A\right) u_0 \, ds}{s^2 - 2s \cos \gamma\pi + 1}.$$

The representation established on the dense set $D(A^2) \subset E$ (32) for the operator function $Y_{0,\alpha}(t; A_\gamma)$ extends by continuity to all E.

Operator functions $Y_{k,\alpha}(t; A_\gamma)$, $L_{0,\alpha}(t; A_\gamma)$, $L_{k,\alpha}(t; A_\gamma)$ are defined respectively by the formulas (11), (20), (21). In particular,

$$L_{0,\alpha}(t; A_\gamma) = I^{\alpha}_{0,t} Y_{0,\alpha}(t; A_\gamma) = \frac{1}{\Gamma(\alpha)} \int_0^t (t-s)^{\alpha-1} Y_{0,\alpha}(s; A_\gamma)\, ds =$$

$$= \frac{\sin\gamma\pi}{\gamma\pi\Gamma(\alpha)} \int_0^t (t-s)^{\alpha-1} \int_0^\infty \frac{C\left(s\eta^{-1/(2\gamma)}; A\right) d\eta}{\eta^2 - 2\eta\cos\gamma\pi + 1} ds =$$

$$= \frac{2\gamma\sin\gamma\pi}{\gamma\pi\Gamma(\alpha)} \int_0^t (t-s)^{\alpha-1} \int_0^\infty \frac{s^{2\gamma}\xi^{2\gamma-1} C(\xi; A)\, d\xi}{s^{4\gamma} - 2(s\xi)^{2\gamma}\cos\gamma\pi + \xi^{4\gamma}} ds =$$

$$= \frac{2\gamma\ \sin\gamma\pi}{\gamma\pi\Gamma(\alpha)} \int_0^\infty \xi^{2\gamma-1} C(\xi; A) \int_0^t \frac{s^{2\gamma}(t-s)^{\alpha-1}\, ds}{s^{4\gamma} - 2(s\xi)^{2\gamma}\cos\gamma\pi + \xi^{4\gamma}}\, d\xi.$$

□

4 Appendix

If A is the generator of an exponentially bounded β times integrated cosine operator of the function $C_\beta(t; A)$, then for

$$0 < \alpha < 1,\ \ \beta \le \frac{1-\alpha}{1+\alpha},\ \ \ \gamma = \frac{(\beta-1)(1+\alpha)}{2} + \alpha \le 0$$

performance for $Y_{0,\alpha}(t; A)$ in Theorem 1 can be written as

$$Y_{0,\alpha}(t; A) = \frac{1}{2\pi i} \int_{\sigma-i\infty}^{\sigma+i\infty} e^{\lambda t}\, \lambda^\alpha \left(\lambda^{\alpha+1} I - A\right)^{-1} d\lambda =$$

$$= \frac{1}{2\pi i} \int_{\sigma-i\infty}^{\sigma+i\infty} e^{\lambda t} \lambda^\gamma \int_0^\infty e^{-\tau\lambda^{(\alpha+1)/2}} C_\beta(\tau; A) d\tau d\lambda =$$

$$= \int_0^\infty C_\beta(\tau; A) \frac{1}{2\pi i} \int_{\sigma-i\infty}^{\sigma+i\infty} \lambda^\gamma e^{\lambda t - \tau\lambda^{(\alpha+1)/2}} d\lambda d\tau = \int_0^\infty C_\beta(\tau; A) I_{0,t}^{-\gamma} f_{\tau,(\alpha+1)/2}(t)\, d\tau, \tag{36}$$

in doing so, we used the introduced in [16, p. 357] function

$$f_{\tau,\gamma}(t) = \begin{cases} \frac{1}{2\pi i} \int\limits_{\sigma-i\infty}^{\sigma+i\infty} \exp\left(tz - \tau z^\gamma\right)\, dz, & t \geq 0, \\ 0, \quad t < 0, \end{cases}$$

where $\sigma > 0,\ \tau > 0,\ 0 < \gamma < 1$.

The function $f_{\tau,\gamma}(t)$ for $t > 0$ is expressed in terms of a Wright-type function ([17, Chap. 1]) $f_{\tau,\gamma}(t) = t^{-1} e_{1,\gamma}^{1,0}\left(-\tau t^{-\gamma}\right)$, where is the function

$$e_{\alpha,\beta}^{\mu,\delta}(z) = \sum_{k=0}^\infty \frac{z^k}{\Gamma(\alpha k + \mu)\, \Gamma(\delta - \beta k)}, \quad \alpha > \max\{0; \beta\}, \quad \mu, z \in C$$

satisfies the assessment

$$e_{1,\beta}^{1,\delta}(-\tau) \leq M_n(\tau) \exp\left(-(1-\beta)\beta^{\beta/(1-\beta)}\tau^{1/(1-\beta)}\right), \tag{37}$$

$$M_n(\tau) = \sum_{m=0}^n \frac{(\beta\tau)^m}{\Gamma(\delta + m(1-\beta))},$$

and the number n is chosen from the condition $\delta + n(1-\beta) \geq 1$.

In the equality (36), the fractional integral $I_{0,t}^{-\gamma} f_{\tau,(\alpha+1)/2}(t)$ is calculated (see formula (1.2.12) in [17]) and we arrive at the equality

$$Y_{0,\alpha}(t; A) = \frac{1}{t^{\gamma+1}} \int_0^\infty C_\beta(\tau; A) e_{1,(\alpha+1)/2}^{1,-\gamma}\left(-\tau t^{-(\alpha+1)/2}\right)\, d\tau. \tag{38}$$

Note that the convergence of the integral in the representation (38) is ensured by the estimate (37).

In the limiting case, when $\alpha = 0$, $\beta = 0$, $\gamma = -1/2$, the formula (36) becomes (14). Indeed, in this particular case we have (see [16, p. 369, formula (32)])

$$f_{\tau,1/2}(t) = \frac{\tau}{2t\sqrt{\pi t}} \exp\left(-\frac{\tau^2}{4t}\right)$$

and, taking into account the integral 2.3.4.1 [12]), we obtain

$$Y_{0,0}(t;A)=\int\limits_0^\infty C(\tau;A)I_{0,t}^{1/2}f_{\tau,(1/2}(t)d\tau=$$

$$=\int\limits_0^\infty C(\tau;A)I_{0,t}^{1/2}\left(\frac{\tau}{2t\sqrt{\pi t}}\exp\left(-\frac{\tau^2}{4t}\right)\right)d\tau=$$

$$=\frac{1}{2\pi}\int\limits_0^\infty \tau C(\tau;A)\int\limits_0^t \frac{1}{s\sqrt{s(t-s)}}\exp\left(-\frac{\tau^2}{4s}\right)dsd\tau=$$

$$=\frac{1}{2\pi\sqrt{t}}\int\limits_0^\infty \tau C(\tau;A)\int\limits_{1/t}^\infty \left(\xi-\frac{1}{t}\right)^{-1/2}\exp\left(-\frac{\xi\tau^2}{4}\right)d\xi d\tau=$$

$$=\frac{1}{\sqrt{\pi t}}\int\limits_0^\infty \exp\left(-\frac{\tau^2}{4t}\right)C(\tau;A)d\tau,$$

which coincides with the representation (14), while, naturally, one should assume that $Y_{0,0}(t;A)=T(t;A)$, $Y_{0,1}(t;A)=C_0(t;A)=C(t;A)$.

References

1. Tersenov SA (1973) Introduction to the theory of equations degenerating on the boundary. Novosibirsk State University, Novosibirsk [in Russian]
2. Glushak AV (2020) Transmutation operators as a solvability concept of abstract singular equations. Trends in mathematics. Transmutation operators and applications. Springer, Birkhauser, pp 379–410
3. Samko SG, Kilbas AA, Marichev OI (1993) Fractional integrals and derivatives: theory and applications. CRC Press, Boca Raton
4. Kilbas AA, Srivastava HM, Trujillo JJ (2006) Theory and applications of fractional differential equations. North Holland mathematics studies, vol 204. Elsevier
5. Glushak AV (1997) The Bessel operator function. Dokl Math 55(1):103–105
6. Glushak AV, Pokruchin OA (2016) Criterion for the solvability of the Cauchy problem for an abstract Euler-Poisson-Darboux equation. Differ Eqs 52(1):39–57
7. Glushak AV (2020) A family of Bessel operator functions. Geometry and mechanics, Itogi Nauki i Tekhniki Ser Modern Mat Its Appl Topic Obz, vol 187, VINITI RAN, Moscow, pp 36–43 [in Russian]
8. Da Prato G, Iannelli M (1980) Linear integro-differential equations in banach spaces. Rend Sem Mat Univ Padova 62:207–219
9. Fox C (1961) The *G* and *H* functions as symmetrical Fourier kernels. Trans Am Math Soc 98:395–429
10. Wright EM (1935) The asymptotic expansion of the generalized hypergeometric function. Lond Math Soc 10:286–293

11. Bajlekova E (2001) Fractional evolution equations in banach spaces. PhD thesis. Eindhoven University of Technology, Eindhoven
12. Prudnikov AP, Brychkov YA, Marichev OI (1981) Integraly i ryady. Elementarnye funktsii (Integrals and series. Elementary functions). Moscow, Nauka [in Russian]
13. Glushak AV (2017) Abstract Cauchy problem for the Bessel-Struve equation. Differ Eqs 53(7):864–878
14. Prudnikov AP, Brychkov YA, Marichev OI (1986) Integraly i ryady. Dopolnitel'nye glavy (Integrals and series: additional chapters). Moscow, Nauka [in Russian]
15. Zhitomirskii YI (1955) Cauchy's problem for systems of linear partial differential equations with differential operators of Bessel type. Mat Sb 36(2):299–310
16. Yosida K (1967) Funktsional'nyy analiz (Functional analysis) (Moscow, Mir) [in Russian]
17. Pskhu AV (2005) Equations in partial derivatives of fractional order. Nauka, Moscow [in Russian]

Local Bifurcations of Periodic Traveling Waves in the Generalized Weakly Dissipative Ginzburg-Landau Equation

Anatoly Kulikov and Dmitry Kulikov

Abstract In this paper we consider a periodic boundary value problem for the generalized Ginzburg-Landau. The generalized version of the weakly dissipative Ginzburg-Landau equation differs from the traditional version by replacing the cubic nonlinearity with nonlinearity of arbitrary odd degree. We will show that the periodic boundary value problem has a countable set of solutions that are single-mode and periodic in the evolutionary variable. We will examine the stability question as well as local bifurcations of such solutions when they change stability. In this case, the two-dimensional attracting invariant tori bifurcate emerges when stability is lost from single-mode solutions. These are non-resonant tori that have appeared in the generic situation. The main results are obtained on the basis and development of methods of the theory of dynamical systems with an infinite-dimensional phase space. These include the method of invariant manifolds and normal forms, as well as the principle of self-similarity. This principle allows us to reduce the problem of bifurcations of a countable set of single-mode solutions to the analysis of the corresponding problem.

Keywords Ginzburg-Landau equation · Periodic boundary conditions · Stability · Bifurcations · Normal forms · Invariant tori

1 Introduction

One of the most famous nonlinear evolutionary equations of mathematical physics can be considered as the corresponding partial differential equation

$$u_t = gu - (d + ic)u|u|^2 + (a + ib)u_{xx}, \tag{1}$$

A. Kulikov · D. Kulikov (✉)
P.G. Demidov Yaroslavl State University, Yaroslavl 150003, Russia
e-mail: kulikov_d_a@mail.ru

A. Kulikov
e-mail: anat_kulikov@mail.ru

V. Vasilyev (ed.), *Differential Equations, Mathematical Modeling and Computational Algorithms*, Springer Proceedings in Mathematics & Statistics 423,
https://doi.org/10.1007/978-3-031-28505-9_3

where $u = u(t, x) = u_1(t, x) + iu_2(t, x), a, b, c, d, g \in R, d > 0, a \geq 0, g > 0$. Note that Eq. (1) is the referred to as the Ginzburg-Landau evolutional complex equation. It appears in several branches of physics as well as chemical kinetics as a mathematical model [1–3]. It is studied together with the periodic boundary conditions [1]

$$u(t, x + 2l) = u(t, x).$$

For chemical kinetics problems, the corresponding boundary conditions of "impenetrability" (homogeneous Neumann boundary conditions) are used

$$u_x(t, 0) = u_x(t, l) = 0.$$

We normalize the variables t, x and the functions $u(t, x)$ as follows:

$$t \to \gamma_1 t,\ x \to \gamma_2 x,\ u \to \gamma_3 u$$

and assume that $l = \pi, d = 1, g = 1$, if these constants are positive. We will study special cases, generalizations and modifications of Eq. (1) and its variations. For instance, if $c = b = 0$, then Eq. (1) is called the variational Ginzburg-Landau equation [1, 4–6]. A variational version of the Ginzburg-Landau equation is found in a section of modern physics as the theory of condensed matter and requires special investigation. Note that if $a = 0$ then we obtain the "weakly dissipative Ginzburg-Landau equation" [7–12]. For this version of the Ginzburg-Landau equation, we also apply the generalized cubic Schrodinger equation [11]. Next observe that if $g = d = a = 0$, then the original version of the Ginzburg-Landau equation is transformed into one of the variations of the nonlinear Schrodinger equation. Analogous to the nonlinear Schrodinger equation, the Ginzburg-Landau equation also occurs in nonlinear optics [8], as well as in some sections of hydrodynamics [2]. In monograph [13] the hypothesis is given that when replacing Eq. (1) with the following

$$u_t = gu - (d + ic)u|u|^4 + (a + ib)u_{xx},$$

according to its authors, a significant change in the dynamics of solutions is possible. In particular, the hard oscillations are possible. In other words the subcritical bifurcations are realized. In this paper, we will consider the generalized weakly dissipative Ginzburg-Landau equation, which includes both variants of the Ginzburg-Landau equation from the introduction.

2 Formulation of the Problem

Our aim is to examine the following boundary value problem

$$u_t = u - (1 + ic)u|u|^{2p} - ibu_{xx}, \tag{2}$$

$$u(t, x + 2\pi) = u(t, x), \tag{3}$$

where $u = u(t, x) = u_1(t, x) + iu_2(t, x), c \in R, b > 0, p \in N$. Note that if $p = 2$ we obtain one of the versions of the equation in monograph [13]. If $p = 1$, we then obtain the initial version of the weakly dissipative Ginzburg-Landau equation.

Next, if we consider the following initial condition for the boundary value problem (2), (3)

$$u(0, x) = f(x), \tag{4}$$

where $f(x) \in H_2$, then via the results from [14, 15] that the initial-boundary value problem (2), (3), (4) is locally well-solvable. Also recall that the inclusion $f(x) \in H_2$ resembles the following characteristics:

(1) $f(x)$ has period 2π;

(2) $f(x)$ has generalized derivatives up to the inclusive second order derivatives $f(x), f'(x), f''(x) \in L_2(-\pi, \pi)$.

This space H_2 is the phase space of solutions to the initial-boundary value problem (2), (3), (4). The nonlinear boundary value problem (2), (3) has a countably family single-mode solutions in the space variable x and periodic in t

$$u = u_n(t, x) = \eta_n \exp(inx + i\sigma_n t), \tag{5}$$

where $n \in Z$ (Z is the set of integer), $|\eta_n| = 1, \sigma_n = bn^2 - c$. Indeed, substitution of the right side of equality (5) into Eq. (2) after elementary simplifications leads to a complex equation for determining η_n, σ_n

$$i\sigma_n = 1 - (1 + ic)|\eta_n|^{2p} + ibn^2.$$

Next notice that along with solution (5), the boundary value problem (2), (3) also has solutions in the corresponding form

$$u_n(t, x, h) = \exp(ih) \exp(inx + i\sigma_n t), \ h \in R.$$

The solutions $u_n(t, x, h)$ form a one-dimensional invariant subspace in the phase space of solutions to the boundary value problem (2), (3). Since h is arbitrary, in further constructions we can assume that $\eta_n = 1$. Replacing an unknown function

$$u(t, x) = \exp(i\omega_n t + inx)v(t, y), \tag{6}$$

where $\omega_n = bn^2$, $y = x + 2bnt$, $n = 0, \pm 1, \pm 2, \ldots$ leads us to the following equation for $v(t, y)$

$$v_t = v - (1 + ic)v|v|^{2p} - ibv_{yy},$$

which should be considered with the corresponding periodic boundary conditions

$$v(t, y + 2\pi) = v(t, y).$$

Notice that substitution (6) transforms the solutions of boundary value problem (2), (3) into solutions of the same boundary value problem. Therefore, the study of the neighborhood of each of the family of solutions (5) can be substituted by a similar problem for one of them: $u_0(t, x) = \exp(i\sigma_0 t)$, where $\sigma_0 = -c$. In physics, the solution $u_0(t, x) = u_0(t)$ is often called a spatially homogeneous cycle (or "thermodynamic" branch, Andronov-Hopf cycle). The remaining solutions of family (5) for $n \neq 0$ are periodic traveling waves and periodically depend on t and x.

3 Stability Analysis of Periodic Traveling Waves

As previously noted, the stability analysis of solutions of $u_n(t, x)$ (stability of one-dimensional manifolds $V_1(u)$) can be reduced by virtue of the principle of self-similarity [9] to the analysis of similar questions for a spatially homogeneous periodic solution of $u_0(t) = \exp(i\sigma_0 t)$, where $\sigma_0 = -c$. In turn, to analyze the stability of the solution $u_0(t)$ by setting

$$u(t, x) = u_0(t)(1 + w(t, x)). \tag{7}$$

For the deviation $w(t, x)$ we obtain the following nonlinear boundary value problem

$$w_t = A(p)w - (1 + ic)F(w, p), \tag{8}$$

$$w(t, x + 2\pi) = w(t, x), \tag{9}$$

where $F(w, p) = F_2(w, p) + F_3(w, p) + F_0(w, p)$ is a two-variable polynomial of degree $2p + 1$. For further constructions, we consider the following terms

$$F_2(w, p) = \frac{1}{2}p\Big((p + 1)w^2 + 2(p + 1)w\overline{w} + (p - 1)\overline{w}^2\Big),$$
$$F_3(w, p) = \frac{1}{6}p\Big((p^2 - 1)w^3 + 3(p^2 + p)w^2\overline{w} + 3(p^2 - 1)w\overline{w}^2 + (p^2 - 3p + 2)\overline{w}^3\Big).$$

$F_0(w, p)$ denotes the terms at zero that have an order of smallness in the variables $w, \overline{w}$ higher than the third. This leads us to $A(p)w = -p(1 + ic)(w + \overline{w}) - ibw_{xx}$.

Next, we will reformulate the linear differential operator A in real form. Instead of the complex-valued function $w = w_1 + iw_2$, we form the vector function $v = colon(w_1, w_2)$. In this case, we rewrite the linear differential operator A as follows

$$Av = \begin{pmatrix} -2p & b\partial^2 \\ -2cp - b\partial^2 & 0 \end{pmatrix} \begin{pmatrix} w_1 \\ w_2 \end{pmatrix}.$$

where we apply the short notation $\partial^2 f = \dfrac{\partial^2 f}{\partial x^2}$. This leads us to the functions in the corresponding form

$$v_k(x) = \exp(ikx) \begin{pmatrix} h_{1k} \\ h_{2k} \end{pmatrix},$$

where $h_{1,k}, h_{2,k}$ are real or complex constants, $k = 0, \pm 1, \pm 2, \ldots$ In this case, the problem of determining the eigenvalues and eigenelements of the linear differential operator A reduces to analyzing the spectrum of the following countable family of matrices

$$A_k = \begin{pmatrix} -2p & -bk^2 \\ bk^2 - 2cp & 0 \end{pmatrix}$$

and to determining of the roots of the family of characteristic equations

$$\lambda^2 + 2p\lambda + q_k = 0,$$

where $k = 0, \pm 1, \pm 2, \ldots, q_k = bk^2(bk^2 - 2cp), p \in Z$. For $k = 0$ we obtain $\lambda_{1,0} = 0, \lambda_{2,0} = -2p < 0$, i.e. for all values of the parameters p and b the linear differential operator A has a zero eigenvalue corresponding to the eigenfunction $v_0(x) = colon(0, 1)$ or $H_0(x) = i$ in the complex record form.

Let now $k \neq 0$. Note that $\lim\limits_{|k| \to \infty} q_k = \infty$. Consequently, the inequalities $q_k > 0$ for all $k \neq 0$ lead to the following inequality

$$Re\lambda_{k,j} \leq -\gamma_0 < 0 \tag{10}$$

for all values of k and $j = 1, 2$. Thus we obtain, $q_k > 0$ for all $k \in Z \backslash \{0\}$, if $b > 2pc$ ($b > 0$ by condition). Otherwise, when $b < 2cp$, the linear differential operator A has at least one of its eigenvalues in the right half-plane of the complex plane (one of the numbers $\lambda_{k,1}$ or $\lambda_{k,2}$ are positive). Finally, for $b = 2pc$ the linear operator A has a triple zero eigenvalue, which in the complex notation corresponds to the following eigenfunctions

$$H_0(x) = i, \; H_1(x) = (-c + i)\cos x, \; H_2(x) = (-c + i)\sin x.$$

The linear differential operator A corresponding to this choice of parameters will be denoted by $A_0 : A_0 w = -p(1+ic)(w+\overline{w}) - 2icpw_{xx}$. Thus, in addition to the zero-equilibrium state, the nonlinear boundary value problem (8), (9) has the following one-parameter family of equilibrium states

$$w_*(t,x) = w_*(h) = \exp(ih) - 1,$$

which is easy to verify indirectly. Next by substituting $w_*(h)$ into formula (7) leads to the following equality

$$u_*(t,x,h) = \exp(i\sigma_0 t)\exp(ih),$$

which is a spatially homogeneous solution of the boundary value problem (2), (3) for all $h \in R$.

Notice that the solutions $w_*(h)$ (family) of equilibrium states form a one-dimensional invariant manifold $M_1(h)$, which exists for all values of the parameters of the boundary value problem (8), (9) and for $h = 0$ we have $w = 0$. Therefore, this one-dimensional invariant manifold is a center manifold in a neighborhood of the zero equilibrium state [16, 17], at least for small $|h|$. This remark and theorems on behavior solutions outside the center manifolds are the base to the assertion.

Theorem 1 *(1) Suppose that $b > 2cp$, then $M_1(h)$ be a local attractor for solutions to the boundary value problem (8), (9). In particular, all equilibria for small $|h|$, including the zero-equilibrium state, are stable but not asymptotically stable. (2) Suppose that $b < 2cp$. Then all the equilibrium states forming $M_1(h)$, including the zero-equilibrium state are unstable (saddle points).*

Remark 1 If $b = 2cp$, then the invariant manifold $M_1(h)$ exists and is formed by a one-parameter family $(w_*(h) = \exp(ih) - 1)$ of equilibrium states. In particular, the equilibrium state $w = 0$ for which the critical case of a threefold zero eigenvalue emerges, belongs to $M_1(h)$. Hence, in this case, an additional analysis of the question of stability of zero state of equilibrium is required. This is due to the fact that the stability theorem with respect to linear approximation cannot be used even in the case of ordinary differential equations.

Corollary 1 *From the previous constructions, when transitioning from the boundary value problem (2), (3) to the auxiliary boundary value problem (8), (9), substitution (7) and from the self-similarity principle, we obtain the following features:*

(1) for $b > 2pc$ all traveling waves $u_n(t,x,h) = \exp(i(nx + \sigma_n t + h))$, where $n = \pm 1, \pm 2, \ldots, h \in R$ and spatially homogeneous solutions $u_0(t,h) = \exp(i(\sigma_0(t+h)))$ are stable;

(2) for $b < 2cp$ they are all unstable;

(3) for $b = 2pc$ $(c = b/(2p))$ the critical case stability problem of solutions $u_n(t,x,h)$ are realized.

Further in the next section, the boundary value problems (2), (3) and (8), (9) will be considered in cases where the threefold zero eigenvalue of the operator A is close to critical. This means that the boundary value problems studied below will be considered if

$$b = 2pc(1 + \gamma\varepsilon), \tag{11}$$

where $\varepsilon \in (0, \varepsilon_0)$, $\gamma = \pm 1$ and 0. Appropriate values of γ will be chosen at the final stage of the analysis of the studied boundary value problems.

Before proceeding to the direct analysis of the bifurcation problem, we introduce some notation and also recall one fairly well-known statement from the theory of linear boundary value problems for ordinary differential equations, which we formulate in a form adapted to our case. Consider the differential operator

$$A(\varepsilon)y = A_0 y + \gamma\varepsilon A_1 y,\ \ A_0 y = -p(1 + ic)(y + \overline{y}) - 2pciy'',$$
$$A_1 y = -2pciy'',\ y(x) = y_1(x) + iy_2(x).$$

In this case $y(x)$ is a sufficiently smooth 2π periodic function.

Remark 2 We will consider the following linear nonhomogeneous boundary value problem

$$A_0 y = f(x),\ y(x + 2\pi) = y(x), \tag{12}$$

where the complex-valued function $f(x) \in L_2(-\pi, \pi)$ and has period 2π. The boundary value problem (12) has a solution if $f(x)$ satisfies the following two conditions:

$$(a)\ Re(a_0(c + i)) = 0,\ \text{where } a_0 = \frac{1}{\pi}\int_{-\pi}^{\pi} f(x)dx;$$

$$(b)\ Im a_1 = Im b_1 = 0,\ \text{where } a_1 = \frac{1}{\pi}\int_{-\pi}^{\pi} f(x)\cos x dx,\ b_1 = \frac{1}{\pi}\int_{-\pi}^{\pi} f(x)\sin x dx.$$

The solution of the boundary value problem (12) is unique, for which the following equalities hold:

$$(a)\ Re(y_0(c + i)) = 0,\ \text{where } y_0 = \frac{1}{\pi}\int_{-\pi}^{\pi} y(x)dx;$$

$$(b)\ Im y_1 = Im z_1 = 0,\ \text{where } y_1 = \frac{1}{\pi}\int_{-\pi}^{\pi} y(x)\cos x dx,\ z_1 = \frac{1}{\pi}\int_{-\pi}^{\pi} y(x)\sin x dx$$

The conditions for solvability when using the complex notation are given. They have a more familiar form. Next, we will consider the corresponding nonhomogeneous boundary value problem

$$A_0 y(x) = f(x),\ y(x+2\pi) = y(x),$$

where $y(x) = colon(y_1(x), y_2(x))$, $f(x) = colon(f_1(x), f_2(x))$. Now A_0 in real form is expressed as

$$A_0 = \begin{pmatrix} -2p & 2pc\partial^2 \\ -2pc - 2pc\partial^2 & 0 \end{pmatrix}.$$

It has a triple zero eigenvalue, which corresponds to three eigenfunctions

$$H_0(x) = \begin{pmatrix} 0 \\ 1 \end{pmatrix},\ H_1(x) = \begin{pmatrix} -c \\ 1 \end{pmatrix} \cos x,\ H_2(x) = \begin{pmatrix} -c \\ 1 \end{pmatrix} \sin x.$$

Conjugate operator

$$A_0^* = \begin{pmatrix} -2p & -2pc - 2pc\partial^2 \\ 2pc\partial^2 & 0 \end{pmatrix}$$

is defined on sufficiently smooth 2π periodic vector functions $z(x) = colon(z_1(x), z_2(x))$. Naturally, it has a triple zero eigenvalue corresponding to the eigenfunctions

$$E_0(x) = \begin{pmatrix} -c \\ 1 \end{pmatrix},\ E_1(x) = \begin{pmatrix} 0 \\ 1 \end{pmatrix} \cos x,\ E_2(x) = \begin{pmatrix} 0 \\ 1 \end{pmatrix} \sin x,$$

The solvability conditions arise in the following form

$$< f, E_j >= 0,\ j = 0, 1, 2,$$

where $< f(x), q(x) >$ denotes the scalar product in the corresponding function space

$$< f(x), q(x) >= \int_{-\pi}^{\pi} (f(x), q(x)) dx,$$

where $q(x) = colon(q_1(x), q_2(x))$, and the brackets $(*, **)$ inside the integral denote the inner product in R^2 [18, 19].

The statements from Remark 2 are known as solvability conditions.

4 Turing—Prigogine Bifurcation

We will focus on the analysis of the nonlinear boundary value problem (2), (3) for the determined values by equality (11). Next, we will transition to a modified version of the boundary value problem (8), (9) with the following substitution

$$u(t,x) = u_0(t)\exp(i\varphi)(1 + w(t,x)), \tag{13}$$

where, as in substitution (7) $u_0(t) = \exp(i\sigma_0 t)$. As a result, now for $w(t,x)$ we obtain a boundary value problem similar to boundary value problem (8), (9)

$$w_t + i\varphi_t(1 + w) = A(\varepsilon)w - (1 + ic)(F_2(w) + F_3(w) + F_0(w)), \tag{14}$$

$$w(t, x + 2\pi) = w(t,x). \tag{15}$$

In this case, $\varphi = \varphi(t,\varepsilon)$ and $\varphi_t(t,0) = 0$, i.e. $\varphi(t,0) = h$ is an arbitrary real constant.

We indicate an essential feature of the boundary value problem (14), (15). We denote $H_{2,even}$ as the subspace of the function space H_2, containing only even functions $f(x)$. In this case, the specificity of the right side of Eq. (14) is such that this subspace is invariant for solutions of the boundary value problem (14), (15). If $w(0,x) \in H_{2,even}$, then its solution is $w(t,x)$ for all t, when it exists, belongs to $H_{2,even}$. In this case, the periodic boundary conditions (15) can be replaced by the homogeneous Neumann boundary conditions

$$w_x(t,0) = w_x(t,\pi) = 0, \tag{16}$$

assuming that $x \in [0,\pi]$. First we restrict ourselves to the analysis of the auxiliary boundary value problem (14), (16). With this choice of boundary conditions, the linear differential operator A_0 has a double zero eigenvalue, whose corresponding eigenfunctions are

$$H_0(x) = i,\ H_1(x) = (-c + i)\cos x.$$

Let us recall some well-known assertions. Denote by $\Omega(r)$ the ball of radius r centered at the zero of the phase space $H_{2,even}$. As is well known (see, for example, [17]), boundary value problem (14), (16) in a neighborhood of the equilibrium $w = 0$ has a smooth two-dimensional invariant manifold $M_2(\varepsilon) \in \Omega(r)$, where r is a sufficiently small positive constant. All solutions of the auxiliary boundary value problem (14), (16) from this neighborhood $\Omega(r)$ approach $M_2(\varepsilon)$ with the exponential rate over time. In this case, solutions to the boundary value problem (14), (16) that belong to $M_2(\varepsilon)$, can be sought in the following form [9, 10, 12]

$$w(t,x,\varepsilon) = \varepsilon^{1/2}Q_1(x,z) + \varepsilon Q_2(x,z) + \varepsilon^{3/2}Q_3(x,z) + \varepsilon Q_4(x,z,\varepsilon). \tag{17}$$

The functions $Q_j(x, z)$, $j = 1, 2, 3$, $Q_4(x, z, \varepsilon)$ reveal the following properties:

(1) they depend on their variables rather smoothly if $|z| < z_0$, $\varepsilon \in (0, \varepsilon_0)$ and, in addition, $Q_4(x, z, 0) = 0$ (z_0, ε_0 – some positive constants);

(2) as a function of x they belong to $W_2^2[0, \pi]$ (the corresponding Sobolev space is denoted by $W_2^2[0, \pi]$) and satisfy the boundary conditions (16).

Further, we assume that the functions $\varphi = \varphi(s, \varepsilon)$, $z = z(s)$ depend on the slow time $s = \varepsilon t$. They satisfy the corresponding system of two ordinary differential equations

$$\varphi_s = \Psi_0(z, \varepsilon),\ z_s = \Psi_1(z, \varepsilon), \tag{18}$$

where the right-hand sides smoothly depend on z, ε, if $|z| < z_0$ and $\varepsilon \in (0, \varepsilon_0)$. The system of differential equations (18) is called the normal form. It can be replaced with a shortened version [20]

$$\varphi_s = \Theta_0(z),\ z_s = \Theta_1(z), \tag{19}$$

where $\Theta_0(z) = \Psi_0(z, 0)$, $\Theta_1(z) = \Psi_1(z, 0)$. Such a variant of system (18), i.e. system (19) is called "truncated normal form". It is this kind of normal form that plays the main role in the analysis of local bifurcations. We substitute the sum (17) into the auxiliary boundary value problem (14), (16) and note that $z_t = z_s\varepsilon$, $\varphi_t = \varphi_s\varepsilon$.

As a result of such a substitution, we obtain a sequence of linear boundary value problems of the terms at equal powers $\varepsilon^{1/2}$. So for $\varepsilon^{1/2}$ we obtain a homogeneous boundary value problem for $Q_1 = Q_1(x, z)$ of the following form

$$A_0 Q_1 = 0,\ Q_{1x}(0, z) = Q_{1x}(\pi, z) = 0,$$

as solutions of which the function can be chosen

$$Q_1(x, z) = zH_1(x) = z(-c + i)\cos x.$$

Collecting the terms at $\varepsilon, \varepsilon^{3/2}$, we obtain two nonhomogeneous boundary value problems. Thus, to determine the function $Q_2 = Q_2(x, z)$, we obtain the following boundary value problem

$$A_0 Q_2 = (1 + ic)\Phi_2(x)z^2 + i\Theta_0(z), \tag{20}$$

$$Q_{2x}(0, z) = Q_{2x}(\pi, z) = 0, \tag{21}$$

where $\Phi_2(x)z^2 = F_2(H_1(x, z))$ and we procure

$$\Phi_2(x)z^2 = \frac{p}{2}z^2\left[(p + 1)(c - i)^2 + 2(p + 1)(c^2 + 1) + (p - 1)(c + i)^2\right]\cos^2 x.$$

These computations assumed that $\varphi_t = \varepsilon\varphi_s$ and therefore $\varphi_t = \varepsilon\Theta_0(z) + o(\varepsilon)$. The boundary value problem (20), (21) has a solution from the specified class of functions if $\Theta_0(z) = pc(c^2+1)z^2$. In this case, the corresponding solution (see solvability conditions)

$$Q_2(x,z) = v(x)z^2 = (v_0 + v_2\cos 2x)z^2,$$
$$v_0 = -\frac{1}{4}(1+ic)((2p+3)c^2+1),\ v_2 = \frac{(2p+1)c^2-1}{12} - i\frac{(4p+5)c^2+1}{24c}.$$

We obtain the corresponding linear nonhomogeneous boundary value problem by collecting the terms at $\varepsilon^{3/2}$, at the third step of the implementation of the algorithm:

$$A_0Q_3 = \Theta_1(z)H_1(x) - z\gamma A_1H_1(x) + i\Theta_0(z)zH_1(x) + \\ +(1+ic)(F_3(Q_1) + \Phi_3(Q_1,Q_2))z^3, \tag{22}$$

$$Q_{3x}(0,z) = Q_{3x}(\pi,z) = 0. \tag{23}$$

In the boundary value problem (22), (23) $Q_3 = Q_3(x,z)$,

$$F_3(Q_1) = F_3(H_1(x)) = \frac{p}{6}(1+ic)\Big((p^2-1)(-c+i)^3 + 3p(p+1)(-c+i)^2(-c-i) + \\ +3(p^2-1)(-c+i)(c+i)^2 + (p-1)(p-2)(-c-i)^3\Big)\cos^3 x,$$
$$\Phi_3(Q_1,Q_2) = p\Big((p+1)H_1(x)v(x) + \\ +(p+1)\Big(H_1(x)\overline{v}(x) + \overline{H}_1(x)v(x)\Big) + (p-1)\overline{H}_1(x)\overline{v}(x)\Big).$$

It follows from the solvability conditions for the nonhomogeneous boundary value problem (22), (23) that in this case one should choose

$$\Theta_1(z) = \nu_p z - l_p z^3,$$

where $\nu_p = -2\gamma c^2 p$, $l_p = \frac{p}{6}((4p^2+22p+4)c^4 + (2p-11)c^2 + 1)$.

Thus, the analysis of the boundary value problem (14), (16) has been reduced to the study of a system of ordinary differential equations (the "shortened" or "truncated" normal form). In our case, it is presented in the following form

$$\varphi_s = pc(c^2+1)z^2, \tag{24}$$

$$z_s = \nu_p z - l_p z^3. \tag{25}$$

Lemma 1 *Differential equation (25), in addition to the zero equilibrium state* $S_0(z=0)$, *has nonzero equilibrium states*

$$S_+ : z_+ = \sqrt{\frac{\nu_p}{l_p}}, \quad S_- : z_- = -\sqrt{\frac{\nu_p}{l_p}},$$

if $\nu_p/l_p > 0$.

For $l_p > 0\,(\nu_p > 0)$, *the equilibrium states* S_+, S_- *are asymptotically stable and they are unstable if* $l_p < 0\,(\nu_p < 0)$. *In turn,* S_0 *is an asymptotically stable equilibrium state if* $\nu_p < 0$ *or* $\nu_p = 0, l_p > 0$.

The proof of Lemma 1 is fairly straight forward. In fact, even in the situation where $\nu_p \neq 0$, one should use the stability theorem in the first (linear) approximation. For $\nu_p = 0$ we get the equation $z_s = -l_p z^3$ and its solution $z(s) \to 0$ for $s \to \infty$, if $l_p > 0$ and $z(s)$ leaves the neighborhood of zero if $l_p < 0$. In our case $l_p > 0$ for any positive integer p and all $c \in R$, due to the discriminant of the square trinomial

$$l_p(\xi) = (4p^2 + 22p + 4)\xi^2 - (2p - 11)\xi + 1$$

is negative.

We choose γ such that $\nu_p > 0$ (for example, $\gamma = -1$). The equilibrium states S_+, S_- of the differential equation (25) correspond to the solutions

$$\varphi_+(s) = (pc(c^2+1)z_+^2)s + h_+, \quad \varphi_-(s) = (pc(c^2+1)z_-^2)s + h_-$$

of differential equation (24). Here $h_+, h_- \in R$ and are arbitrary. Transitioning to a more complete system (18) in this case gives us

$$\varphi_\pm(s) = (pc(c^2+1)z_\pm^2 + O(\varepsilon))s.$$

It follows from the results of [21, 22] and previous constructions that the assertion is true.

Lemma 2 *There exists a constant* $\varepsilon_p > 0$, *such that for all* $\varepsilon \in (0, \varepsilon_p)$ *there are two sets of functions*

$$\{\varphi_+(t,\varepsilon), w_+(x,\varepsilon); \varphi_-(t,\varepsilon), w_-(x,\varepsilon)\},$$

satisfying the nonhomogeneous boundary value problem (14), (16). For such functions, the following asymptotic formulas are valid

$$w_\pm(x,\varepsilon) = \varepsilon^{1/2} z_\pm(-c+i)\cos x + \varepsilon z_\pm^2(v_0 + v_2\cos 2x) + o(\varepsilon),$$
$$\varphi_\pm(t,\varepsilon) = (pc(c^2+1)z_\pm^2 + O(\varepsilon))\varepsilon t.$$

Also observe that these functions satisfy the boundary value problem (14), (15). Moreover, due to the translational invariance for the solutions of the boundary value problem (14), (15), it also has the following pairs of solutions

$$(w_+(x+h_+,\varepsilon), \varphi_+(t,\varepsilon)), \quad (w_-(x+h_-,\varepsilon), \varphi_-(t,\varepsilon)).$$

Next note that the boundary value problem (14), (16) is invariant under the change

$$x \to \pi - x, \ z_+ \to z_-,$$

then for the boundary value problem (14), (15) there remains only one set

$$(w_+(x + h_+, \varepsilon), \varphi_+(t, \varepsilon)),$$

which includes all the corresponding solutions by choosing an appropriate shift h_+. All these constructions and remarks allow us to formulate the main result, which refers to the boundary value problem (2), (3).

Theorem 2 *There exists $\varepsilon_p > 0$, such that for all $\varepsilon \in (0, \varepsilon_p)$ the nonlinear boundary value problem (2), (3) for $b_p = 2cp(1-\varepsilon)$ $(\gamma = -1)$ has a two-parameter family of the periodic in t solutions $V_0(h_0, h)$*

$$u_0(t, x, \varepsilon) = \exp(i\sigma_0 t + i\varphi_+(t, \varepsilon) + ih_0)(1 + w_+(x + h, \varepsilon)),$$

where $\varphi_+(t, \varepsilon) = \Big(pc(c^2 + 1)z_+^2 + O(\varepsilon)\Big)\varepsilon t,\ z_+ = \sqrt{\nu_p / l_p},$

$$w_+(x + h, \varepsilon) = \varepsilon^{1/2} z_+(-c + i)\cos(x + h) + \varepsilon z_+^2(v_0 + v_2 \cos 2(x + h)) + o(\varepsilon),$$

where $h_0, h \in R$ and are arbitrary, the constants v_0, v_2 were specified earlier. The two-dimensional invariant manifold $V_0(h_0, h)$ is a local attractor.

The validity of the assertion follows from the principle of self-similarity from Eq. (6). The following assertion is corollary from Theorem 2.

Corollary 2 *Boundary value problem (2), (3) has a countable set of two-dimensional attracting invariant manifolds $V_n(h_0, h)$, generated by the following solutions*

$$\begin{aligned} u_n(t, x, \varepsilon) = \exp(i\sigma_n t + inx + i\varphi_+(t, \varepsilon) + ih_0)\times \\ \times\Big(1 + w_+(x + 4npc(1 - \varepsilon)t + h, \varepsilon)\Big), \end{aligned} \tag{26}$$

where $n = \pm 1, \pm 2, \ldots, \sigma_n = -c + 2pc(1 - \varepsilon)n^2$, and the functions $\varphi_+(t, \varepsilon)$ and $w_+(x, \varepsilon)$ were found earlier in the process of implementing the modified Krylov-Bogolyubov algorithm (see formula (17) and boundary value problems (20), (21) and (22), (23)) and using the principle of self-similarity.

From the asymptotic formulas and the method of constructing solutions $u_0(t, x, \varepsilon), u_n(t, x, \varepsilon)$, it follows that $V_n(h_0, h)$ for all $n \in Z$ are two-dimensional invariant tori. Moreover, the torus $V_0(h_0, h)$ is filled with solutions that are periodic in t, and the solutions that form $V_n(h_0, h)$ as $n \neq 0$ are almost periodic functions of the variable t with a non-resonant set of eigenfrequencies. We emphasize that the solutions that form these two-dimensional tori are stable but cannot be asymptotically stable as in the neighborhood of each of these solutions there is always one more representative of the corresponding family.

5 Conclusions

The aim of this work was to generalize the results of works [9, 10], where particular cases of the boundary value problem (2), (3) for $p = 1, 2$ were considered. In this work we were able to show the following characteristics. Qualitatively, the results for all p are fairly close. In all boundary value problems with different values of p there exists a countable set of traveling waves that are periodic in t. When they lose stability, two-dimensional invariant tori, which are attracting invariant manifolds, bifurcate from each of them. For $n = 0$ the torus $V_0(h_0, h)$ is filled with periodic solutions, and the tori $V_n(h_0, h)$ are non-resonant in the generic situation. Thus, the hypothesis that the replacement of the cubic nonlinearity by the fifth-degree nonlinearity leads to subcritical bifurcations in the vicinity of traveling waves turned out to be not completely consistent. In any case, it is of paramount interest to consider the weakly dissipative versions of Ginzburg-Landau equation.

However, for the basic and generalized versions of the weakly dissipative version of the Ginzburg-Landau equation, the dynamics can be quite complex. The periodic boundary value problem (2), (3), with an appropriate choice of the coefficients of the equation, can have a countable set of local attractors, each of which is a two-dimensional invariant torus. The torus with number n ($n = 0, \pm 1, \dots$) is formed by solutions (26) whose norm in the phase space (that is, in H_2) tends to infinity if $|n| \to \infty$. At the same time, the norm of all these solutions in the space $L_2(-\pi, \pi)$ is close to $\sqrt{2\pi}$.

This work was carried out within the framework of a development programme for the Regional Scientific and Educational Mathematical Center of the Yaroslavl State University with financial support from the Ministry of Science and Higher Education of the Russian Federation (agreement on provision of subsidy from the federal budget No. 075-02-2022-886).

References

1. Aranson IS, Kramer L (2002) The world of the complex Ginzburg-Landau equation. Rev Modern Phys 74:99–143. https://doi.org/10.1103/RevModPhys.74.99
2. Landau LD, Lifshitz EM (1986) Theoretical physics. Hydrodynamics, Nauka, Moscow (in Russian)
3. Kuramoto Y (1984) Chemical oscillations, waves, and turbulence. Springer, Berlin Heidelberg. https://doi.org/10.1007/978-3-642-69689-3

4. Kulikov AN, Rudy AS (2003) States of equilibrium of considered matter with Ginzburg-Landau ψ^4-model. Chaos, Solitons Fractals 15:75–85. https://doi.org/10.1016/S0960-0779(02)00091-7
5. Kulikov AN, Kulikov DA (2017) Equilibrium states of the variational Ginzburg-Landau equation. Vestnik of National Research Nuclear University MEPhI 6:496–502. https://doi.org/10.1134/S2304487X17060074 (in Russian)
6. Kulikov DA (2020) Generalized variant of the variational Ginzburg–Landau equation. Vestnik of National Research Nuclear University MEPhI 9:329–337. https://doi.org/10.1134/S2304487X20040045 (in Russian)
7. Deissler RJ (1987) Turbulent birth, spots and sludge in a generalized Ginzburg-Landau equation. Phys Lett A 120:334–340. https://doi.org/10.1016/0375-9601(87)90726-2
8. Witham GB (1974) Linear and nonlinear waves, and turbulence. Wiley, London
9. Kulikov AN, Kulikov DA (2010) Local bifurcations of plane running waves for the generalized cubic Schrodinger equation. Differ Eqs 46:1299–1308. https://doi.org/10.1134/S0012266110090065
10. Kulikov AN, Kulikov DA (2009) Postcritical and subcritical bifurcations of traveling waves of the modified Ginzburg-Landau equation. Vestnik Udmurtskogo Univ 4:71–78. https://doi.org/10.20537/vm090407 (in Russian)
11. Kolesov AY, Kulikov AN, Rozov NK (2006) Cylindrical traveling waves for the generalized cubical Schrodinger equation. Dokl Math 73:125–128. https://doi.org/10.1134/S1064562406010340
12. Kulikov AN, Kulikov DA (2017) Local bifurcations in the periodic boundary value problem for the generalized Kuramoto-Sivashinsky equation. Autom Remote Control 78:1955–1966. https://doi.org/10.1134/S0005117917110029
13. Malinetskiy GG, Potapov AB, Podlazov AV (2006) Nonlinear dynamics: approaches, results, hopes. Komkniga, Moscow (in Russian)
14. Segal I (1963) Nonlinear semigroups. Ann Math 78:339–364. https://doi.org/10.2307/1970347
15. Yakubov SY (1970) Solvability of the Cauchy problem for abstract second-order quasilinear hyperbolic equations and their applications. Tr MMO 23:37–60. (in Russian)
16. Kulikov AN (1976) On smooth invariant manifolds of the semigroup of nonlinear operators in a Banach space. Study Stab Theory Oscil 114–129. (in Russian)
17. Marsden J, McCracken M (1982) Hopf bifurcation and its applications. Springer, New York
18. Naimark MA (1959) Linear differential operators. Nauka, Moscow
19. Functional analysis (1972) Reference math library. Nauka, Moscow. (in Russian)
20. Guckenheimer J, Holmes P (1983) Nonlinear oscillations, dynamical systems, and bifurcations of vector fields. Springer Science+Business Media, New York. https://doi.org/10.1007/978-1-4612-1140-2
21. Kolesov AY, Kulikov AN, Rozov NK (2003) Invariant tori of a class of point mappings: the annulus principle. Differ Eqs 39:614–631. https://doi.org/10.1023/A:1026133701786
22. Kolesov AY, Kulikov AN, Rozov NK (2003) Invariant tori of a class of point transformations: preservation of an invariant torus under perturbations. Differ Eqs 39:775–790. https://doi.org/10.1023/B:DIEQ.0000008405.33370.ff

Towards Discrete Octonionic Analysis

Rolf Sören Kraußhar, Anastasiia Legatiuk, and Dmitrii Legatiuk

Abstract In recent years, there is a growing interest in the studying octonions, which are 8-dimensional hypercomplex numbers forming the biggest normed division algebras over the real numbers. In particular, various tools of the classical complex function theory have been extended to the octonionic setting in recent years. However not so many results related to a discrete octonionic analysis, which is relevant for various applications in quantum mechanics, have been presented so far. Therefore, in this paper, we present first ideas towards discrete octonionic analysis. In particular, we discuss several approaches to a discretisation of octonionic analysis and present several discrete octonionic Stokes' formulae.

Keywords Octonions · Discrete Clifford analysis · Discrete operators · Stokes' formula · Discrete octonions

1 Introduction

As very well-known, complex analysis provides a very powerful toolkit to study numerous boundary value problems arising in classical harmonic analysis in the two-dimensional case. Motivated by modern problems of engineering and physics, there has been a rapidly growing interest in developing higher-dimensional versions of complex function theory to extend the classical toolkit for a successful treatment of higher-dimensional problems. While engineering mainly focusses on three-dimensional settings, modern physics, for example particle physics, also require tools in the context of dimensions $n > 3$. Einstein's relativity theory already requires four dimensions, including time. Also the standard model of particle physics of electroweak action requires four dimensions. An enormous challenge in modern physics

R. Sören Kraußhar · D. Legatiuk (✉)
Chair of Mathematics, Universität Erfurt, Erfurt, Germany
e-mail: dmitrii.legatiuk@uni-erfurt.de

A. Legatiuk
Chair of Applied Mathematics, Bauhaus-Universität Weimar, Weimar, Germany

V. Vasilyev (ed.), *Differential Equations, Mathematical Modeling and Computational Algorithms*, Springer Proceedings in Mathematics & Statistics 423,
https://doi.org/10.1007/978-3-031-28505-9_4

however, is to understand how gravity can be incorporated on the level of particle physics. Studying of this problems leads to the consideration of even higher dimensional settings, such as string and super-string theory, where the latter requires 12 dimensions. More recent models of a generalised standard model give stronger indications to use an eight dimensional model, see for example works [4, 15, 22] which shall provide one motivation for this paper from the physical point of view.

From the mathematical point of view, there are several possibilities to extend complex numbers and complex analysis to higher dimensions. One approach is to work with associative Clifford algebras leading to several function theories that consider functions defined on open subsets of an arbitrary dimensional vector space $\mathbb{R}^{n+1}$ that take values in a 2^n-dimensional Clifford algebra $\mathbb{R}^n$. On this way, a higher-dimensional version of the Cauchy-Riemann operator $D := \sum_{i=0}^{n} \partial_{x_i} e_i$ factorising the $n+1$-dimensional Laplacian in the form of an elliptic first order differential operator is considered. Its function theory is widely known as *Clifford analysis*, see for instance [2], and offers many powerful generalisations of complex function theory, such as a Cauchy integral formula, Taylor and Laurent series expansions, a residue theory and a toolkit to study operators of Calderon-Zygmund type on strongly Lipschitz surfaces. A series of textbooks, see for example [17], presents a toolkit of related integral operators that can be used to tackle associated boundary value problems. Recently a lot of progress has been made in also elaborating discrete versions of Clifford analysis which also opened the door to apply these function theoretic tool numerically in bounded and unbounded domains, see [3, 5–8, 11–13, 16] among others.

However, besides the use of associative Clifford algebras, there are also other possibilities of generalising complex function theory to higher dimensions. If the Cayley-Dickson duplication process to the complex numbers is applied, then we first arrive at the four-dimensional Hamiltonian quaternions, which, however, still is a Clifford algebra; and after applying it once more, we obtain a new algebra, namely the *octonions*, see [1]. Octonions are not any more associative – so, they are neither a Clifford algebra nor representable with matrices. However, they still form a normed non-associative division algebra having no zero-divisors. From the recent viewpoint of generalised particle physics, see again for example [4, 15, 22], octonions seem to offer a more adequate model for a unified description of particle physics including gravity, see also [18]. However, there is still a lack of results on the level of octonionic function theory.

According to our knowledge, the first contribution to introduce an octonionic generalisation of complex function theory was provided by P. Dentoni and M. Sce in 1973 in [10], where a Cauchy integral formula for null-solutions to the octonionic Cauchy-Riemann operator has been presented. Later, a lot of fundamental contributions were provided by K. Nono [23] in 1988, and the school of Xingmin-Li, Li-Zhong Peng and their co-authors starting with 2000 up to now, see [25–27, 29]. In these papers, for instance, generalisations of a Cauchy integral formula together with Plemelj projection formulas and with some basic applications to Calderon-Zygmund type operators [30] including a generalisation of the three-line theorem from J. Peetre [28], as well as Taylor and Laurent series expansions have intensively been studied [26]. More recently, J. Kauhanen and H.Orelma started to look more intensively at some elementary octonionic boundary value problems and analysed more precisely the algebraic structure of the set of octonionic null-solutions of Cauchy-Riemann operators, see for example [19–21]. As also mentioned by J. Kauhanen and H. Orelma, in contrast to Clifford analysis, octonionic monogenic functions do not form $\mathbb{O}$-modules but only $\mathbb{R}$-modules. This fact has a strong influence on the study of generalised Hilbert spaces in the octonionic setting, which is a topic of very recent research, see for example [9, 14, 24]. For solving related boundary values in practice, it is necessary to apply discretised versions of the related octonionic operators.

Although a discretisation of octonionic analysis is important for practical use of function theoretic tools, to the best of our knowledge, no results related to a discrete octonionic analysis have been presented so far. Therefore, the aim of this short paper is presenting some first results in this direction. In particular, we introduce discretised versions of the octonionic Cauchy-Riemann operators and establish a generalised version of the Stokes' formula. As it will be clearly seen, already at this level, we encounter substantial differences to the classical discrete Clifford analysis: because of the non-associativity of octonionic multiplication, discretisation of octonionic analysis needs to be discussed more carefully. Additionally, we will also indicate the difference to the continuous case, which appear due to working with forward and backward Cauchy-Riemann operators. Thus, this paper serves as a first step for developing discrete octonionic analysis, and results presented here will be further extended in future work.

2 Preliminaries and Notations

2.1 *Continuous Octonionic Analysis*

Before introducing discrete constructions, let us briefly recall some basic information about octonions $\mathbb{O}$ and continuous octonionic analysis. Let us consider 8-dimensional Euclidean space $\mathbb{R}^8$ with the basis unit vectors e_k, $k = 0, 1, \ldots, 7$ and points $\mathbf{x} = (x_0, x_1, \ldots, x_7)$. Then in real coordinates, octonions are expressed in the form

$$x = x_0e_0 + x_1e_1 + x_2e_2 + x_3e_3 + x_4e_4 + x_5e_5 + x_6e_6 + x_7e_7,$$

where $e_4 = e_1e_2$, $e_5 = e_1e_3$, $e_6 = e_2e_3$ and $e_7 = e_4e_3 = (e_1e_2)e_3$. Additionally we have $e_i^2 = -1$ and $e_0e_i = e_ie_0$ for all $i = 1, \ldots, 7$, and $e_ie_j = -e_je_i$ for all mutual distinct $i, j \in \{1, \ldots, 7\}$. Table 1 shows multiplication rules for real octonions. As it can be clearly seen from this table, multiplication of octonions is not associative, precisely we have $(e_ie_j)e_k = -e_i(e_je_k)$.

There are several possibilities to extend the classical function theory to octonions. One way consists of the Riemann-approach, following the line of investigation of P.

Table 1 Multiplication table for real octonions $\mathbb{O}$

	e_0	e_1	e_2	e_3	e_4	e_5	e_6	e_7
e_0	1	e_1	e_2	e_3	e_4	e_5	e_6	e_7
e_1	e_1	-1	e_4	e_5	$-e_2$	$-e_3$	$-e_7$	e_6
e_2	e_2	$-e_4$	-1	e_6	e_1	e_7	$-e_3$	$-e_5$
e_3	e_3	$-e_5$	$-e_6$	-1	$-e_7$	e_1	e_2	e_4
e_4	e_4	e_2	$-e_1$	e_7	-1	$-e_6$	e_5	$-e_3$
e_5	e_5	e_3	$-e_7$	$-e_1$	e_6	-1	$-e_4$	e_2
e_6	e_6	e_7	e_3	$-e_2$	$-e_5$	e_4	-1	$-e_1$
e_7	e_7	$-e_6$	e_5	$-e_4$	e_3	$-e_2$	e_1	-1

Dentoni and M. Sce [10], K. Nono [23], the school of Xingmin-Li and Zhong Peng, see for instance [25, 26] and others. In their spirit one may introduce.

Definition 1 (*Octonionic monogenicity*) Let $U \subseteq \mathbb{O}$ be open. A function $f : U \rightarrow \mathbb{O}$ is called left (right) octonionic monogenic if $\mathcal{D}f = 0$ (esp. $f\mathcal{D} = 0$). Here, $\mathcal{D} := \frac{\partial}{\partial x_0} + \sum_{i=1}^{7} e_i \frac{\partial}{\partial x_i}$ is the octonionic first order Cauchy-Riemann operator. If f satisfies $\overline{\mathcal{D}}f = 0$ (resp. $f\overline{\mathcal{D}} = 0$), then we call f left (right) octonionic anti-monogenic.

In contrast to Clifford analysis, where one considers null-solutions to the Cauchy-Riemann operator defined on the paravector space $\mathbb{R} \oplus \mathbb{R}^7$ with values in the Clifford algebra $\mathcal{C}\ell_7$, which is a real vector spaces isomorphic to $\mathbb{R}^{128}$, the octonionic approach really considers maps from $\mathbb{R}^8$ into $\mathbb{R}^8$. Another essential difference is the fact that left (right) octonionic monogenic functions do neither form a right nor a left $\mathbb{O}$-module. Following for instance J. Kauhanen and H. Orelma in [21], one can take as a very simple counterexample: the function $f(x) := x_1 - x_2 e_4$. Then we have $\mathcal{D}[f(x)] = e_1 - e_2 e_4 = e_1 - e_1 = 0$. However, $g(x) := (f(x)) \cdot e_3 = (x_1 - x_2 e_4)e_3 = x_1 e_3 - x_2 e_7$ satisfies $\mathcal{D}[g(x)] = e_1 e_3 - e_2 e_7 = e_5 - (-e_5) = 2e_5 \neq 0$.

As already mentioned in the classical paper [25], the lack of associativity prevents us from getting a direct analogue of Stokes' formula in the octonionic setting. Even if both $\mathcal{D}f = 0$ and $g\mathcal{D} = 0$, we do not have in general

$$\int_{\partial G} g(x)\ (d\sigma(x) f(x)) = 0 \text{ nor } \int_{\partial G} (g(x) d\sigma(x))\ f(x) = 0.$$

Instead, quoting from [30], we obtain the following relation

$$\int_{\partial G} g(x)\ (d\sigma(x) f(x)) = \int_{G} \Big(g(x)(\mathcal{D}f(x)) + (g(x)\mathcal{D}) f(x) - \sum_{j=0}^{7} [e_j, \mathcal{D}g_j(x), f(x)] \Big) dV, \tag{1}$$

where $[a, b, c] := (ab)c - a(bc)$ is the so-called *associator* (which would vanish in the cases of associativity). Although the associator appears in most of octonionic constructions, it is nonetheless possible to introduce specific structures, where the associator would vanish. For example, it has been pointed out in [25], that considering the two functions being octonionic monogenic and Stein-Weiss conjugate harmonics, i.e. $\frac{\partial g_j}{\partial x_i} = \frac{\partial g_i}{\partial x_j}$ for all $0 \leq i < j \leq 7$, the associator will vanish.

Moreover, it is still possible to obtain a generalisation of the Cauchy's integral formula to octonionic setting [23, 27]:

Proposition 1 (Cauchy's integral formula) *Let $U \subseteq \mathbb{O}$ be open and $G \subseteq U$ be an 8-D compact oriented manifold with a strongly Lipschitz boundary ∂G. If $f : U \to \mathbb{O}$ is left octonionic monogenic, then for all $x \in G$*

$$f(x) = \frac{3}{\pi^4} \int_{\partial G} q_0(y-x)\Big(d\sigma(y) f(y)\Big).$$

However, we have to emphasise carefully on the fact that putting the parenthesis differently, leads to the different formula

$$\frac{3}{\pi^4} \int_{\partial G} \Big(q_0(y-x) d\sigma(y)\Big) f(y) = f(x) + \int_G \sum_{i=0}^{7} \Big[q_0(y-x), \mathcal{D} f_i(y), e_i\Big] dy_0 \cdots dy_7,$$

involving the associator again.

2.2 *Discretisation of Octonionic Analysis*

Let us now introduce a discrete setting for octonions. Consider the unbounded uniform lattice $h\mathbb{Z}^8$ with the lattice constant $h > 0$, which is defined in the classical way as follows

$$h\mathbb{Z}^8 := \left\{ \mathbf{x} \in \mathbb{R}^8 \,|\, \mathbf{x} = (m_1 h, m_2 h, \ldots, m_8 h), m_j \in \mathbb{Z}, j = 1, 2, \ldots, 8 \right\}.$$

Next, we define the classical forward and backward differences $\partial_h^{\pm j}$ as

$$\begin{aligned} \partial_h^{+j} f(mh) &:= h^{-1}(f(mh + e_j h) - f(mh)), \\ \partial_h^{-j} f(mh) &:= h^{-1}(f(mh) - f(mh - e_j h)), \end{aligned} \tag{2}$$

for discrete functions $f(mh)$ with $mh \in h\mathbb{Z}^n$. In the sequel, we consider functions defined on $\Omega_h \subset h\mathbb{Z}^8$ and taking values in octonions $\mathbb{O}$. As usual, all important properties such as, l^p-summability $(1 \leq p < \infty)$, are defined component-wisely.

Next step is to introduce discretisation of the Cauchy-Riemann operators in octonions. Several approaches to the discretisation of the Cauchy-Riemann (and Dirac) operators have been presented in recent years. In particular, the discrete Clifford analysis is generally based on the idea of splitting each basis element e_k, $k = 0, 1, \dots, 7$, into two basis elements e_k^+ and e_k^-, $k = 0, 1, \dots, 7$, i.e., $e_k = e_k^+ + e_k^-$, corresponding to the forward and backward directions, respectively, see [3, 13] for the details. A typical choice for such a basis is one satisfying the relations:

$$\begin{cases} e_j^- e_k^- + e_k^- e_j^- = & 0, \\ e_j^+ e_k^+ + e_k^+ e_j^+ = & 0, \\ e_j^+ e_k^- + e_k^- e_j^+ = & -\delta_{jk}, \end{cases}$$

where δ_{jk} is the Kronecker delta. This approach has several advantages and, in particular, it leads to a canonical factorisation of a star-Laplacian Δ_h by a pair of discrete Dirac operators. Unfortunately, this approach is not so well suited for working in the octonionic setting, because it is not so easy to respect the non-associativity of octonionic multiplication.

Another way of working with discrete Cauchy-Riemann and Dirac operators is to represent these operators by help of matrices containing finite difference approximations of partial derivatives, see for example [5, 11, 16] and references therein. Similar to the first approach, using matrix-based discretisation for discretising the octonionic analysis will be difficult because of non-associativity, which is not respected by the classical matrix multiplication.

For proposing a discretisation of octonionic analysis respecting the non-associativity of octonionic multiplication, we will work with the approach presented in [12] and consisting in a direct discretisation of the continuous Dirac operators by forward and backward finite difference operators. Thus, by help of the finite difference operators (2), we introduce *discrete forward Cauchy-Riemann operator* $D^+ \colon l^p(\Omega_h, \mathbb{O}) \to l^p(\Omega_h, \mathbb{O})$ and *discrete backward Cauchy-Riemann operators* $D^- \colon l^p(\Omega_h, \mathbb{O}) \to l^p(\Omega_h, \mathbb{O})$ as follows

$$D_h^+ := \sum_{j=0}^{7} e_j \partial_h^{+j}, \quad D_h^- := \sum_{j=0}^{7} e_j \partial_h^{-j}. \tag{3}$$

A small disadvantage of this approach is related to the factorisation of the star-Laplacian, which is not just a composition of the Cauchy-Riemann operator and its conjugated operator, but requires a more complicated combination. It is easy to show by direct computations, that the star-Laplacian Δ_h can be represented as follows:

$$\Delta_h = \frac{1}{2}\left(D_h^+ \overline{D_h^-} + D_h^- \overline{D_h^+}\right) \text{ with } \Delta_h := \sum_{j=0}^{7} \partial_h^{+j} \partial_h^{-j},$$

where $\overline{D_h^-}$ and $\overline{D_h^+}$ represent conjugated operators:

$$\overline{D_h^-} = \partial_h^{-0} - \sum_{j=1}^{7} e_j \partial_h^{-j}, \quad \overline{D_h^+} = \partial_h^{+0} - \sum_{j=1}^{7} e_j \partial_h^{+j}.$$

In the rest of the paper, we will work with the discrete Cauchy-Riemann operators (3), because this discretisation clearly respects the non-associativity of octonionic multiplication.

3 The Discrete Stokes' Formula in Octonions

In this section, we introduce the discrete Stokes' formula in octonionic setting. Additionally, we will underline the difference between octonionic constructions and the classical discrete Clifford analysis. Moreover, for keeping notations shorter, we will omit the lattice constant h in the argument of discrete functions for the proof of discrete Stokes' formula, i.e. notations $f(m)$ or $f(m_1, m_2, m_3)$ will be used instead of $f(mh)$ or $f(m_1h, m_2h, m_3h)$, respectively.

The following theorem presents the discrete octonionic Stokes' formula for the whole space:

Theorem 1 *The discrete Stokes' formula for the whole space with the lattice $h\mathbb{Z}^8$ is given by*

$$\sum_{m\in\mathbb{Z}^8} \left\{\left[g(mh)D_h^+\right] f(mh) - g(mh)\left[D_h^- f(mh)\right]\right\} h^8 = 0 \tag{4}$$

for all discrete functions f and g such that the series converge.

Proof To underline clearly the effect of non-associativity of the octonionic multiplication, the proof will be presented with all explicit calculations. We start the proof by working with the first term on the left-hand side in (4):

$$\sum_{m\in\mathbb{Z}^8} \left[g(m)D_h^+\right] f(m)h^8 = \sum_{m\in\mathbb{Z}^8} \sum_{j=0}^{7} \left[\partial^{+j} g(m)e_j\right] f(m)h^8$$

$$= \sum_{m\in\mathbb{Z}^8} \sum_{j=0}^{7} \sum_{i=0}^{7} \sum_{k=0}^{7} \left[\partial^{+j} g_i(m)e_ie_j\right] f_k(m)e_kh^8.$$

Next, using the relation $(e_ie_j)e_k = -e_i(e_je_k)$ and the definition of D_h^+ leads to the following expression

$$\sum_{m\in\mathbb{Z}^8}\sum_{j=0}^{7}\sum_{i=0}^{7}\sum_{k=0}^{7}\left[-\partial^{+j}g_i f_k(m)e_i(e_je_k)\right]h^8$$

$$=\sum_{m\in\mathbb{Z}^8}\sum_{j=0}^{7}\sum_{i=0}^{7}\sum_{k=0}^{7}\left[-\left(g_i(m+e_j)-g_i(m)\right)f_k(m)e_i(e_je_k)\right]h^8$$

$$=\sum_{m\in\mathbb{Z}^8}\sum_{j=0}^{7}\sum_{i=0}^{7}\sum_{k=0}^{7}\left[-g_i(m+e_j)e_if_k(m)+g_i(m)e_if_k(m)\right](e_je_k)h^8.$$

Performing change of variables in the last expression, we get

$$\sum_{m\in\mathbb{Z}^8}\sum_{j=0}^{7}\sum_{i=0}^{7}\sum_{k=0}^{7}\left[-g_i(m)e_if_k(m-e_j)+g_i(m)e_if_k(m)\right](e_je_k)h^8$$

$$=\sum_{m\in\mathbb{Z}^8}\sum_{j=0}^{7}\sum_{i=0}^{7}\sum_{k=0}^{7}\left[g_i(m)e_i\left(f_k(m-e_j)+f_k(m)\right)\right](e_je_k)h^8$$

$$=\sum_{m\in\mathbb{Z}^8}\sum_{j=0}^{7}\sum_{i=0}^{7}\sum_{k=0}^{7}g_i(m)e_i\partial^{-j}f_k(e_je_k)h^8$$

$$=\sum_{m\in\mathbb{Z}^8}\sum_{j=0}^{7}\sum_{i=0}^{7}\sum_{k=0}^{7}g_i(m)e_i\left(\partial^{-j}e_jf_ke_k\right)h^8=\sum_{m\in\mathbb{Z}^8}g(m)\left[D_h^-f(m)\right]h^8.$$

Thus, the statement of the theorem is proved.

As we see from this theorem, the discrete Stokes' formula does not contain the associator in contrast to the continuous case (1). This is an interesting result, and a possible reason for vanishing of the associator could be the fact, that the discrete octonionic Stokes' formula contains two different differential operators: forward and backward Cauchy-Riemann operators, while in the continuous case both operators are the same. Additionally, it is worth to underline that the non-associativity affect the sign of the second summand in (4), which is not the case in the discrete Clifford analysis [6, 8].

Next, we consider the case of the upper half-lattice, defined as follows

$$h\mathbb{Z}^8_+ := \left\{(h\underline{m}, hm_7)\colon \underline{m}\in\mathbb{Z}^7, m_7\in\mathbb{Z}_+\right\}.$$

The discrete octonionic Stokes' formula for the upper half-lattice is provided by the following theorem:

Theorem 2 *The discrete Stokes' formula for the upper half-lattice $h\mathbb{Z}^8_+$ is given by*

$$\sum_{m\in\mathbb{Z}^8_+} \left\{\left[g(mh)D_h^+\right] f(mh) - g(mh)\left[D_h^- f(mh)\right]\right\} h^8$$
$$= \sum_{\underline{m}\in\mathbb{Z}^7} e_7\left(g(\underline{m},1) f_k(\underline{m},0)\right) h^8 \tag{5}$$

for all discrete functions f and g such that the series converge.

Proof The proof of this theorem is similar to the proof of the discrete Stokes' formula for the whole space. Nonetheless, it is necessary to address the fact, that the discrete Cauchy-Riemann operators can be applied only for points with $m_7 \geq 1$. We start the proof by working with the first term on the left-hand side in (5):

$$\sum_{m\in\mathbb{Z}^8_+} \left[g(m)D_h^+\right] f(m)h^8 = \sum_{m\in\mathbb{Z}^8_+}\sum_{j=0}^{6}\left[\partial^{+j} g(m)e_j\right] f(m)h^8$$

$$+ \sum_{m\in\mathbb{Z}^8_+} \left[\partial^{+7} g(m)e_7\right] f(m)h^8 = \sum_{m\in\mathbb{Z}^8_+}\sum_{j=0}^{6}\sum_{i=0}^{7}\sum_{k=0}^{7}\left[\partial^{+j} g_i(m)e_ie_j\right] f_k(m)e_kh^8$$

$$+ \sum_{\underline{m}\in\mathbb{Z}^7}\left\{\sum_{m_7\geq 1}\sum_{i=0}^{7}\sum_{k=0}^{7}\left[\left(g_i(m+e_7)f_k(m) - g_i(m)f_k(m)\right)e_7e_i\right]e_kh^8\right\}.$$

Next, we will work with the second sum. By using the relation $(e_ie_j)e_k = -e_i(e_je_k)$ and performing change of variables, we get the following expression

$$\sum_{\underline{m}\in\mathbb{Z}^7}\left\{\sum_{m_7\geq 1}\sum_{i=0}^{7}\sum_{k=0}^{7}\left[\left(-g_i(m+e_7)f_k(m) + g_i(m)f_k(m)\right)e_7\right]e_ie_kh^8\right\}$$

$$= \sum_{\underline{m}\in\mathbb{Z}^7}\left\{\sum_{m_7\geq 1}\sum_{i=0}^{7}\sum_{k=0}^{7}\left[\left(g_i(m)f_k(m) - g_i(m)f_k(m-e_7)\right)e_7\right]e_ie_kh^8\right\}$$

$$= \sum_{\underline{m}\in\mathbb{Z}^7}\left\{\sum_{m_7\geq 1}\sum_{i=0}^{7}\sum_{k=0}^{7} g_i(m)f_k(m)e_7\left(e_ie_k\right)h^8\right.$$

$$\left.- \sum_{m_7\geq 2}\sum_{i=0}^{7}\sum_{k=0}^{7} g_i(m)f_k(m-e_7)e_7\left(e_ie_k\right)h^8\right\}$$

$$= \sum_{\underline{m}\in\mathbb{Z}^7} \left\{ \sum_{m_7\geq 1} \sum_{i=0}^{7} \sum_{k=0}^{7} g_i(m) f_k(m) e_7 (e_i e_k) h^8 \right.$$

$$- \sum_{m_7\geq 1} \sum_{i=0}^{7} \sum_{k=0}^{7} g_i(m) f_k(m - e_7) e_7 (e_i e_k) h^8$$

$$\left. + \sum_{i=0}^{7} \sum_{k=0}^{7} g_i(\underline{m}, 1) f_k(\underline{m}, 0) e_7 (e_i e_k) h^8 \right\}.$$

Combining this result with the first sum of the original expression, we finally get the following equality

$$\sum_{m\in\mathbb{Z}^8_+} \left[g(m) D_h^+\right] f(m) h^8 = \sum_{m\in\mathbb{Z}^8_+} g(m) \left[D_h^- f(m)\right] h^8$$

$$+ \sum_{\underline{m}\in\mathbb{Z}^7} e_7 \left(g(\underline{m}, 1) f_k(\underline{m}, 0)\right) h^8,$$

which proofs the assertion of the theorem.

Similarly, a discrete Stokes' formula can be established for the lower half-lattice, defined as follows

$$h\mathbb{Z}^8_- := \left\{(h\underline{m}, hm_7) \colon \underline{m} \in \mathbb{Z}^7, m_7 \in \mathbb{Z}_-\right\}.$$

We have then the following theorem:

Theorem 3 *The discrete Stokes' formula for the lower half-lattice $h\mathbb{Z}^8_-$ is given by*

$$\sum_{m\in\mathbb{Z}^8_-} \left\{\left[g(mh) D_h^+\right] f(mh) - g(mh) \left[D_h^- f(mh)\right]\right\} h^8$$
$$= - \sum_{\underline{m}\in\mathbb{Z}^7} e_7 \left(g(\underline{m}, 0) f_k(\underline{m}, -1)\right) h^8 \tag{6}$$

for all discrete functions f and g such that the series converge.

Proof The proof of this theorem is analogue to the previous proof, and, therefore, we will present a shorter version of the proof. Hence, we have:

$$\sum_{m\in\mathbb{Z}^8_-}\left[g(m)D_h^+\right]f(m)h^8=\sum_{m\in\mathbb{Z}^8_-}\sum_{j=0}^{6}\sum_{i=0}^{7}\sum_{k=0}^{7}\left[\partial^{+j}g_i(m)e_ie_j\right]f_k(m)e_kh^8$$

$$+\sum_{\underline{m}\in\mathbb{Z}^7}\left\{\sum_{m_7\leq -1}\sum_{i=0}^{7}\sum_{k=0}^{7}\left[\left(g_i(m+e_7)f_k(m)-g_i(m)f_k(m)\right)e_7e_i\right]e_kh^8\right\}.$$

Working with the second sum, we get

$$\sum_{\underline{m}\in\mathbb{Z}^7}\left\{\sum_{m_7\leq -1}\sum_{i=0}^{7}\sum_{k=0}^{7}\left[\left(g_i(m+e_7)f_k(m)-g_i(m)f_k(m)\right)e_7e_i\right]e_kh^8\right\}$$

$$=\sum_{\underline{m}\in\mathbb{Z}^7}\left\{\sum_{m_7\leq -1}\sum_{i=0}^{7}\sum_{k=0}^{7}g_i(m)f_k(m)e_7\left(e_ie_k\right)h^8\right.$$

$$\left.-\sum_{m_7\leq 0}\sum_{i=0}^{7}\sum_{k=0}^{7}g_i(m)f_k(m-e_7)e_7\left(e_ie_k\right)h^8\right\}$$

$$=\sum_{\underline{m}\in\mathbb{Z}^7}\left\{\sum_{m_7\leq -1}\sum_{i=0}^{7}\sum_{k=0}^{7}g_i(m)f_k(m)e_7\left(e_ie_k\right)h^8\right.$$

$$-\sum_{m_7\leq -1}\sum_{i=0}^{7}\sum_{k=0}^{7}g_i(m)f_k(m-e_7)e_7\left(e_ie_k\right)h^8$$

$$\left.-\sum_{i=0}^{7}\sum_{k=0}^{7}g_i(\underline{m},0)f_k(\underline{m},-1)e_7\left(e_ie_k\right)h^8\right\}.$$

Combining this result with the first sum of the original expression, we obtain the assertion of the theorem.

4 Summary

While a lot of results in the continuous octonionic analysis have been presented in recent years, construction of a discrete counterpart of the continuous theory is still missing. Therefore, in this short paper, we discussed first ideas towards developing a discrete octonionic analysis. In particular, we discuss several approaches to a discretisation of octonionic analysis, and underlined, that because of non-associativity of octonions not all approaches common in the discrete Clifford analysis are applicable in the octonionic setting. After that, we presented several discrete octonionic Stokes' formulae: for the whole spaces, upper-half lattice, and lower-half lattice. The results presented in this paper will be further extended in future work.

References

1. Baez J (2002) The octonions. Bull Am Math Soc 39:145–205
2. Brackx F, Delanghe R, Sommen F (1982) Clifford analysis. Pitman research notes in mathematics, vol 76. Boston
3. Brackx F, De Schepper H, Sommen F, Van de Voorde L (2009) Discrete Clifford analysis: a germ of function theory. Hypercomplex analysis. Birkhäuser, Basel, pp 37–53
4. Burdik C, Catto S, Gürcan Y, Khalfan A, Kurt L, La Kato V (2019) $SO(9,1)$ group and examples of analytic functions. J Phys Conf Ser 1194:012016
5. Cerejeiras P, Faustino N, Vieira N (2008) Numerical Clifford analysis for nonlinear Schrödinger problem. Numer Methods Partial Differ Equ 24(4):1181–1202
6. Cerejeiras P, Kähler U, Ku M, Sommen F (2014) Discrete Hardy spaces. J Fourier Anal Appl 20(4):715–750
7. Cerejeiras P, Kähler U, Legatiuk A, Legatiuk D (2020) Boundary values of discrete monogenic functions over bounded domains in $\mathbb{R}^3$. In: D. Alpay M, Vajiac (eds) Linear systems, signal processing and hypercomplex analysis. Operator theory: advances and applications, vol 275. Birkhäuser, pp 149–165
8. Cerejeiras P, Kähler U, Legatiuk A, Legatiuk D (2021) Discrete Hardy spaces for bounded domains in $\mathbb{R}^n$. Complex Anal Oper Theory 15:4
9. Constales D, Kraußhar RS (2021) Octonionic Kerzman-Stein operators. Complex Anal Oper Theory 15:104
10. Dentoni P, Sce M (1973) Funzioni regolari nell'algebra di Cayley. Rend Sem Mat Univ Padova 50:251–267
11. Faustino N, Gürlebeck K, Hommel A, Kähler U (2006) Difference potentials for the Navier-Stokes equations in unbounded domains. J Differ Equ Appl 12(6):577–595
12. Faustino N, Kähler U (2007) Fischer decomposition for difference Dirac operators. Adv Appl Clifford Algebr 17:37–58
13. Faustino N, Kähler U, Sommen F (2007) Discrete Dirac operators in Clifford analysis. Adv Appl Clifford Algebr 17(3):451–467
14. Frenod E, Ludkowski SV (2017) Integral operator approach over octonions to solution of nonlinear PDE. Far East J Math Sci
15. Gogberashvili M (2016) Octonionic geometry and conformal transformations. Int J Geom Methods Mod Phys 13(7):1650092
16. Gürlebeck K, Hommel A (2001) On finite difference Dirac operators and their fundamental solutions. Adv Appl Clifford Algebr 11:89–106
17. Gürlebeck K, Sprößig W (1997) Quaternionic and Clifford calculus for physicists and engineers. John Wiley & Sons, Chichester-New York
18. Gürsey F, Tze H (1996) On the role of division and Jordan algebras in particle physics. World Scientific, Singapore
19. Kauhanen J, Orelma H (2017) Some theoretical remarks of octonionic analysis. AIP Conf Proc 1907:0300056
20. Kauhanen J, Orelma H (2018) Cauchy-Riemann operators in octonionic analysis. Adv Appl Clifford Algebr 28:1
21. Kauhanen J, Orelma H (2019) On the structure of octonion regular functions. Adv Appl Clifford Algebr 29:77

22. Najarbashi G, Seifi B, Mirzaei S (2016) Two-and three-qubit geometry, quaternionic and octonionic conformal maps, and intertwining stereographic projection. Quantum Inf Process 15:509–528
23. Nono K (1988) On the octonionic linearization of Laplacian and octonionic function theory. Bull Fukuoka Univ Ed Part III(37):1–15
24. Huo Q, Ren Q (2022) Structure of octonionic Hilbert spaces with applications in the Parseval equality and Cayley-Dickson algebras. J Math Phys 63(4):042101
25. Li X-M, Peng L-Z (2000) On Stein-Weiss conjugate harmonic function and octonion analytic function. Approx Theory Appl 16:28–36
26. Li X-M, Zhao K, Peng L-Z (2001) The Laurent series on the octonions. Adv Appl Clifford Algebr 11:205–217
27. Li X-M, Peng L-Z (2002) The Cauchy integral formulas on the octonions. Bull Belg Math Soc 9:47–62
28. Li X-M, Peng L-Z (2004) Three-line theorems on the octonions. Acta Math Sin 20(3):483–490
29. Li X-M, Kai Z, Peng L-Z (2005) Characterization of octonionic analytic functions. Complex Var 50(13):1031–1040
30. Li X-M, Peng L-Z, Qian T (2008) Cauchy integrals on Lipschitz surfaces in octonionic space. J Math Anal Appl 343:763–777

Axiomatic Method for Constructing a Generalized Solution of a Mixed Problem for a Telegraph Equation

Igor S. Lomov

Abstract The paper presents an algorithm for constructing a rapidly converging series representing a generalized solution of a mixed problem for a telegraphic equation considered in a half-band. Reviewed the case of an essentially non-self-adjoint operator in a spatial variable. The system of root functions of a differential operator, in addition to its eigenfunctions, contains an infinite number of associated functions. The constructed series can be considered as a generalized d'Alembert formula. A new axiomatic A.P. Khromov's method is applied to construct the solution. The proposed approach superseds the traditional method of separating variables for solving mixed problems, which usually results in to slowly converging series. For the problem under consideration, in general, the method of separating variables is not applicable, since the coefficient of the equation depends both on the spatial variable and on time.

Keywords Telegraph equation · Mixed problem · Generalized d'Alembert formula · Fourier method · Non-self-adjoint operator · Divergent series

1 Introduction

A number of mathematical models used in problems of sound theory (elasticity), light, electricity and magnetism, contain the so-called telegraph equation $u_{tt}(x, t) = u_{xx}(x, t) - qu(x, t)$. A mixed problem is posed. Consider the case when the potential q can also depend on time, $q = q(x, t)$. To construct a solution to a generalized mixed problem, we use the recently developed axiomatic method of A.P. Khromov [1]. Previously, he developed a sequential method for constructing a generalized solution to the problem under consideration [2, 3]. The advantage of these methods over the methods used earlier consist in the fact that minimum requirements are imposed on the initial data of the problem, the justification of the result attracts a

I. S. Lomov (✉)
Moscow State University, Moscow, Russia
e-mail: lomov@cs.msu.ru

V. Vasilyev (ed.), *Differential Equations, Mathematical Modeling and Computational Algorithms*, Springer Proceedings in Mathematics & Statistics 423,
https://doi.org/10.1007/978-3-031-28505-9_5

minimum number of additional statements, and the solution is given in the form of a rapidly converging functional series.

Let's consider four problems sequentially, for which we will find generalized solutions.

2 A Mixed Problem for a Homogeneous Wave Equation with a Nonzero Initial Deviation

Consider the following problem

$$u_{tt}(x,t) = u_{xx}(x,t), \quad (x,t) \in (0,1) \times (0,+\infty), \tag{1}$$

$$u(0,t) = 0, \quad u_x(0,t) = u_x(1,t), \quad t \geq 0, \tag{2}$$

$$u(x,0) = \varphi(x), \; u_t(x,0) = 0, \quad x \in [0,1], \tag{3}$$

$\varphi(x)$—complex-valued, integrable on (0, 1) functions, $\varphi(x) \in \mathcal{L}(0,1)$. We use the notation derivatives $u_x = \frac{\partial u}{\partial x}$, etc.

The peculiarity of the problem (1)–(3) is due to the fact that the corresponding Sturm–Liouville operator $L_0: \quad ly = -y''(x),\ x \in (0,1),\ y(0) = 0,\ y'(0) = y'(1)$, is essentially non-self-adjoint (according to Ilyin)—the system of root functions of this operator, in addition to its eigenfunctions, contains an infinite number of associated functions (the Samarsky-Ionkin problem). Let's write out this system.

Denote by ϱ_k the square roots of the eigenvalues operator, $\{u_k(x)\}$—system of eigen and associated operator functions, moreover, $u_{2k-1}(x)$—eigenfunctions, $u_{2k}(x)$—associated functions, $k \geq 1$, $\{v_k(x)\}$—biorthogonally conjugate system of functions, $(u_k, v_n) = \delta_{kn} = \begin{cases} 1, k = n, \\ 0, k \neq n \end{cases}$, where $(u_k, v_n) = \int_0^1 u_k(x) v_n(x) dx$.

Then $\varrho_0 = 0,\ \varrho_{2k-1} = \varrho_{2k} = 2\pi k,\ k \geq 1, \quad u_0(x) = x,\ v_0(x) = 2,\ u_{2k-1}(x) = \sin 2\pi k x,\ v_{2k-1}(x) = 4(1-x)\sin 2\pi k x, \qquad u_{2k}(x) = -\frac{x}{4\pi k}\cos 2\pi k x,\ v_{2k}(x) = -16\pi k \cos 2\pi k x$. So the chosen system $\{u_k(x)\}$ of root functions of the operator forms unconditional basis in the space $\mathcal{L}^2(0,1)$. System $\{v_k(x)\}$ also forms an unconditional basis in this space.

The formal solution of the problem (1)–(3) by the Fourier method is

$$\begin{aligned}
&u(x,t) = \tfrac{1}{2}\big\{2(x+t)(1,\varphi)+\\
&+4\sum_{n=1}^{\infty}\big[(\varphi(\tau),(1-\tau)\sin 2\pi n\tau)\sin 2\pi n(x+t)+\\
&+(\varphi(\tau),\cos 2\pi n\tau)(x+t)\cos 2\pi n(x+t)\big]+\\
&+2(x-t)(1,\varphi)+4\sum_{n=1}^{\infty}\big[(\varphi(\tau),(1-\tau)\sin 2\pi n\tau)\sin 2\pi n(x-t)+\\
&+(\varphi(\tau),\cos 2\pi n\tau)(x-t)\cos 2\pi n(x-t)\big]\big\}.
\end{aligned} \tag{4}$$

Definition 1 By the classical solution (almost everywhere solution) of the problem (1)–(3) we mean the function $u(x, t)$ continuous and continuously differentiable with respect to x and t in half-strip $[0, 1] \times [0, \infty)$, and the functions $u_x(x, t), u_t(x, t)$ are absolutely continuous in $x \in [0, 1]$ and $t \in [0, \infty)$, respectively, satisfying the conditions (2), (3) and almost everywhere in x and t the Eq. (1).

Let us present a uniqueness theorem for the classical solution of the problem (1)–(3). Fix an arbitrary number $T > 0$, let Q_T—rectangle, $Q_T = [0, 1] \times [0, T]$, denoted by Q is the class of functions integrable on Q_T, $f \in Q \Leftrightarrow f(x, t) \in \mathcal{L}(Q_T)$.

Theorem 1 *If $u(x, t)$ is a classical solution to the problem (1)–(3) with condition $u_{tt}(x, t) \in Q$ $(\forall T > 0)$, then it is unique and can be found by the formula (4), in which the series on the right for any fixed $t > 0$ converge absolutely and uniformly in $x \in [0, 1]$.*

The proof of the theorem follows the scheme described in [4] and does not depends on specific boundary conditions.

Note that the series (4) makes sense for any function $\varphi(x) \in \mathcal{L}(0, 1)$, although now it can also be divergent. Nevertheless, we will assume that it is a *formal solution* of the problem (1)–(3), but now understood *purely formally*. This problem (1)–(3) will be called the *generalized mixed problem*. Finding a solution to a generalized mixed problem means finding the "sum" of, generally speaking, a divergent series. "Sum" in quotes means that this is the sum of a divergent (generally) series (see [5, p. 101], [6, p. 6, 19]).

Finding a solution to the generalized mixed problem (1)–(3) means finding the "sum" of the divergent series (4).

In addition to the three axioms about divergent series [6, p. 19], following A.P. Khromov, we will also use the following integration rule for a divergent series:

$$\int \sum = \sum \int, \tag{5}$$

where $\int$ is a definite integral.

Let's go back to the row (4). Before transforming it, let us write the formal expansion of the function $\varphi(x)$ into a series in terms of the root system functions of the operator L_0:

$$\begin{aligned}\varphi(x) \sim 2x(1, \varphi) + 4 \sum_{n=1}^{\infty} [(\varphi(\tau), (1 - \tau) \sin 2\pi n\tau) \sin 2\pi nx + \\ +(\varphi(\tau), \cos 2\pi n\tau) x \cos 2\pi nx].\end{aligned} \tag{6}$$

The series (4) can be represented as

$$u(x, t) = \sum\nolimits_{+} + \sum\nolimits_{-}, \tag{7}$$

where $\sum_{\pm} = \sum_{n=1}^{\infty} \ldots (x \pm t)$. Comparing (6), (7), we conclude that to find the "sum" of the series (4), we need to find the "sum" of the series (6) .

Let the "sum" of the series (6) for $x \in [0, 1]$ be some function $g(x) \in \mathcal{L}(0, 1)$. Then, in accordance with rule (5), we have

$$\begin{aligned} &\int_0^x g(\eta)d\eta = 2(1, \varphi)\int_0^x \eta d\eta + \\ &+4\sum_{n=1}^{\infty}\big[(\varphi(\tau), (1-\tau)\sin 2\pi n\tau)\int_0^x \sin 2\pi n\eta d\eta + \\ &+(\varphi(\tau), \cos 2\pi n\tau)\int_0^x \eta\cos 2\pi n\eta d\eta\big], \quad x \in [0, 1]. \end{aligned} \tag{8}$$

The following generalization to the considered system $\{u_k(x)\}$ of Lebesgue's theorem on term-by-term integration of the trigonometric Fourier series takes place.

Theorem 2 *Let a function $\varphi(x) \in \mathcal{L}(0, 1)$ be given that has the series (6) as its biorthogonal expansion in the system $\{u_k(x)\}$. If the segment is $[A, B] \subseteq [0, 1]$, then*

$$\begin{aligned} &\int_A^B \varphi(x)dx = \int_A^B 2x(1, \varphi)dx + \sum_{n=1}^{\infty}\int_A^B [4(\varphi(\tau), (1-\tau)\sin 2\pi n\tau)\sin 2\pi nx + \\ &+4(\varphi(\tau), \cos 2\pi n\tau)x\cos 2\pi nx]dx. \end{aligned}$$

Those, the biorthogonal series (6) can be integrated term-by-term, the resulting series converges and its sum is equal to $\int_A^B \varphi(x)dx$. In this case, the series (6) itself may not converge.

The proof of Theorem 2 is carried out in Sect. 5.

According to Theorem 2, the sum of the series (8), the usual sum, is the function $\int_0^x \varphi(\eta)d\eta$. But then, $\int_0^x g(\eta)d\eta = \int_0^x \varphi(\eta)d\eta$, i.e. $g(x) = \varphi(x)$ is true almost everywhere on the interval [0, 1], we have found the "sum" of the series (6), which can also be divergent.

The formal series (6) is defined for all values of $x \in \mathbf{R}$. Denote by $\widetilde{\varphi}(x)$ the "sum" of the series (6) for all values of $x \in \mathbf{R}$. By virtue of (6) and (7) we conclude that the "sum" $u(x, t)$ of the series (4) is a function

$$u(x, t) = \frac{1}{2}[\widetilde{\varphi}(x + t) + \widetilde{\varphi}(x - t)]. \tag{9}$$

Proven

Theorem 3 *The solution of the generalized mixed problem (1)–(3) is the function $u(x, t)$ from the class Q defined by the formula (9).*

Let us find an algorithm for extending the function $\widetilde{\varphi}(x)$ from the segment $[0, 1]$, where $\widetilde{\varphi}(x) = \varphi(x)$, to the whole number line. Assuming that $\widetilde{\varphi}(x)$ is a smooth function, we substitute the relation (9) into the boundary conditions (2). We obtain two equalities: $\widetilde{\varphi}(x) = -\widetilde{\varphi}(-x),\ x \in \mathbf{R}$, i.e., the function $\widetilde{\varphi}(x)$—odd, and

$$\widetilde{\varphi}'(1+x) = 2\widetilde{\varphi}'(x) - \widetilde{\varphi}'(1-x),\ x \in \mathbf{R}, \tag{10}$$

where it is taken into account that $\widetilde{\varphi}'(x)$—is an even function. We integrate the equality (10) over the interval $[0, x]$, and we get

$$\widetilde{\varphi}(1+x) = 2\widetilde{\varphi}(x) + \widetilde{\varphi}(1-x),\ x > 0. \tag{11}$$

The relation (11) allows us to extend the function $\widetilde{\varphi}(x) = \varphi(x),\ x \in [0, 1]$, from the segment $[0, 1]$ to the semiaxis $x > 0$, then we continue the function to the semiaxis $x < 0$ as an odd function.

3 Mixed Problem for an Inhomogeneous Wave Equation with Zero Initial Deviation

Consider the following generalized mixed problem

$$u_{tt}(x,t) = u_{xx}(x,t) + f(x,t), \quad (x,t) \in (0,1) \times (0,+\infty), \tag{12}$$

$$u(0,t) = 0, \quad u_x(0,t) = u_x(1,t), \quad t \geq 0, \tag{13}$$

$$u(x,0) = u_t(x,0) = 0, \quad x \in [0,1], \tag{14}$$

where $f(x,t)$ is a function of class Q.

The formal solution of the problem (12)–(14) by the Fourier method is

$$\begin{aligned}
&u(x,t) = \tfrac{1}{2}\int\limits_0^t d\tau \int\limits_0^{t-\tau} \{2(x+\eta)(1, f(\xi,\tau)) + \\
&+4\sum_{n=1}^{\infty} \left[(f(\xi,\tau), (1-\xi)\sin 2\pi n\xi)\sin 2\pi n(x+\eta) + \right.\\
&\left.+(f(\xi,\tau), \cos 2\pi n\xi)(x+\eta)\cos 2\pi n(x+\eta)\right] + \\
&+2(x-\eta)(1, f(\xi,\tau)) + 4\sum_{n=1}^{\infty} \left[(f(\xi,\tau), (1-\xi)\sin 2\pi n\xi)\sin 2\pi n(x-\eta) + \right.\\
&\left.+(f(\xi,\tau), \cos 2\pi n\xi)(x-\eta)\cos 2\pi n(x-\eta)\right]\}d\eta,
\end{aligned}$$

we used the rule (5) and took the integrals out of the signs of the sums. Let's combine terms with arguments $(x+\eta)$ and $(x-\eta)$, we get

$$u(x,t) = \tfrac{1}{2}\int\limits_0^t d\tau \int\limits_{x-t+\tau}^{x+t-\tau} \{2\eta(1, f(\xi,\tau))+$$
$$+4\sum_{n=1}^{\infty}\left[(f(\xi,\tau),(1-\xi)\sin 2\pi n\xi)\sin 2\pi n\eta+\right. \tag{15}$$
$$\left.+(f(\xi,\tau),\cos 2\pi n\xi)\eta\cos 2\pi n\eta\right]\}d\eta = \tfrac{1}{2}\int\limits_0^t d\tau \int\limits_{x-t+\tau}^{x+t-\tau} \widetilde{f}(\eta,\tau)d\eta,$$

the last equality is explained by the fact that the bracketed expression $\{\cdot\}$ in (15), as it follows from the formula (6), has the "sum" $\widetilde{f}(\eta,\tau)$, where $\widetilde{f}(\eta,\tau)$ is the extension of the function $f(\eta,\tau)$ along τ to the entire real axis using the same formulas, which is for the function $\varphi(x)$.

Thus, fair

Theorem 4 *The solution $u(x,t)$ of the generalized mixed problem (12)–(14) is a function of class Q defined by the formula*

$$u(x,t) = \frac{1}{2}\int\limits_0^t d\tau \int\limits_{x-t+\tau}^{x+t-\tau} \widetilde{f}(\eta,\tau)d\eta. \tag{16}$$

From the formula (16), using the continuation formulas, we obtain the estimate

$$\|u(x,t)\|_{\mathcal{L}(Q_T)} \le c_T \|f(x,t)\|_{\mathcal{L}(Q_T)}, \quad \forall T>0,\ c_T = \text{const} > 0,$$

this confirms that $u(x,t)$ is a function of class Q.

4 A Mixed Problem for an Inhomogeneous Wave Equation with a Nonzero Initial Deviation

Consider a generalized mixed problem

$$u_{tt}(x,t) = u_{xx}(x,t) + f(x,t), \quad (x,t)\in(0,1)\times(0,+\infty), \tag{17}$$

$$u(0,t)=0, \quad u_x(0,t)=u_x(1,t), \quad t\ge 0, \tag{18}$$

$$u(x,0)=\varphi(x), \ \ u_t(x,0)=0, \quad x\in[0,1], \tag{19}$$

where $f(x,t)$ is a function of class Q, $\varphi(x)\in\mathcal{L}(0,1)$.

The formal solution of the problem (17)–(19) by the Fourier method is $u(x,t) = u_0(x,t)+u_1(x,t)$, where $u_0(x,t)$ is the series (4) and $u_1(x,t)$ is the series (15). Therefore, based on Sects. 2 and 3, we get

Theorem 5 *Generalized mixed problem (17)–(19) has a solution $u(x,t)$ of class Q defined by the formula*

$$u(x,t) = \frac{1}{2}[\widetilde{\varphi}(x+t) + \widetilde{\varphi}(x-t)] + \frac{1}{2}\int_0^t d\tau \int_{x-t+\tau}^{x+t-\tau} \widetilde{f}(\eta,\tau)d\eta. \quad (20)$$

5 Mixed Problem for the Telegraph Equation

We use the results of Sects. 2, 3 and 4 to solve the following problem:

$$u_{tt}(x,t) = u_{xx}(x,t) - q(x,t)u(x,t), \quad (x,t) \in (0,1) \times (0,+\infty), \quad (21)$$

$$u(0,t) = 0, \quad u_x(0,t) = u_x(1,t), \quad t \geq 0, \quad (22)$$

$$u(x,0) = \varphi(x), \; u_t(x,0) = 0, \quad x \in [0,1], \quad (23)$$

where $\varphi(x) \in \mathcal{L}(0,1)$, the function $q(x,t)$ is such that there is a function $q_0(x) \in \mathcal{L}(0,1)$, such that $|q(x,t)| \leq q_0(x)$, the function $q(x,t)u(x,t)$ is a function of class Q.

From Theorem 5 we obtain that finding a solution to the problem (21)–(23) in the class Q reduces to finding in this class the solution of the integral equation

$$u(x,t) = \frac{1}{2}[\widetilde{\varphi}(x+t) + \widetilde{\varphi}(x-t)] - \frac{1}{2}\int_0^t d\tau \int_{x-t+\tau}^{x+t-\tau} \widetilde{q(\eta,\tau)u(\eta,\tau)}d\eta, \quad (24)$$

where $\widetilde{q(\eta,\tau)u(\eta,\tau)}$ is the extension along η to the entire real axis from the interval $[0,1]$ for each τ of the function $q(\eta,\tau)u(\eta,\tau)$ by the same formulas as the function $\varphi(x)$.

The integral equation has a unique solution in the class Q obtained by the method of successive substitutions. This solution is given by the formula

$$u(x,t) = A(x,t) = \sum_{n=0}^{\infty} a_n(x,t), \quad (25)$$

where

$$a_0(x,t) = \tfrac{1}{2}[\widetilde{\varphi}(x+t) + \widetilde{\varphi}(x-t)],$$
$$a_n(x,t) = \tfrac{1}{2}\int_0^t d\tau \int_{x-t+\tau}^{x+t-\tau} \widetilde{f}_{n-1}(\eta,\tau)d\eta, \quad n = 1,2,\ldots,$$

where $\widetilde{f}_n(\eta,\tau) = f_n(\eta,\tau) = -q(\eta,\tau)a_n(\eta,\tau)$ for $\eta \in [0,1]$, $n = 0, 1, \ldots$, $f_n(\eta,\tau)$ extends over the variable η from [0, 1] to the whole line in the same way as the function $\varphi(x)$, $\widetilde{f}_n(\eta,\tau) = \widetilde{-q(\eta,\tau)a_n}(\eta,\tau)$.

The formula (25) can be called the generalized d'Alembert formula.

Theorem 6 *If $\varphi(x) \in \mathcal{L}(0,1)$ then the $A(x,t)$ (25) converges absolutely and uniformly (with exponential speed) in the rectangle Q_T for any $T > 0$.*

The proof of the theorem follows directly from the following estimate for the common term of the series (25).

Lemma 1 *Let $\varphi(x) \in \mathcal{L}(0,1)$, T—arbitrary positive number. Then the estimates hold*

$$\|a_n(x,t)\|_{C(Q_T)} \le c_T^{n+1}\|q_0\|_1^n\|\varphi\|_1\frac{T^{n-1}}{(n-1)!}, \quad n \in \mathbf{N},\ c_T = \text{const} > 0.$$

The proof of the lemma is carried out using the method of mathematical induction.

6 The Term-by-Term Integration Theorem

Here we justify Theorem 2 on the term-by-term integration of the biorthogonal expansion with respect to the system $\{u_k(x)\}$ integrable on the interval [0, 1] functions. We adhere to the well-known scheme of proving the Lebesgue theorem, with the correction that now the expansion in a series is not carried out according to orthonormal system, but biorthogonal system. Let us rename $\varphi(x)$ in Theorem 2 by $f(x)$.

So, let a function $f(x) \in \mathcal{L}(0,1)$ be given, which has as its biorthogonal expansion in the $\{u_k(x), v_k(x)\}$ system

$$2x(1,f) + 4\sum_{n=1}^{\infty}[(f(\tau),(1-\tau)\sin 2\pi n\tau)\sin 2\pi nx + \\ +(f(\tau),\cos 2\pi n\tau)x\cos 2\pi nx]. \tag{26}$$

Let $[A,B] \subseteq [0,1]$, then it is required to prove that

$$\int_A^B f(x)dx = 2\int_A^B x(1,f)dx + 4\sum_{n=1}^{\infty}\int_A^B\big[(f(\tau),(1-\tau)\sin 2\pi n\tau)\sin 2\pi nx + \\ +(f(\tau),\cos 2\pi n\tau)x\cos 2\pi nx\big]dx,$$

those, the series (26) can be integrated term by term, the resulting series converges and its sum is equal to $\int_A^B f(x)dx$. In this case, the series itself (26) may diverge.

Consider the function

$$\varphi(x) = \begin{cases} 1, & x \in [A, B], \\ 0, & x \in [0, 1] \setminus [A, B]. \end{cases}$$

Each of the systems $\{u_k(x)\}$, $\{v_k(x)\}$, forms an unconditional basis in the space $\mathcal{L}^2(0, 1)$. Let us expand the function $\varphi(x)$ into a series in the system $\{v_k(x), u_k(x)\}$, and call it the conjugate series:

$$\begin{aligned} &\varphi(x) \sim 2\alpha_0 + 4\sum_{k=1}^{\infty}[\alpha_k(1-x)\sin 2\pi kx + \beta_k \cos 2\pi kx] = \\ &= 2(\varphi(\tau), \tau) + 4\sum_{k=1}^{\infty}[(\varphi(\tau), \sin 2\pi k\tau)(1-x)\ sin2\pi kx + \\ &+(\varphi(\tau), \tau\cos 2\pi k\tau)\cos 2\pi kx]. \end{aligned} \tag{27}$$

Let us calculate the coefficients $\alpha_0, \alpha_k, \beta_k$, $k \geq 1$, of the series (27). We have

$$\alpha_0 = (\varphi(\tau), \tau) = \int_A^B \tau d\tau = \tfrac{1}{2}(B^2 - A^2),$$

$$\alpha_k = (\varphi(\tau), \sin 2\pi k\tau) = \int_A^B \sin 2\pi k\tau d\tau = \tfrac{1}{2\pi k}(\cos 2\pi kA - \cos 2\pi kB),$$

$$\begin{aligned} &\beta_k = (\varphi(\tau), \tau\cos 2\pi k\tau) = \int_A^B \tau\cos 2\pi k\tau d\tau = \tfrac{1}{2\pi k}[B\sin 2\pi kB - A\sin 2\pi kA + \\ &+\tfrac{1}{2\pi k}(\cos 2\pi kA - \cos 2\pi kB)]. \end{aligned}$$

Let us substitute the obtained relations for the coefficients into the partial sum $S_n(x)$ of the series (27):

$$\begin{aligned} &S_n(x) = B^2 - A^2 + 4\sum_{k=1}^{n}\Big[\tfrac{1}{2\pi k}(\cos 2\pi kA - \cos 2\pi kB)(1-x)\sin 2\pi kx + \\ &+\tfrac{1}{2\pi k}(B\sin 2\pi kB - A\sin 2\pi kA)\cos 2\pi kx + \tfrac{1}{4\pi^2k^2}(\cos 2\pi kA - \\ &-\cos 2\pi kB)\cos 2\pi kx\Big]. \end{aligned}$$

Let us prove that (1) the sequence $\{S_n(x)\}$ converges $\forall x \in [0, 1]$, (2) the sequence $\{S_n(x)\}$ is uniformly bounded in n and x to $[0, 1]$.

(1). To prove the convergence of the series (27), we apply the Dirichlet-Abel test and the comparison test for numerical series. We transform the products of trigonometric functions into sums and group terms. We will receive

$$S_n(x) = B^2 - A^2 + \tfrac{1-x-A}{\pi} \sum_{k=1}^{n} \tfrac{\sin 2\pi k(A+x)}{k} - \tfrac{1-x+A}{\pi} \sum_{k=1}^{n} \tfrac{\sin 2\pi k(A-x)}{k} +$$
$$+ \tfrac{x-1+B}{\pi} \sum_{k=1}^{n} \tfrac{\sin 2\pi k(B+x)}{k} + \tfrac{1-x+B}{\pi} \sum_{k=1}^{n} \tfrac{\sin 2\pi k(B-x)}{k} + \tag{28}$$
$$+ \tfrac{1}{\pi^2} \sum_{k=1}^{n} \tfrac{1}{k^2} (\cos 2\pi k B - \cos 2\pi k A) \cos 2\pi k x.$$

According to the usual scheme, we obtain the estimates

$$\Big|\sum_{k=1}^{n} \sin 2\pi k(A \pm x)\Big| \le \frac{1}{|\sin \pi(A \pm x)|}, \quad \forall n, \ \forall x \in [0, 1],$$

$A \pm x \neq 0, A + x \neq 1$. If $A \pm x = 0$ or $A + x = 1$, then the corresponding sums are equal to zero;

$$\Big|\sum_{k=1}^{n} \sin 2\pi k(B \pm x)\Big| \le \frac{1}{|\sin \pi(B \pm x)|}, \quad \forall n, \ \forall x \in [0, 1],$$

$B - x \neq 0, B \pm x \neq 1, 2$. If $B \pm x = 1, 2$ or $B - x = 0$, then the corresponding sums are equal to zero.

Thus, the sums of sines in the first four partial sums in (28) are bounded in absolute value for all values of n and $x \in [0, 1]$. Consequently, the series corresponding to these sums converge in every point $x \in [0, 1]$. The series corresponding to the last sum in (28) converges absolutely and uniformly on the set [0, 1].

Thus, the sequence $\{S_n(x)\}$ converges at every point $x \in [0, 1]$, i.e. the series (27) converges on [0, 1].

(2). Let us prove that there is a constant $c > 0$ such that $|S_n(x)| \le c, \forall n, \ \forall x \in [0, 1]$. To do this, we prove that each of the sums on the right-hand side (28) is uniformly bounded.

Let us use the well-known estimate ([7, p. 318])

$$\Big|\sum_{k=1}^{n} \frac{\sin kt}{k}\Big| \le 2\sqrt{\pi}, \quad \forall n, \ \forall t \in \mathbf{R}.$$

Putting in the first sum in (28) $t = 2\pi(A + x)$, we obtain

$$\Big|\sum_{k=1}^{n} \frac{\sin 2\pi k(A + x)}{k}\Big| = \Big|\sum_{k=1}^{n} \frac{\sin kt}{k}\Big| \le 2\sqrt{\pi}, \quad \forall n, \ \forall x \in [0, 1].$$

Similarly, we evaluate the next three sums in (28). For the last sum in (28), we obtain an upper bound in terms of the constant $c = 4$, $\forall n$, $\forall x \in [0, 1]$, since

$\sum_{k=1}^{n} \frac{1}{k^2} < 2,\ \forall n$. For the sum $S_n(x)$, we obtain an estimate uniform in n and $x \in [0, 1]$ in terms of the constant $c_1 = 1 + \frac{24}{\sqrt{\pi}} + \frac{4}{pi^2}$:

$$|S_n(x)| \leq c_1, \quad \forall n \geq 1,\ \forall x \in [0, 1]. \tag{29}$$

The results obtained in (1), (2) make it possible to apply the Lebesgue theorem on passing to the limit ([7, p. 139]):

$$\int_0^1 f(x)\varphi(x)dx = \lim_{n\to\infty} \int_0^1 f(x)S_n(x)dx,$$

or, use the relation (27),

$$\int_A^B f(x)dx = 2\alpha_0 \int_0^1 f(x)dx + 4\sum_{k=1}^{\infty} \Big[\alpha_k \int_0^1 f(x)(1-x)\sin 2\pi kxdx +$$

$$+\beta_k \int_0^1 f(x)\cos 2\pi kxdx\Big] = 2(1, f)\int_A^B xdx +$$

$$+4\sum_{k=1}^{\infty} \Big[(f(\tau), (1-\tau)\sin 2\pi k\tau)\int_A^B \sin 2\pi kxdx +$$

$$+(f(\tau), \cos 2\pi k\tau)\int_A^B x\cos 2\pi kxdx\Big],$$

those, we get the required formula. Theorem 2 is proved.

The author is grateful to A.P. Khromov for helpful discussions of the results of this work.

References

1. Khromov AP (2022) Divergent series and a generalized mixed problem for the wave equation Modern problems of the theory of functions and their applications. Saratov University, Saratov, Issue (21), pp 319–324 (in Russian)
2. Khromov AP, Kornev VV (2019) Classical and generalized solutions of a mixed problem for an inhomogeneous wave equation. Comput Math Math Phys **59**(2):286–300 (in Russian). https://doi.org/10.1134/S0965542516020135
3. Lomov IS (2021) Efficient application of the Fourier technique for constructing a solution to a mixed problem for a telegraph equation. Mosc Univ Comput Math Cybern **45**:168–173 (in Russian). https://doi.org/10.3103/S0278641921040038

4. Khromov AP (2019) Necessary and sufficient conditions for the existence of a classical solution of a mixed problem for a homogeneous wave equation in the case of an integrable potential. Differ Equ **55**(5):717–731 (in Russian). https://doi.org/10.1134/S0012266119050112
5. Euler L (1949) Differential calculus. Moscow (in Russian)
6. Hardy G (2006) Divergent series. Moscow (in Russian)
7. Natanson IP (1957) The theory of functions of a real variable. Moscow (in Russian)

Non-local Substitutions for Liouville Equations with Three and Four Independent Variables

Aleksey Mironov and Lyubov Mironova

Abstract We obtained the non-local transformations of the Cole—Hopf type, which translate the Liouville equations with three and four independent variables into the Bianchi equations. The solutions with arbitrary functions of these Liouville equations are constructed.

Keywords Liouville equation · Bianchi equation · Non-local transformation

1 On the Group Properties of Bianchi Equations

Consider a homogeneous equation with a dominant partial derivative with variable coefficients (Bianchi equation)

$$u_{xyz} + au_{xy} + bu_{yz} + cu_{xz} + du_x + eu_y + fu_z + gu = 0. \tag{1}$$

In the paper [1] some group properties of this equation have been considered. It is known that the set of equivalence transformations for (1)

$$\overline{x} = \alpha(x), \quad \overline{y} = \beta(y), \quad \overline{z} = \gamma(z), \quad u = \omega(x, y, z)\overline{u}. \tag{2}$$

Two equations of the form (1) are called equivalent in function [2, p 117], if they pass into each other during transformations (2), in which

$$\alpha(x) = x, \quad \beta(y) = y, \quad \gamma(z) = z.$$

A. Mironov · L. Mironova (✉)
Kazan (Volga Region) Federal University, Kazan, Russia
e-mail: lbmironova@yandex.ru

V. Vasilyev (ed.), *Differential Equations, Mathematical Modeling and Computational Algorithms*, Springer Proceedings in Mathematics & Statistics 423,
https://doi.org/10.1007/978-3-031-28505-9_6

In the paper [3] it was shown that two equations of the form (1) are equivalent in function if and only if the Laplace invariants

$$
\begin{gathered}
H_1 = a_y + ac - d,\ H_2 = a_x + ab - e,\ H_3 = c_x + bc - f,\\
H_4 = b_z + ab - e,\ H_5 = b_y + bc - f,\ H_6 = c_z + ac - d,\\
H_7 = a_{xy} + bd + ce + af - 2abc - g,\\
H_8 = b_{yz} + bd + ce + af - 2abc - g,\\
H_9 = c_{xz} + bd + ce + af - 2abc - g
\end{gathered} \tag{3}
$$

are the same for both equations.

If we look for the operator allowed by the Eq. (1)

$$\alpha\partial_x + \beta\partial_y + \gamma\partial_z + \tau\partial_u,$$

then it turns out that part of the system of defining equations will be

$$\partial_u\alpha = \partial_u\beta = \partial_u\gamma = 0, \quad \partial_u^2\tau = 0.$$

It is known [2, pp. 99–100] that in this case the Lie algebra of the Eq. (1) there is $L = L^r \oplus L^\infty$, where the algebra L^r of dimension r is formed by operators of the form

$$X = \xi^1(x,y,z)\partial_x + \xi^2(x,y,z)\partial_y + \xi^3(x,y,z)\partial_z + \sigma(x,y,z)u\partial_u, \tag{4}$$

and L^∞ is an Abelian subalgebra typical of linear equations with the operator $\omega(x,y,z)\partial_u$, where ω is the solution of the Eq. (1). It is clear that the operator $u\partial_u$ is allowed by any Eq. (1), therefore, this operator can be included in L^∞ and assume that $\sigma(x,y,z)$ is defined in (4) up to a constant summand.

To construct the defining equations we use the third continuation of the operator (4)

$$
\begin{gathered}
X_3 = \xi^1\partial_x + \xi^2\partial_y + \xi^3\partial_z + \sigma u\partial_u + \tau^1\partial_{u_1} + \tau^2\partial_{u_2} + \tau^3\partial_{u_3} +\\
+\tau^{11}\partial_{u_{11}} + \tau^{12}\partial_{u_{12}} + \tau^{13}\partial_{u_{13}} + \tau^{22}\partial_{u_{22}} + \tau^{23}\partial_{u_{23}} + \tau^{33}\partial_{u_{33}} +\\
+\tau^{111}\partial_{u_{111}} + \tau^{112}\partial_{u_{112}} + \tau^{113}\partial_{u_{113}} + \tau^{122}\partial_{u_{122}} + \tau^{123}\partial_{u_{123}} +\\
+\tau^{133}\partial_{u_{133}} + \tau^{222}\partial_{u_{222}} + \tau^{223}\partial_{u_{223}} + \tau^{233}\partial_{u_{233}} + \tau^{333}\partial_{u_{333}}.
\end{gathered}
$$

The notation used here is $u_1 = u_x$, $u_2 = u_x$,..., $u_{12} = u_{xy}$,..., $u_{333} = u_{zzz}$. We get

$$\begin{aligned}
\tau^1 &= \sigma_x u + (\sigma - \xi^1_x)u_1 - \xi^2_x u_2 - \xi^3_x u_3,\\
\tau^2 &= \sigma_y u - \xi^1_y u_1 + (\sigma - \xi^2_y)u_2 - \xi^3_y u_3,\\
\tau^3 &= \sigma_z u - \xi^1_z u_1 - \xi^2_z u_2 + (\sigma - \xi^3_z)u_3,\\
\tau^{12} &= \sigma_{xy} u + (\sigma_y - \xi^1_{xy})u_1 + (\sigma_x - \xi^2_{xy})u_2 - \xi^3_{xy}u_3 -\\
&-\xi^1_y u_{11} + (\sigma - \xi^1_x - \xi^2_y)u_{12} - \xi^3_y u_{13} - \xi^2_x u_{22} - \xi^3_x u_{23},\\
\tau^{13} &= \sigma_{xz} u + (\sigma_z - \xi^1_{xz})u_1 - \xi^2_{xz}u_2 + (\sigma_x - \xi^3_{xz})u_3 -\\
&-\xi^1_z u_{11} - \xi^2_z u_{12} + (\sigma - \xi^1_x - \xi^3_z)u_{13} - \xi^2_x u_{23} - \xi^3_x u_{33},\\
\tau^{23} &= \sigma_{yz} u - \xi^1_{yz}u_1 + (\sigma_z - \xi^2_{yz})u_2 + (\sigma_y - \xi^3_{yz})u_3 -\\
&-\xi^1_z u_{12} - \xi^1_y u_{13} - \xi^2_z u_{22} + (\sigma - \xi^1_y - \xi^3_z)u_{23} - \xi^3_y u_{33},\\
\tau^{123} &= \sigma_{xyz} u + (\sigma_{yz} - \xi^1_{xyz})u_1 + (\sigma_{xz} - \xi^2_{xyz})u_2 + (\sigma_{xy} - \xi^3_{xyz})u_3 -\\
&-\xi^1_{yz}u_{11} + (\sigma_z - \xi^2_{yz} - \xi^1_{xz})u_{12} + (\sigma_y - \xi^1_{xy} - \xi^3_{yz})u_{13} -\\
&-\xi^2_{xz}u_{22} + (\sigma_x - \xi^3_{xz} - \xi^2_{xy})u_{23} - \xi^3_{xy}u_{33} -\\
&-\xi^1_z u_{112} - \xi^1_y u_{113} - \xi^2_z u_{122} + (\sigma - \xi^1_x - \xi^2_y - \xi^3_z)u_{123} -\\
&-\xi^3_y u_{133} - \xi^2_x u_{223} - \xi^3_x u_{233}.
\end{aligned}$$

By applying the operator X_3 to the Eq. (1), we obtain the defining equations

$$\begin{gathered}
\xi^1_y = \xi^1_z = \xi^2_x = \xi^2_z = \xi^3_x = \xi^3_y = 0,\\
\sigma_x + (b\xi^1)_x + b_y\xi^2 + b_z\xi^3 = 0,\\
\sigma_y + c_x\xi^1 + (c\xi^2)_y + c_z\xi^3 = 0,\\
\sigma_z + a_x\xi^1 + a_y\xi^2 + (a\xi^3)_z = 0,\\
\sigma_{xy} + c\sigma_x + b\sigma_y + (f\xi^1)_x + (f\xi^2)_y + f_z\xi^3 = 0,\\
\sigma_{xz} + a\sigma_x + b\sigma_z + (e\xi^1)_x + e_y\xi^2 + (e\xi^3)_z = 0,\\
\sigma_{yz} + a\sigma_y + c\sigma_z + d_x\xi^1 + (d\xi^2)_y + (d\xi^3)_z = 0,\\
\sigma_{xyz} + a\sigma_{xy} + b\sigma_{yz} + c\sigma_{xz} + d\sigma_x + e\sigma_y + f\sigma_z +\\
+(g\xi^1)_x + (g\xi^2)_y + (g\xi^3)_z = 0.
\end{gathered} \tag{5}$$

Defining Eq. (5) can be written using Laplace invariants (3) in the form

$$\begin{gathered}
\xi^1_y = \xi^1_z = \xi^2_x = \xi^2_z = \xi^3_x = \xi^3_y = 0,\\
(\sigma + b\xi^1 + c\xi^2 + a\xi^3)_x = (H_3 - H_5)\xi^2 + (H_2 - H_4)\xi^3,\\
(\sigma + b\xi^1 + c\xi^2 + a\xi^3)_y = (H_5 - H_3)\xi^1 + (H_1 - H_6)\xi^3,\\
(\sigma + b\xi^1 + c\xi^2 + a\xi^3)_z = (H_4 - H_2)\xi^1 + (H_6 - H_1)\xi^2,\\
H_{1x}\xi^1 + (H_1\xi^2)_y + (H_1\xi^3)_z = 0,\\
H_{6x}\xi^1 + (H_6\xi^2)_y + (H_6\xi^3)_z = 0,\\
(H_2\xi^1)_x + H_{2y}\xi^2 + (H_2\xi^3)_z = 0,\\
(H_4\xi^1)_x + H_{4y}\xi^2 + (H_4\xi^3)_z = 0,\\
(H_3\xi^1)_x + (H_3\xi^2)_y + H_{3z}\xi^3 = 0,\\
(H_5\xi^1)_x + (H_5\xi^2)_y + H_{5z}\xi^3 = 0,\\
(H_7\xi^1)_x + (H_7\xi^2)_y + (H_7\xi^3)_z = 0,\\
(H_8\xi^1)_x + (H_8\xi^2)_y + (H_8\xi^3)_z = 0,\\
(H_9\xi^1)_x + (H_9\xi^2)_y + (H_9\xi^3)_z = 0.
\end{gathered} \tag{6}$$

The first row in (6) shows that

$$\xi^i = \xi^i(x_i), \quad i = \overline{1,3}.$$

The second, third and fourth rows from (6) are differential equations for determining the function σ, after ξ^1, ξ^2, ξ^3 have been obtained. The equations starting from the fifth row are responsible for the results of the group classification.

Some consequences can be deduced directly from the defining equations in the form (6). If all $H_i, i = \overline{1,9}$, are identically equal to zero, then the Eq. (1) is equivalent to the equation $u_{xyz} = 0$ and admits an infinite-dimensional Lie algebra of operators of the form

$$\xi^1(x)\partial_x + \xi^2(y)\partial_y + \xi^3(z)\partial_z$$

with arbitrary $\xi^1(x)$, $\xi^2(y)$, $\xi^3(z)$.

Let's introduce the relations into consideration

$$p_{12} = \frac{H_3}{H_5}, \quad p_{13} = \frac{H_2}{H_4}, \quad p_{23} = \frac{H_1}{H_6}, \tag{7}$$

$$\begin{aligned} q_1 &= \frac{(\ln H_1)_{yz}}{H_1}, \quad q_2 = \frac{(\ln H_2)_{xz}}{H_2}, \quad q_3 = \frac{(\ln H_3)_{xy}}{H_3}, \\ q_4 &= \frac{(\ln H_4)_{xz}}{H_4}, \quad q_5 = \frac{(\ln H_5)_{xy}}{H_5}, \quad q_6 = \frac{(\ln H_6)_{yz}}{H_6}, \\ q_i &= \frac{(\ln H_i)_{xyz}}{H_i}, \quad i = 7, 8, 9. \end{aligned} \tag{8}$$

Substitute $H_1 = p_{23}H_6$, $H_6 \neq 0$, in the fifth row (6)

$$p_{23}(H_{6x}\xi^1 + (H_6\xi^2)_y + (H_6\xi^3)_z) + p_{23x}H_6\xi^1 + p_{23y}H_6\xi^2 + p_{23z}H_6\xi^3 = 0.$$

Since the term in parentheses vanishes, it follows

$$\xi^1 p_{23x} + \xi^2 p_{23y} + \xi^3 p_{23z} = 0. \tag{9}$$

The identity (9) means that either $p_{23} = const$ or p_{23} is an invariant of the group G with the operator (4).

If $p_{23} = const$, then from the fifth and sixth rows (6) we get

$$\xi^1(\ln H_6)_x + \xi^2(\ln H_6)_y + \xi^3(\ln H_6)_z + \xi^2_y + \xi^3_z = 0. \tag{10}$$

Differentiating by y, z we get

$$\xi^1\frac{((\ln H_6)_{yz})_x}{(\ln H_6)_{yz}} + \xi^2\frac{((\ln H_6)_{yz})_y}{(\ln H_6)_{yz}} + \xi^3\frac{((\ln H_6)_{yz})_z}{(\ln H_6)_{yz}} + \xi^2_y + \xi^3_z = 0. \tag{11}$$

Subtracting (10) from (11) and then multiplying by $(\ln H_6)_{yz}/H_6$, we get

$$\xi^1 q_{6x} + \xi^2 q_{6y} + \xi^3 q_{6z} = 0.$$

Thus, again either $q_6 = const$ or q_6 is an invariant of the group G with the operator (4).

Then similar identities can be obtained for $p_{12}, p_{13}, q_i, i = \overline{1,5}$.

Similar identities can be obtained for relations

$$P_1 = \frac{H_7}{H_8}, \quad P_2 = \frac{H_7}{H_9}, \quad P_3 = \frac{H_8}{H_9}.$$

For example, considering the relation P_1, we come to the identity

$$\xi^1 P_{1x} + \xi^2 P_{1y} + \xi^3 P_{1z} = 0.$$

Again, either $P_1 = const$, or P_1 is an invariant of the group G with the operator (4). If $P_1 = const$, then row 12 from (6) gives

$$\xi^1 (\ln H_8)_x + \xi^2 (\ln H_8)_y + \xi^3 (\ln H_8)_z + \xi^1_x + \xi^2_y + \xi^3_z = 0. \tag{12}$$

Differentiating by x, y, z we get

$$\xi^1 \frac{((\ln H_8)_{xyz})_x}{(\ln H_8)_{xyz}} + \xi^2 \frac{((\ln H_8)_{xyz})_y}{(\ln H_8)_{xyz}} + \xi^3 \frac{((\ln H_8)_{xyz})_z}{(\ln H_8)_{xyz}} + \xi^1_x + \xi^2_y + \xi^3_z = 0. \tag{13}$$

Subtracting (12) from (13) and multiplying by $(\ln H_8)_{xyz}/H_8$, we get

$$\xi^1 q_{8x} + \xi^2 q_{8y} + \xi^3 q_{8z} = 0.$$

Thus, either $q_8 = const$ or q_8 is an invariant of the group G with the operator (4).

Based on the above statements, classes of equations of the form (1) admitting Lie algebras of the largest dimensions were listed in the work [1].

In the case when $q_i = const, i = \overline{1,6}$, the invariant H_i is a solution of the Liouville equation (this follows from (8)), the formula of the general solution of which is known [2, p 123]. Similarly, if any of the constructions q_i, $i = \overline{7,9}$, is constant, then the corresponding invariant H_i is the solution of the equation

$$(\ln H_i)_{xyz} = q_i H_i.$$

In this regard, the task of constructing is of interest exact solutions of the three-dimensional analogue of the Liouville equation

$$u_{xyz} = e^u. \tag{14}$$

We can propose the following method of constructing an exact solution based on the application of Lie groups of point transformations.

The usual algorithm for calculating the group of point transformations allowed by the Eq. (14) leads to the Lie algebra of operators

$$X = \xi(x)\partial_x + \eta(y)\partial_y + \zeta(z)\partial_z - (\xi'(x) + \eta'(y) + \zeta'(z))\partial_u,$$

where $\xi(x)$, $\eta(y)$, $\zeta(z)$ are arbitrary functions.

To determine the invariants of the group allowed by the Eq. (14), we obtain the system

$$\frac{dx}{\xi(x)} = \frac{dy}{\eta(y)} = \frac{dz}{\zeta(z)} = \frac{du}{-\xi'(x) - \eta'(y) - \zeta'(z)}. \tag{15}$$

The first integrals of the system (15) have the form

$$u + \ln|\xi(x)\eta(y)\zeta(z)| = C_1,$$

$$\varphi(x) - \psi(y) = C_2, \quad \varphi(x) - \chi(z) = C_3,$$

$$\varphi'(x) = \frac{1}{\xi(x)}, \quad \psi'(y) = \frac{1}{\eta(y)}, \quad \chi'(z) = \frac{1}{\zeta(z)}.$$

Let's introduce new variables

$$v = u + \ln|\xi(x)\eta(y)\zeta(z)|, \quad t = \varphi(x) - \psi(y), \quad \tau = \varphi(x) - \chi(z).$$

Invariant with respect to the group of point transformations allowed by the Eq. (14) , the solution has the form $v = w(t, \tau)$. As a result, we come to the equation for determining the function w

$$w_{tt\tau} + w_{t\tau\tau} = e^w. \tag{16}$$

The Eq. (16) has a solution

$$w = \ln\frac{-12}{(t+\tau)^3}.$$

Then (here $\xi(x)\eta(y)\zeta(z) > 0$)

$$u = -\ln(\xi(x)\eta(y)\zeta(z)) + \ln\frac{-12}{(2\varphi(x) - \psi(y) - \chi(z))^3} =$$

$$= \ln\frac{-12\frac{1}{\xi(x)}\frac{1}{\eta(y)}\frac{1}{\zeta(z)}}{(2\varphi(x) - \psi(y) - \chi(z))^3}.$$

Denoting $\lambda(x) = 2\varphi(x)$, $\mu(y) = -\psi(y)$, $\nu(z) = -\chi(z)$, we obtain an exact solution of the Eq. (14), depending on three arbitrary functions

$$u = \ln \frac{-6\lambda'(x)\mu'(y)\nu'(z)}{(\lambda(x) + \mu(y) + \nu(z))^3}.$$

In [4, 5] some group properties of the fourth-order Bianchi equation were considered. The homogeneous Bianchi equation of the fourth order is

$$\begin{gathered} u_{x_1x_2x_3x_4} + a_1 u_{x_2x_3x_4} + a_2 u_{x_1x_3x_4} + a_3 u_{x_1x_2x_4} + a_4 u_{x_1x_2x_3} + \\ + a_{12} u_{x_3x_4} + a_{13} u_{x_2x_4} + a_{14} u_{x_2x_3} + a_{23} u_{x_1x_4} + a_{24} u_{x_1x_3} + a_{34} u_{x_1x_2} + \\ + a_{123} u_{x_4} + a_{124} u_{x_3} + a_{134} u_{x_2} + a_{234} u_{x_1} + a_{1234} u = 0. \end{gathered} \tag{17}$$

It is implied here that the coefficients are variable.

The Laplace invariants for this equation have the form

$$\begin{gathered} h_{i,j} = a_{ix_j} + a_i a_j - a_{ij}, \\ h_{i,jk} = a_{ix_jx_k} + a_i a_{jk} + a_j a_{ik} + a_k a_{ij} - 2a_i a_j a_k - a_{ijk}, \\ h_{i,jkl} = a_{ix_jx_kx_l} + a_i a_{jkl} + a_j a_{ikl} + a_k a_{ijl} + a_l a_{ijk} + \\ + a_{ij} a_{kl} + a_{ik} a_{jl} + a_{il} a_{jk} - 2a_i a_j a_{kl} - 2a_i a_k a_{jl} - \\ -2a_i a_l a_{jk} - 2a_j a_k a_{il} - 2a_j a_l a_{ik} - 2a_k a_l a_{ij} + \\ +6a_i a_j a_k a_l - a_{ijkl}, \quad \{i, j, k, l\} = \{1, 2, 3, 4\}, \quad j < k < l. \end{gathered}$$

Here we consider coefficients that differ in the order of the indices to be equal (for example, $a_{123} = a_{231}$). There are a total of 28 Laplace invariants for this equation. Two equations of the form (17) are equivalent in function if and only if they have all the corresponding Laplace invariants equal.

Note that if all Laplace invariants are identically zero, then the Eq. (17) is equivalent to the equation $u_{x_1x_2x_3x_4} = 0$ and admits an infinite-dimensional Lie algebra of operators of the form

$$\xi^1(x_1)\partial_{x_1} + \xi^2(x_2)\partial_{x_2} + \xi^3(x_3)\partial_{x_3} + \xi^4(x_4)\partial_{x_4}$$

with arbitrary $\xi^i(x_i)$.

Similarly to the case of the third-order Bianchi equation, we can introduce into consideration the constructions

$$p_{ij} = \frac{h_{j,i}}{h_{i,j}}, \quad q_{ij} = \frac{(\ln h_{i,j})_{x_ix_j}}{h_{i,j}}, \quad i, j = \overline{1,4};$$

$$p^l_{ijk} = \frac{h_{l,l_1l_2}}{h_{i,jk}}, \quad q_{ijk} = \frac{(\ln h_{i,jk})_{x_ix_jx_k}}{h_{i,jk}}, \quad \{l, l_1, l_2\} = \{i, j, k\};$$

$$p^n_{ijkl} = \frac{h_{n,n_1n_2n_3}}{h_{i,jkl}}, \quad q_{ijkl} = \frac{(\ln h_{i,jkl})_{x_1x_2x_3x_4}}{h_{i,jkl}}, \quad \{n, n_1, n_2, n_3\} = \{i, j, k, l\}.$$

These constructions are used in [5] to obtain classes of fourth-order Bianchi equations with certain group properties.

It is easy to notice that for constants $q_{ij}, q_{ijk}, q_{ijkl}$ the Laplace invariants are again solutions of the Liouville equation and its three-dimensional and four-dimensional analogues.

2 Three-Dimensional Analogue of the Liouville Equation

Let us consider an approach to the problem of constructing exact solutions to non-linear equations based on non-local transformations of variables. Equation

$$u_{xyz} = \lambda e^u \tag{18}$$

is a three-dimensional analogue of the Liouville equation

$$u_{xy} = \lambda e^u. \tag{19}$$

Equation (19), in particular, plays a key role in the problem of group classification of second-order hyperbolic equations [2, pp. 116–125]

$$v_{xy} + a(x, y)v_x + b(x, y)v_y + c(x, y)v = 0.$$

The general solution of the Eq. (19) is well known and can be constructed in various ways [2, p. 123], [6, pp. 239–240]. As noted earlier, the Eq. (18) is used in the study of the group properties of the third-order Bianchi Eq. (1).

Here a non-local transformation (such as the Cole—Hopf substitution [7]) is constructed, translating the Eq. (18) into the simplest Bianchi equation

$$v_{xyz} = 0, \tag{20}$$

which has a general solution with three arbitrary functions

$$v = \alpha(x, y) + \beta(x, z) + \gamma(y, z). \tag{21}$$

In this case, an algorithm based on the use of group methods is used [6, pp. 237–241].

Equation (18) admits the Lie algebra of operators

$$X = \xi(x)\partial_x + \eta(y)\partial_y + \zeta(z)\partial_z - (\dot{\xi}(x) + \dot{\eta}(y) + \dot{\zeta}(z))\partial_u,$$

where $\xi(x)$, $\eta(y)$, $\zeta(z)$ are arbitrary functions [1].

On the other hand, the Eq. (20) admits the Lie algebra of operators

$$X_0 = \xi(x)\partial_x + \eta(y)\partial_y + \zeta(z)\partial_z,$$

where $\xi(x), \eta(y), \zeta(z)$ are also arbitrary. In addition, like any linear equation, Eq. (20) admits a stretching operator

$$Y = v\partial_v.$$

In this regard, assume that there is a non-local transformation

$$u = \varphi(v, v_x, v_y, v_z) \tag{22}$$

such that the system of Eqs. (18), (20), (22) admits the Lie algebra of operators

$$\begin{gathered} X = \xi(x)\partial_x + \eta(y)\partial_y + \zeta(z)\partial_z - (\dot{\xi}(x) + \dot{\eta}(y) + \dot{\zeta}(z))\partial_u, \\ Y = v\partial_v. \end{gathered}$$

We find the first continuations of operators

$$\begin{gathered} X_1 = \xi(x)\partial_x + \eta(y)\partial_y + \zeta(z)\partial_z - (\dot{\xi}(x) + \dot{\eta}(y) + \dot{\zeta}(z))\partial_u - \\ -(\ddot{\xi}(x) - \dot{\xi}(x)u_x)\partial_{u_x} - (\ddot{\eta}(y) - \dot{\eta}(y)u_y)\partial_{u_y} - (\ddot{\zeta}(z) - \dot{\zeta}(z)u_z)\partial_{u_z} + \\ +\dot{\xi}(x)v_x\partial_{v_x} + \dot{\eta}(y)v_y\partial_{v_y} + \dot{\zeta}(z)v_z\partial_{v_z}, \end{gathered}$$

$$Y_1 = v\partial_v + v_x\partial_{v_x} + v_y\partial_{v_y} + v_z\partial_{v_z}.$$

We get relations

$$Y_1(u - \varphi)|_{u=\varphi} = v\varphi_v + v_x\varphi_{v_x} + v_y\varphi_{v_y} + v_z\varphi_{v_z} = 0, \tag{23}$$

$$X_1(u - \varphi)|_{u=\varphi} = -(\dot{\xi} + \dot{\eta} + \dot{\zeta}) + \dot{\xi}(x)v_x\varphi_{v_x} + \dot{\eta}(y)v_y\varphi_{v_y} + \dot{\zeta}(z)v_z\varphi_{v_z} = 0. \tag{24}$$

Since the function v has the form (21), from (23) and (24) we get the system

$$\begin{gathered} (\alpha + \beta + \gamma)\varphi_v + (\alpha_x + \beta_x)\varphi_{v_x} + (\alpha_y + \gamma_y)\varphi_{v_y} + (\beta_z + \gamma_z)\varphi_{v_z} = 0, \\ -(\dot{\xi} + \dot{\eta} + \dot{\zeta}) + \dot{\xi}(x)(\alpha_x + \beta_x)\varphi_{v_x} + \dot{\eta}(y)(\alpha_y + \gamma_y)\varphi_{v_y} + \dot{\zeta}(z)(\beta_z + \gamma_z)\varphi_{v_z} = 0. \end{gathered} \tag{25}$$

The system (25) is satisfied by the relation

$$u = \varphi(v, v_x, v_y, v_z) = \ln\frac{cv_xv_yv_z}{v^3} = \ln c + \ln v_x + \ln v_y + \ln v_z - 3\ln v. \tag{26}$$

Substituting (26) into the Eq. (18) taking into account (21) leads to a formula defining a class of solutions to the Eq. (18) depending on three arbitrary functions

$$u = \ln\left(-\frac{6}{\lambda}\frac{f_1'(x)f_2'(y)f_3'(z)}{(f_1(x) + f_2(y) + f_3(z))^3}\right). \tag{27}$$

Here $f_1(x)$, $f_2(y)$, $f_3(z)$—arbitrary continuously differentiable functions.

3 Fourth-Order Analogue of the Liouville Equation

Now consider the equation

$$u_{xyzt} = \lambda e^u, \tag{28}$$

related to the fourth-order linear Bianchi equation, whose group properties are considered in [4, 5].

Similarly to the case of the Eq. (18), we construct a non-local transformation that translates the Eq. (28) into the equation

$$v_{xyzt} = 0, \tag{29}$$

the general solution of which

$$v = \alpha(x, y, z) + \beta(x, y, t) + \gamma(x, z, t) + \delta(y, z, t). \tag{30}$$

Equation (28) admits a Lie algebra of operators

$$X = \xi(x)\partial_x + \eta(y)\partial_y + \zeta(z)\partial_z + \tau(t)\partial_t - (\dot{\xi}(x) + \dot{\eta}(y) + \dot{\zeta}(z) + \dot{\tau}(t))\partial_u,$$

where $\xi(x)$, $\eta(y)$, $\zeta(z)$, $\tau(t)$ are arbitrary functions [5].

On the other hand, the Eq. (29) admits the Lie algebra of operators

$$X_0 = \xi(x)\partial_x + \eta(y)\partial_y + \zeta(z)\partial_z + \tau(t)\partial_t,$$

as well as the stretching operator

$$Y = v\partial_v.$$

Looking for a non-local transformation

$$u = \varphi(v, v_x, v_y, v_z, v_t) \tag{31}$$

such that the system of Eqs. (28), (29), (31) admits the Lie algebra of operators

$$X = \xi(x)\partial_x + \eta(y)\partial_y + \zeta(z)\partial_z + \tau(t)\partial_t - (\dot{\xi}(x) + \dot{\eta}(y) + \dot{\zeta}(z) + \dot{\tau}(t))\partial_u,$$
$$Y = v\partial_v.$$

We calculate the first continuations of operators

$$X_1 = \xi(x)\partial_x + \eta(y)\partial_y + \zeta(z)\partial_z + \tau(t)\partial_t - (\dot{\xi}(x) + \dot{\eta}(y) + \dot{\zeta}(z) + \dot{\tau}(t))\partial_u -$$
$$-(\ddot{\xi}(x) - \dot{\xi}(x)u_x)\partial_{u_x} - (\ddot{\eta}(y) - \dot{\eta}(y)u_y)\partial_{u_y} - (\ddot{\zeta}(z) - \dot{\zeta}(z)u_z)\partial_{u_z} -$$
$$-(\ddot{\tau}(t) - \dot{\tau}(t)u_t)\partial_{u_t} + \dot{\xi}(x)v_x\partial_{v_x} + \dot{\eta}(y)v_y\partial_{v_y} + \dot{\zeta}(z)v_z\partial_{v_z} + \dot{\tau}(t)v_t\partial_{v_t},$$

$$Y_1 = v\partial_v + v_x\partial_{v_x} + v_y\partial_{v_y} + v_z\partial_{v_z} + v_t\partial_{v_t}$$

and we write down the ratios

$$Y_1(u-\varphi)|_{u=\varphi} = v\varphi_v + v_x\varphi_{v_x} + v_y\varphi_{v_y} + v_z\varphi_{v_z} + v_t\varphi_{v_t} = 0, \tag{32}$$

$$X_1(u-\varphi)|_{u=\varphi} = -(\dot{\xi}+\dot{\eta}+\dot{\zeta}+\dot{\tau}) + \dot{\xi}v_x\varphi_{v_x} + \dot{\eta}v_y\varphi_{v_y} + \dot{\zeta}v_z\varphi_{v_z} + \dot{\tau}v_t\varphi_{v_t} = 0. \tag{33}$$

The function v has the form (30), therefore from (32)–(33) we get the system

$$\begin{array}{c} (\alpha+\beta+\gamma+\delta)\varphi_v + (\alpha_x+\beta_x+\gamma_x)\varphi_{v_x} + (\alpha_y+\beta_y+\delta_y)\varphi_{v_y} + \\ +(\alpha_z+\gamma_z+\delta_z)\varphi_{v_z} + (\beta_t+\gamma_t+\delta_t)\varphi_{v_t} = 0, \\ -(\dot{\xi}+\dot{\eta}+\dot{\zeta}+\dot{\delta}) + \dot{\xi}(\alpha_x+\beta_x+\gamma_x)\varphi_{v_x} + \dot{\eta}(\alpha_y+\beta_y+\delta_y)\varphi_{v_y} + \\ +\dot{\zeta}(\alpha_z+\gamma_z+\delta_z)\varphi_{v_z} + \dot{\delta}(\beta_t+\gamma_t+\delta_t)\varphi_{v_t} = 0. \end{array} \tag{34}$$

The system (34) is satisfied by the relation

$$u = \ln \frac{cv_x v_y v_z v_t}{v^4}. \tag{35}$$

Substituting (35) into (28) and taking into account (30), we get the solution of the Eq. (28)

$$u = \ln\left(\frac{24}{\lambda}\frac{f_1'(x)f_2'(y)f_3'(z)f_4'(t)}{(f_1(x)+f_2(y)+f_3(z)+f_4(t))^4}\right),$$

where $f_1(x)$, $f_2(y)$, $f_3(z)$, $f_4(t)$—arbitrary continuously differentiable functions.

This paper has been supported by the Kazan Federal University Strategic Academic Leadership Program (Priority—2030).

References

1. Mironov AN (2013) Classes of Bianchi equations of third order. Math Notes 94:369–378. https://doi.org/10.1134/S0001434613090083
2. Ovsyannikov LV (1978) Group analysis of differential equations. Nauka, Moscow [in Russian]
3. Dzhokhadze OM (2004) Laplace invariants for some classes of linear partial differential equations. Differ Equ 40:63–74. https://doi.org/10.1023/B:DIEQ.0000028714.62481.2d
4. Mironov AN (2009) On the Laplace invariants of a fourth-order equation. Differ Equ 45:1168–1173. https://doi.org/10.1134/S0012266109080084
5. Mironov AN (2013) On some classes of fourth-order Bianchi equations with constant ratios of Laplace invariants. Differ Equ 49:1524–1533. https://doi.org/10.1134/S0012266113120070
6. Fushchich VI, Shtelen' VM, Serov NI (1989) Symmetric analysis and exact solutions of nonlinear equations of mathematical physics. Naukova dumka, Kyiv [in Russian]
7. Cole JD (1951) On a quasi-linear parabolic equation occurring in aerodynamics. Quart Appl Math 9:225–236. https://doi.org/10.1090/qam/42889

Convergence Rates of a Finite Difference Method for the Fractional Subdiffusion Equations

Li Liu, Zhenbin Fan, Gang Li, and Sergey Piskarev

Abstract We consider the convergence of an effective numerical method of the subdiffusion equation with the Caputo fractional derivative in time. We investigate an implicit difference scheme and an explicit difference scheme by using the projection method in space and a finite difference method which was proposed by Ashyralyev in time. Combining the method of functional analysis and the technique of numerical analysis, we utilize the idea of layering in temporal direction to obtain that the local truncation error is $O(n^{-\alpha})$. Then we prove that the implicit and explicit numerical methods converge at a rate of $O(\tau^{\alpha})$ in time. Finally, a numerical experiment is given to confirm the α-th order accuracy.

Keywords Fractional subdiffusion equations · Weak regularity · Resolvent family · Discretization methods · Error estimate

1 Introduction

In recent years, many scholars devoted to study the linear partial differential equation

$$ {}^{c}\partial_t^{\alpha} u(x,t) = (Bu)(x,t) + f(x,t), \quad (x,t) \in E \times (0,T], \tag{1} $$

where ${}^{c}\partial_t^{\alpha} u(x,t)$ denotes the Caputo time fractional derivative of u with order α, $0 < \alpha < 1$, B is a symmetric uniformly elliptic operator which generates an analytic α-resolvent family, f is a given source term in $E \times (0,T]$, $E \subset \mathbb{R}^n$ ($n \in \mathbb{N}^+$) is a

L. Liu · Z. Fan · G. Li
School of Mathematical Sciences, Yangzhou University, Jiangsu 225002, China
e-mail: zbfan@yzu.edu.cn

S. Piskarev (✉)
Scientific Research Computer Center, Lomonosov Moscow State University, Leninskie Gory, Moscow 119899, Russia
e-mail: piskarev@gmail.com

Mari State University, Lenin Square 1, Yoshkar-Ola City 424000, Russia

V. Vasilyev (ed.), *Differential Equations, Mathematical Modeling and Computational Algorithms*, Springer Proceedings in Mathematics & Statistics 423,
https://doi.org/10.1007/978-3-031-28505-9_7

bounded convex polygonal domain with sufficiently smooth boundary ∂E. This kind of equation has recently attracted increasing interest in the physical, chemical and engineering fields due to its excellent modeling capability. It is known that nature often does not follow the Gaussian predictions and violates the Gaussian universality mirrored in experimental results. In contrast to Gaussian diffusion, thermal diffusion and anomalous diffusion both are involved fields with intriguing subtleties. Fractional diffusion equations can account for the typical anomalous characters which are observed in many phenomena. For example, system (1) can be used to accurately describe the physical phenomena of the thermal diffusion in media with fractal geometry and the anomalous diffusion in highly heterogeneous aquifers. Because of higher degrees of freedom, the models involving fractional derivatives are more successful in applications in situations where non-Gaussian and non-Markovian processes occur. For the more physical interpretations of problem (1), one can see [1, 14, 17, 23, 36, 37, 39] and the references therein. Numerical method is one of the important tools to deal with fractional differential equations. Many numerical schemes for the discretization of Caputo fractional subdiffusion problem (1) have been devised (see, e.g., [2, 3, 5, 6, 16, 24, 33, 38]). Meanwhile, the convergence rates of existing schemes for fractional subdiffusion equations also have been the subjects of numerous studies. However, the discussion on convergence rates of the effective and robust discretization techniques for systems involving Caputo fractional derivative is relatively scarce.

In this paper, we investigate the convergence rates of two novel numerical schemes by virtue of the idea of layering in temporal direction. Under the consideration of weakly regular solution, we find that the order of convergence of fractional difference algorithm can reach $O(n^{-\alpha})$, which means the local truncation error converges at αth order accuracy far away from initial time. Furthermore, we establish the error equations and obtain that the error is of order $O(\tau^{\alpha})$. Throughout this paper, the fractional subdiffusion equation (1) is subjected to the initial and boundary conditions

$$u(x,0) = \phi(x), \quad x \in E, \tag{2}$$

$$u(x,t) = 0, \quad x \in \partial E, \quad t \in (0,T]. \tag{3}$$

Next, let us review some existing results on numerical analysis of fractional differential equations in the following paragraphs. In general, there are three kinds of representative methods to deal with the approximation of spatial direction, which are the spectral method, the finite element method and the finite difference method. The finite difference method, among others, is the most conventional method in the early stage. Stynes et al. [42] discreted the second-order spatial derivative by using the standard center finite difference algorithm, which converges at a rate $O(h^2)$. Jin et al. [21] employed the standard Galerkin finite element method and obtained the second-order accuracy of the space semidiscrete scheme in the L^2 norm. Lin and Xu [33] established a numerical scheme of order $O(h^2)$ based on the Legendre spectral methods in L^2 space for sufficiently smooth solutions. In Eq. (1), the operator

B could be the Laplacian in the domain E, i.e., $Bu = \Delta_x u$, or be a second-order strongly elliptic partial differential operator in some spatial variables,

$$Bu = \sum_{i,j=1}^{n} a^{ij} u_{x_i x_j} + \sum_{j=1}^{n} b^j u_{x_j} + cu.$$

In [44], Vainikko gave a functional-analytical treatment of discretization methods, which is often referred to as the projection method. As a general approach to establish the semidiscrete approximation in space direction, it covers the quadrature formula method and the difference method. Actually, some authors have used the projection methods in various articles and they applied it to obtain many excellent results in different senses (see, e.g., [4, 7, 10, 26, 27, 29–31, 40, 41]). In subsequent study, we use the projection methods to approximate the operator B in the system (1).

To date, there are many difference schemes used for discretization fractional derivative in time. The numerical methods include the L1, L2, L2C methods (see, e.g., [16, 21, 33]), the convolution quadrature method (see, e.g., [11, 46]), the fractional rectangular formula, fractional trapezoidal formula and fractional Newton-Cotes formula derived from the polynomial interpolation technique, and some higher order methods which are based on the explicit expression of the Jacobi polynomials (see, e.g., [9, 15, 35]). Under the assumption of sufficiently smooth solutions, some authors have shown that the high order convergence in temporal direction for various discretization methods to approximate the fractional differential equation; e.g., the schemes in [16, 30, 33, 45] have been proved to converge at a rate of order $O(\tau)$, $O(\tau^{2-\alpha})$, or even $O(\tau^{3-\alpha})$, respectively. In many papers, the high order convergence of some methods resulting from Taylor expansion requires excessive smoothness on the solution. Hence, these results are not robust with respect to the regularity of solutions. In fact, it is impossible for the solutions of the fractional subdiffusion systems to be too smooth no matter how smooth the source terms take. For example, in [31], for initial data $\phi_s \in D(B_s^{\ell+1})$ with the smallest integer ℓ such that $(\ell+1)\alpha \geq 2$, $f_s \equiv 0$, the solutions of the spatially discrete scheme of homogeneous differential equation (7) could be represented as

$$\begin{aligned} u_s(t) &= S_\alpha(t, B_s)\phi_s \\ &= \phi_s + \frac{t^\alpha B_s \phi_s}{\Gamma(\alpha+1)} + \cdots + \frac{t^{\ell\alpha} B_s^\ell \phi_s}{\Gamma(\ell\alpha+1)} + (g_{(\ell+1)\alpha-1} * S_\alpha)(t, B_s) B_s^{\ell+1} \phi_s, \end{aligned} \tag{4}$$

where $S_\alpha(\cdot, B_s)$ is an α-resolvent family generated by the operator B_s, $\Gamma(\cdot)$ is the Gamma function, $g_\alpha(t) = t^{\alpha-1}/\Gamma(\alpha)$. It follows that, for any $t \in (0, T]$,

$$\|\partial_t u_s(t)\| \leq C(\alpha)(1 + t^{\alpha-1}) m_\ell(\phi_s), \quad \|\partial_t^2 u_s(t)\| \leq C(\alpha)(1 + t^{\alpha-2}) m_\ell(\phi_s), \tag{5}$$

where $m_\ell(\phi_s) = \max_{0\le k\le \ell+2} \|B_s^k\phi_s\|$. So, the solutions are only weakly regular near $t = 0$. For another instance, Sakamoto and Yamamoto [43] obtain an estimate when $f = 0$ and $\phi \in L_2(E)$:

$$\|u(\cdot,t)\|_{H^2(E)} + \|^c\partial_t^\alpha u(\cdot,t)\|_{L_2(E)} \le Ct^{-\alpha}\|\phi\|_{L_2(E)} \quad t \in (0,T].$$

We can infer that the αth-order derivative of u is at least unbounded near $t = 0$. Hence, it is necessary to reconsider numerical methods for approximation of Caputo fractional derivative concerning the limited regularity of solution. Needless to say, the addition of singular behavior will generally deteriorate the convergence.

Recently, the fact that the derivative of solution exhibits the weak singularity at $t = 0$ has been recognized in error analysis of the fractional difference schemes: L1 scheme [21, 31, 42], convolution quadrature [20–22]; see also [11, 12, 18, 43]. The existing ways to reach desired higher-order convergence rate include: employing the graded part of the fitted mesh to handle the inherent weak regularity of the solution at $t = 0$ [13, 42]; designing a starting term by virtue of starting weights to capture all leading singularities [20, 24]; making a proper correction at the first step in temporal direction with the help of approximation strategies [22]. Jin et al. [22] focus on seeking the approximation method with higher accuracy from the perspectives of whether the initial data is smooth and what compatibility conditions the source term satisfies. Stynes et al. [42] improve the accuracy and weaken the influence of the singularity at initial time by substituting the fitted mesh for the uniform mesh. All these excellent results are obtained by means of the regularity estimates such as the analogues of inequality (5) to some extent.

In the present paper, we concentrate on the convergence rates in temporal direction for the schemes, which are constructed by projection method in space and discretization in time proposed by Ashyralyev [3], which was theoretically obtained by virtue of the connection of fractional derivative with fractional powers of positive operators. In [3] neither stability nor convergence rate were obtained for this scheme. First, we prove the stability of such schemes. Unlike the common techniques on error estimates in time, we utilize the idea of layering in temporal direction to obtain the local truncation error. In addition, we use an analogues of formula (5) that derived from the resolvent theory to describe and deal with the weak regularity near the initial time $t = 0$ in subsequent estimates. Then we obtain the error equations and error bounds by using the expressions of solutions of the fully discrete difference schemes. Finally, we obtain that the temporal global errors converge at a rate $O(\tau^\alpha)$.

The rest of the paper is organized as follows. In the next section, we give some basic definitions, the spatial semidiscrete difference scheme, two fully discrete difference schemes and some known results that we need in subsequent derivation. The convergence rate of the truncation error is given in Sect. 3. In Sect. 4, we present a rigorous analysis of the orders of convergence for both implicit difference scheme and explicit difference scheme. Finally, we confirm our estimates by some numerical experiments in Sect. 5.

2 Discretization Methods

In this section, we present two fully discrete difference schemes of problem (1). We start from the semidiscrete difference scheme in space, which is obtained by projection method. Then we discrete the time fractional derivative and get two fully discrete difference schemes. Throughout, let C denote a generic constant which is always independent of the step sizes s, τ, and it may change at each occurrence. First, we recall the definition of the Caputo derivative in Eq. (1).

Definition 1 Let $\alpha \in (0, 1)$. Then, we define the operator ${}^c\partial^{\alpha}_{0,t}$ for functions by

$$ {}^c\partial^{\alpha}_{0,t}\psi(t) := J^{1-\alpha}_{0,t}\psi'(t) = \int_0^t \frac{(t-s)^{-\alpha}}{\Gamma(1-\alpha)}\psi'(s)ds. \tag{6} $$

The operator ${}^c\partial^{\alpha}_{0,t}$ is called the Caputo differential operator of order α.

In what follows, we abbreviate the notation ${}^c\partial^{\alpha}_{0,t}$ as ${}^c\partial^{\alpha}_{t}$.

2.1 Spatial Semidiscrete Schemes

Here, we employ a functional-analytical treatment of discretization in spatial direction and present a general semidiscrete approximation scheme; that is, we use the projection method to characterize the spatial semidiscrete difference schemes. We now introduce the framework of the approximation scheme obtained by projection method (see also [44] for more details). Let Ω and Ω_s with $s \in \mathbb{N}^+$ be Banach spaces. Let $\mathcal{B}(\Omega, \Omega_s)$ denote the space of all continuous linear mappings from Ω to Ω_s and let $\mathcal{B}(\Omega)$ denote $\mathcal{B}(\Omega, \Omega)$. Take $B \in \mathcal{B}(\Omega)$, $B_s \in \mathcal{B}(\Omega_s)$ and $p_s \in \mathcal{B}(\Omega, \Omega_s)$ such that $\|p_s x\|_{\Omega_s} \to \|x\|_{\Omega}$ when s tends to infinity, for each $x \in \Omega$.

Definition 2 The sequence $\{x_s\} \subset \Omega_s$ is $\mathcal{P}$-converging to x belonging to Ω if

$$ \lim_{s\to\infty} \|x_s - p_s x\|_{\Omega_s} = 0. $$

It is also written by $x_s \xrightarrow{\mathcal{P}} x$.

Definition 3 The operator family $\{B_s\} \subset \mathcal{B}(\Omega_s)$ is $\mathcal{PP}$-converging to B belonging to $\mathcal{B}(\Omega)$ if for arbitrary sequence $\{x_s\} \subset \Omega_s$, $x_s \xrightarrow{\mathcal{P}} x \in \Omega$ implies $B_s x_s \xrightarrow{\mathcal{P}} Bx$. It is also denoted by $B_s \xrightarrow{\mathcal{PP}} B$.

If $\Omega_s = \Omega$ and p_s is the identity operator on Ω for each $s \in \mathbb{N}^+$, then Definition 3 leads to the traditional pointwise convergence of bounded linear operators. In partial differential equations, the infinitesimal generators of analytic α-resolvent families are unbounded operators in general. So, before we present the semidiscrete difference schemes, we have to introduce the notion of compatibility property of operators.

Definition 4 The sequence of closed linear operators $\{B_s\}$, $s \in \mathbb{N}^+$, is said to be compatible with a closed linear operator B, if for arbitrary $x \in D(B)$ there exists a sequence $\{x_s\}$, $x_s \in D(B_s) \subseteq \Omega_s$, such that $x_s \xrightarrow{\mathcal{P}} x$ and $B_s x_s \xrightarrow{\mathcal{P}} Bx$. We write (B_s, B) are compatible.

One version of the Trotter-Kato theorem in [29], which is essential in the investigation of the approximation theory to differential equations, is shown as follows.

Theorem 1 (Theorem ABC [29]) *Suppose that $0 < \alpha \leq 2$ and B, B_s generate exponentially bounded analytic α-times resolution families $S_\alpha(\cdot, B)$, $S_\alpha(\cdot, B_s)$ in Banach spaces Ω, Ω_s, respectively. The following conditions (A) and (B) are equivalent to condition (C).*

(A) Consistency. There exists $\lambda \in \rho(B) \bigcap \cap_s \rho(B_s)$ such that the resolvents converge, i.e.,

$$(\lambda I_s - B_s)^{-1} \xrightarrow{\mathcal{PP}} (\lambda I - B)^{-1}.$$

(B) Stability. There are some constants $M \geq 1$, $\theta \in (0, \pi/2)$ and ω which are independent of s, such that the sector $\omega + \Sigma_{\theta+\pi/2}$ is included in $\rho(B_s)$ and

$$\sup_{\lambda \in \omega + \Sigma_{\beta+\pi/2}} \left\| \lambda^{\alpha-1}(\lambda^\alpha I_s - B_s)^{-1} \right\|_{B(\Omega_s)} \leq \frac{M}{|\lambda - \omega|}$$

for any $s \in \mathbb{N}$ and for any $0 < \beta < \theta$.

(C) Convergence. For some finite $\omega_1 > 0$ and $\theta \in (0, \pi/2)$, one has

$$\sup_{z \in \Sigma_\beta} e^{-\omega_1 \operatorname{Re} z} \|S_\alpha(z, B_s) x_s - p_s S_\alpha(z, B)x\|_{\Omega_s} \to 0 \quad as \quad s \to \infty$$

whenever $x_s \xrightarrow{\mathcal{P}} x$ for any $x_s \in \Omega_s$, $x \in \Omega$ and for any $0 < \beta < \theta$.

We assume that the infinitesimal generator of analytic α-resolvent family satisfies conditions (A) and (B) in Theorem 1. According to the above convergence definition of the projection method, we can obtain the spatial semidiscrete difference schemes of (1) as follows:

$$\begin{cases} {}^c\partial_t^\alpha u_s(t) = B_s u_s(t) + f_s(t), & t \in (0, T], \\ u_s(0) = \phi_s, \end{cases} \tag{7}$$

where the operators B_s generate analytic C_0-semigroups, B_s and B are compatible, $\phi_s \xrightarrow{\mathcal{P}} \phi$, and $f_s(\cdot) \xrightarrow{\mathcal{P}} f(\cdot)$ in appropriate sense. For the abstract semidiscrete difference schemes, Piskarev and Siegmund [41] have proved the convergence of the solutions due to the Trotter-Kato's theorem and theory of resolvent family. Next, we are going to describe the discretization techniques of the semidiscrete problem (7) in temporal direction.

2.2 Fully Discrete Schemes

In order to obtain the fully discrete schemes, we now consider the temporal discretization of Eq. (7). We divide the interval $[0, T]$ into a uniform grid $Q_\tau = \{t_n = n\tau, n = 0, 1, \ldots, N\}$, where $\tau = T/N$ is the time step size. Suppose $u_s(\cdot)$ is a grid function on Q_τ, where $u_s^n = u_s(t_n)$. We prove the convergence of the schemes under the maximal norm,

$$\|u_s^n\|_{\Omega_s \times Q_\tau} = \sup_{1 \le j \le n} \|u_s(j\tau)\|_{\Omega_s}.$$

In terms of the relationship between the Riemann-Liouville derivative and Caputo derivative, we discretize the Caputo time fractional derivative using the fractional difference algorithm proposed in [3],

$$D_{t_n}^{\alpha} u_s(\cdot) = \sum_{j=0}^{n-1} b_{j+1}^{(n)} \frac{u_s(t_{j+1}) - u_s(t_j)}{\tau^\alpha}, \tag{8}$$

where $b_{j+1}^{(n)} = \Gamma(n - j - \alpha)/\big(\Gamma(1-\alpha)\Gamma(n-j)\big)$, $j = 0, 1, \ldots, n-1$.

Using (8) to approximate the fractional time derivative ${}^c\partial_t^\alpha u_s(t)$, we obtain two fully discrete difference schemes, which are an implicit finite difference scheme and an explicit finite difference scheme. The problem (7) could be approximated by the following implicit difference scheme (see also [26])

$$\begin{cases} D_{t_n}^{\alpha} \widetilde{u}_s(\cdot) = B_s \widetilde{u}_s^n + f_s^n, \\ \widetilde{u}_s(0) = \phi_s. \end{cases} \tag{9}$$

It is worth mentioning that in different spaces, the approximative ways of the functions may be various. For instance, $f_s^n = f_s(t_n)$ in $C(\Omega_s)$, $f_s^n = \frac{1}{\tau}\int_{t_{k-1}}^{t_k} f_s(\theta)d\theta$ in $L^1(\Omega_s)$. In the present paper, we use the first way to define f_s^n. In [26], we have proved that the above implicit difference scheme is uniquely solvable, unconditionally stable and well posed in infinite norm. We also obtained the expressions of the solutions of the implicit difference scheme (9).

Theorem 2 ([26]) *For the fully discrete implicit difference scheme* (9)*, we obtain the following expressions of the solutions,*

$$\widetilde{u}_s^n = \sum_{m=1}^{n} c_m^{(n)} R_s^m \phi_s + \tau^\alpha \sum_{j=1}^{n} \sum_{m=1}^{n-j+1} d_{m,j}^{(n)} R_s^m f_s^j,$$

where $R_s = (I_s - \tau^\alpha B_s)^{-1}$; $c_1^{(n)} = b_1^{(n)}$,

$$c_m^{(n)} = \sum_{j=m}^{n} \left(b_j^{(n)} - b_{j-1}^{(n)}\right) c_{m-1}^{(j-1)}, \quad m \in \{2, 3, \ldots, n\};$$

and $d_{1,j}^{(n)} = 0$, $d_{1,n}^{(n)} = 1$ *and for* $j \in \{1, \ldots, n-1\}$, $m \in \{2, \ldots, n-j+1\}$,

$$d_{m,j}^{(n)} = \sum_{k=m+j-1}^{n} \left(b_k^{(n)} - b_{k-1}^{(n)}\right) d_{m-1,j}^{(k-1)}, \quad n \geq 1.$$

Remark 1 ([27]) For coefficients and the resolvent in Theorem 2, we have

$$c_m^{(n)} > 0, \quad \|R_s^m\|_{\Omega_s \to \Omega_s} = \|(I_s - \tau^\alpha B_s)^{-m}\|_{\Omega_s \to \Omega_s} \leq M, \quad m \in \{1, 2, \ldots, n\},$$

and $d_{m,j}^{(n)} \geq 0$, $m \in \{1, 2, \ldots, n-j+1\}$, $j \in \{1, 2, \ldots, n\}$, $n \geq 1$. Furthermore,

$$\sum_{j=1}^{n} \sum_{m=1}^{n-j+1} d_{m,j}^{(n)} = \frac{(1+\alpha)(2+\alpha)\cdots(n-1+\alpha)}{(n-1)!}, \quad n \geq 2.$$

Lemma 1 ([27]) *Let* $S_n = \sum_{j=1}^{n} \sum_{m=1}^{n-j+1} d_{m,j}^{(n)}$. *Then, for each* $n \geq 2$, *the inequality* $S_n \leq \exp(\alpha)(n-1)^\alpha$ *holds.*

Moreover, we can approximate the semidiscrete fractional difference scheme (7) by the following explicit difference scheme,

$$\begin{cases} D_{t_n}^\alpha \hat{u}_s(\cdot) = B_s \hat{u}_s^{n-1} + f_s^{n-1}, \\ \hat{u}_s(0) = \phi_s, \end{cases} \tag{10}$$

where $D_{t_n}^\alpha$, f_s^{n-1} and ϕ_s are defined as above. The expression of the solution for this explicit difference scheme has been obtained in [27].

Theorem 3 ([27]) *For the fully discrete explicit difference scheme* (10), *the solution is given by*

$$\hat{u}_s^n = \sum_{j=0}^{n} \gamma_j^{(n)} \hat{R}_s^j \phi_s + \tau^\alpha \sum_{j=0}^{n-1} \sum_{m=0}^{n-j-1} \delta_{m,j}^{(n)} \hat{R}_s^m f_s^j, \tag{11}$$

where $\gamma_0^{(1)} = b_0^{(1)}$, $\gamma_1^{(1)} = 1 - b_0^{(1)}$, $\gamma_0^{(n)} = b_1^{(N)} + \sum_{m=1}^{n-2} (b_{m+1}^{(n)} - b_m^{(n)}) \gamma_0^{(m)}$,

$$\gamma_j^{(n)} = (1 - b_{n-1}^{(n)}) \gamma_{j-1}^{(n-1)} + \sum_{m=j}^{n-2} (b_{m+1}^{(n)} - b_m^{(n)}) \gamma_j^{(m)}, \quad j \in \{1, \ldots, n\}, \quad n \geq 2,$$

and $\hat{R}_s = I_s + \alpha^{-1} \tau^\alpha B_s$; $\delta_{0,n-1}^{(n)} = 1$, $n \geq 1$, $\delta_{0,n-2}^{(n)} = 0$, $\delta_{1,n-2}^{(n)} = 1 - b_{n-1}^{(n)}$, $n \geq 2$,

$$\delta_{m,j}^{(n)} = (1 - b_{n-1}^{(n)})\delta_{m-1,j}^{(n-1)} + \sum_{l=m+j+1}^{n-2} (b_{l+1}^{(n)} - b_l^{(n)})\delta_{m,j}^{(l)},$$

for $j \in \{0, \ldots, n-3\}$, $m \in \{0, \ldots, n-j-1\}$, $n \geq 3$. *In addition, for any* $n \in \mathbb{N}^+$, *one has*

$$\sum_{j=0}^{n} \gamma_j^{(n)} = 1, \quad \sum_{j=0}^{n-1} \sum_{m=0}^{n-j-1} \delta_{m,j}^{(n)} b_{n-m-j}^{(n-m)} = 1. \tag{12}$$

Meanwhile, the following relationship between the coefficients $\delta_{m,j}^{(n)}$ in the above formula and the coefficients $d_{m,j}^{(n)}$ which are in the expression of the solution of the implicit difference scheme (9) is as follows.

Lemma 2 ([27]) *For any* $n \geq 1$, *the coefficients* $d_{m,j}^{(n)}$ *and* $\delta_{m,j}^{(n)}$ *satisfy the following equality,*

$$\sum_{m=1}^{n-j} d_{m,j+1}^{(n)} = \sum_{m=0}^{n-j-1} \delta_{m,j}^{(n)}, \quad j \in \{0, 1, \ldots, n-1\}.$$

Lemma 3 ([27]) *Let condition* (B) *with* $\omega = 0$ *be satisfied. Assume that there exists a number* $M_1 \geq 1$ *satisfying* $\|\tau^\alpha B_s\|_{\Omega_s \to \Omega_s} < \alpha/(M_1 + 2)$. *Then,* $\|\hat{R}_s^j\|_{\Omega_s \to \Omega_s} \leq M$, $j \in \{0, 1, \ldots, N\}$, *where* M *is a constant which is independent of the step size* τ.

2.3 *Regularity of the Solutions*

Now, our interest returns to the problem of whether there is the weak regularity of the solution of (7) near the initial time. We want to know whether there is a natural way to estimate the norm of the first-order derivative of the solution. We take the easy way out here. We here present the estimate, inductively, by using the representation of solution and some properties with respect to the solution operator. To begin, choose some auxiliary lemmas and conclusions given in [28, 32] that involved the resolvent operator and the expression of the solution for problem (7).

Definition 5 ([28]) $\{S_\alpha(t, B_s)\}_{t \geq 0} \subset \mathcal{B}(\Omega_s)$ is said to be an α-resolvent family generated by the operator B_s if the following assertions are fulfilled:

(i) $S_\alpha(t, B_s)\phi_s \in C([0, +\infty); \Omega_s)$ and $S_\alpha(0, B_s) = I_s$, for every $\phi_s \in \Omega_s$;

(ii) $S_\alpha(t, B_s)D(B_s) \subset D(B_s)$ and $B_s S_\alpha(t, B_s)\phi_s = S_\alpha(t, B_s)B_s\phi_s$ for any $\phi_s \in D(B_s)$, $t \geq 0$;

(iii) $S_\alpha(t, B_s)\phi_s = \phi_s + (g_\alpha * S_\alpha)(t, B_s)B_s\phi_s$ for all $\phi_s \in D(B_s)$, where $g_\alpha(t) = \frac{t^{\alpha-1}}{\Gamma(\alpha)}$ for $t > 0$.

Lemma 4 ([28]) *Let B_s generates an α-resolvent family $\{S_\alpha(t, B_s)\}_{t\geq 0} \subset \mathcal{B}(\Omega_s)$ and $f_s(\cdot) \in C([0, T]; \Omega_s)$. If u_s is a mild solution of (7), then it has the following form*

$$u_s(t) = S_\alpha(t, B_s)\phi_s + (P_\alpha * f_s)(t), \quad t \in [0, T],$$

where $S_\alpha(t, B_s) = (g_{1-\alpha} * P_\alpha)(t)$, $P_\alpha(t) = \frac{d}{dt}(g_\alpha * S_\alpha)(t)$.

Then, choose the following important property of analytic solution operator, which is obtained by repeated application of (iii) in above Definition 5.

Lemma 5 ([25]) *If B_s generates an α-resolvent family $\{S_\alpha(t, B_s)\}_{t\geq 0}$, then for $\phi_s \in D(B_s^\ell)$ with $\ell\alpha \geq 1$, $S_\alpha(t, B_s)\phi_s$ is differentiable and*

$$\frac{d}{dt}\big(S_\alpha(t, B_s)\phi_s\big) = \sum_{k=1}^{\ell-1} g_{k\alpha}(t)B_s^k\phi_s + \big(g_{\ell\alpha-1} * S_\alpha\big)(t, B_s)B_s^\ell\phi_s, \quad 0 < t \leq T.$$

This lemma leads to an estimate (5). Let us split $u_s(\cdot)$ into two parts. That is, $u_s(\cdot) = w_s(\cdot) + v_s(\cdot)$, where $w_s(t) = S_\alpha(t, B_s)\phi_s$ and $v_s(t) = (P_\alpha * f_s)(t)$ for $0 \leq t \leq T$. According to Lemma 5, it suffices to consider $\|v_s'(t)\|_{\Omega_s}$. Since the source term is smooth, i.e., $f_s(\cdot) \in C^1([0, T]; \Omega_s)$, for $0 < t \leq T$ we have

$$\begin{aligned}
\|v_s'(t)\|_{\Omega_s} &= \|\frac{d}{dt}(P_\alpha * f_s)(t)\|_{\Omega_s} \\
&\leq \|\int_0^t P_\alpha(\theta) f_s'(t-\theta)d\theta\|_{\Omega_s} + \|P_\alpha(t)\|_{\Omega_s\to\Omega_s}\|f_s\|_{C(\Omega_s)} \\
&\leq \int_0^t \|P_\alpha(\theta)\|_{\Omega_s\to\Omega_s}d\theta\|f_s'\|_{C(\Omega_s)} + \|P_\alpha(t)\|_{\Omega_s\to\Omega_s}\|f_s\|_{C(\Omega_s)}. \qquad (13)
\end{aligned}$$

Furthermore, with regard to the bound of the operator P_α, one has the following result.

Lemma 6 ([8, 28]) *Assume that $0 < \alpha < 1$ and the hypotheses of Lemma 5 are true. Then there exists a constant C satisfying $\|P_\alpha(t)\|_{\Omega_s\to\Omega_s} \leq Ce^{\omega t}(1 + t^{\alpha-1})$ for $0 < t \leq T$.*

And then, from (13), we can deduce an estimate about the solution of the nonhomogeneous semidiscrete difference scheme (7).

Theorem 4 *Let the condition (B) hold with $\omega = 0$ and $0 < \alpha < 1$. Assume the operators B_s generate analytic C_0-semigroups, $\phi_s \in D(B_s^\ell)$ with $\ell\alpha \geq 1$ and $f_s(\cdot) \in C^1([0, T]; \Omega_s)$. Then, the solution of Eq. (7) satisfies*

$$\|u_s'(t)\|_{\Omega_s} \leq C(1 + t^{\alpha-1}), \quad \textit{for all} \quad t \in (0, T].$$

Proof According to inequality (13) and Lemma 4, we can obtain

$$
\begin{aligned}
\|v_s'(t)\|_{\Omega_s} &\le \int_0^t \|P_\alpha(\theta)\|_{\Omega_s\to\Omega_s} d\theta \|f_s'\|_{C(\Omega_s)} + \|P_\alpha(t)\|_{\Omega_s\to\Omega_s}\|f_s\|_{C(\Omega_s)} \\
&\le C\int_0^t (1+\theta^{\alpha-1})d\theta\|f_s'\|_{C(\Omega_s)} + C(1+t^{\alpha-1})\|f_s\|_{C(\Omega_s)} \\
&\le Ct^\alpha + Ct + C + Ct^{\alpha-1} \\
&\le CT^\alpha + CT + C + Ct^{\alpha-1} \\
&\le C(1+t^{\alpha-1}), \qquad t \in (0, T]. \qquad (14)
\end{aligned}
$$

Hence, we have $\|u_s'(t)\|_{\Omega_s} \le \|w_s'(t)\|_{\Omega_s} + \|v_s'(t)\|_{\Omega_s} \le C(1+t^{\alpha-1})$ for all $0 < t \le T$.

Actually, no matter how smooth the initial data takes, the solutions of the problem (7) will exhibit weak singularity at the initial time $t = 0$.

3 Truncation Errors

In this section, we show that the approximation approach (8) can achieve αth-order accuracy of convergence in time. To achieve this, we first need to bound the truncation error $({}^c\partial_t^\alpha u_s)(t_n) - D_{t_n}^\alpha u_s(\cdot)$. According to the representations of Caputo fractional derivative and its discretization scheme, we have

$$
\begin{aligned}
({}^c\partial_t^\alpha u_s)(t_n) - D_{t_n}^\alpha u_s(\cdot) &= \int_0^{t_n} \frac{(t_n - z)^{-\alpha}}{\Gamma(1-\alpha)} u_s'(z)dz - \sum_{j=0}^{n-1} b_{j+1}^{(n)} \frac{u_s^{j+1} - u_s^j}{\tau^\alpha} \\
&= \sum_{j=0}^{n-1} \int_{t_j}^{t_{j+1}} \left(\frac{(t_n - z)^{-\alpha}}{\Gamma(1-\alpha)} - b_{j+1}^{(n)}\tau^{-\alpha} \right) u_s'(z)dz \\
&= \sum_{j=0}^{n-1} \Bigg[\int_{j\tau}^{(j+1)\tau} \frac{(n\tau - z)^{-\alpha}}{\Gamma(1-\alpha)} u_s'(z)dz \\
&\quad - \int_{j\tau}^{(j+1)\tau} \frac{\Gamma(n-j-\alpha)}{\Gamma(1-\alpha)\Gamma(n-j)} \tau^{-\alpha} u_s'(z)dz \Bigg] := \sum_{j=0}^{n-1} K_{n,j}.
\end{aligned}
\qquad (15)
$$

In many works (see, e.g., [14, 39]), the results of Caputo derivatives of certain important functions have been provided. With the help of these conclusions, we present a characterization of the smoothness properties of a function in the following lemma.

Lemma 7 *Let $V(t) = (t + \xi - 1)^\alpha$, $\xi \geq 2$ is an arbitrary integer and $\alpha \in (0, 1)$. Then the Caputo fractional derivative of function $V(\cdot)$ at $t = 1$ satisfies*

$$\left({}^c\partial_t^\alpha V\right)(1) = O\left((\xi - 1)^{\alpha-2}\right).$$

Proof For function $V(t) = (t + \xi - 1)^\alpha$, it follows from [14] that

$$\left({}^c\partial_t^\alpha V\right)(t) = \frac{\Gamma(\alpha+1)}{\Gamma(\alpha)} \frac{(\xi-1)^{\alpha-2} t^{1-\alpha}}{\Gamma(2-\alpha)} {}_2F_1\left(1, 1-\alpha; 2-\alpha; -\frac{t}{\xi-1}\right),$$

where ${}_2F_1(a, b; c; d)$ denotes the Gauss' hypergeometric function, which is defined by

$${}_2F_1(a, b; c; z) = \frac{\Gamma(c)}{\Gamma(a)\Gamma(b)} \sum_{k=0}^{\infty} \frac{\Gamma(a+k)\Gamma(b+k)}{\Gamma(c+k)k!} z^k, \quad (a, b \in \mathbb{R}, -c \notin \mathbb{N}).$$

Then, let $t = 1$, we have

$$\begin{aligned}
({}^c\partial_t^\alpha V)(1) &= \frac{\Gamma(\alpha+1)}{\Gamma(\alpha)} \frac{(\xi-1)^{\alpha-2} 1^{1-\alpha}}{\Gamma(2-\alpha)} {}_2F_1\left(1, 1-\alpha; 2-\alpha; -\frac{1}{\xi-1}\right) \\
&= \frac{\alpha}{\Gamma(2-\alpha)} \sum_{k=0}^{\infty} \frac{(1-\alpha)\Gamma(1+k)}{(k+1-\alpha)k!} (-1)^k (\xi-1)^{\alpha-2-k} \\
&= \frac{\alpha}{\Gamma(1-\alpha)} \sum_{k=0}^{\infty} \frac{(-1)^k}{(k+1-\alpha)} (\xi-1)^{\alpha-2-k} \\
&= O\left((\xi-1)^{\alpha-2}\right).
\end{aligned}$$

The proof is completed.

Theorem 5 *Assume the hypothesis of Theorem 4 holds. Let $u_s(\cdot)$ be the solution of the general semidiscrete difference scheme (7). Then we have the following local truncation error estimate*

$$\|({}^c\partial_t^\alpha u_s)(t_n) - D_{t_n}^\alpha u_s(\cdot)\|_{\Omega_s} \leq Cn^{-\alpha}. \tag{16}$$

That is, the algorithm (8) of approximation for Caputo fractional derivative converges at a rate $O(\tau^\alpha t_n^{-\alpha})$, $n = 1, 2, \ldots, N$.

Proof In terms of the weak regularity near the initial time $t = 0$, we split the summation (15) as follows and deal with it separately.

$$\sum_{j=0}^{n-1} K_{n,j} = \begin{cases} K_{1,0}, & n = 1, \\ K_{n,0} + K_{n,n-1} + \sum_{j=1}^{n-2} K_{n,j}, & n \geq 2. \end{cases}$$

Note that when $n = 2$, the third term on the right-hand side vanishes. Next, we estimate each of the three parts separately. For $K_{n,0}$, when $n = 1$, we have

$$\begin{aligned}
\|K_{1,0}\|_{\Omega_s} &= \|({}^c\partial_t^\alpha u_s)(t_1) - D_{t_1}^\alpha u_s(\cdot)\|_{\Omega_s} \\
&= \left\| \int_0^\tau \frac{(\tau - z)^{-\alpha}}{\Gamma(1-\alpha)} u_s'(z)dz - b_1^{(1)}\tau^{-\alpha}\big(u_s(t_1) - u_s(t_0)\big) \right\|_{\Omega_s} \\
&\le \frac{1}{\Gamma(1-\alpha)} \int_0^\tau (\tau - z)^{-\alpha} C z^{\alpha-1} dz + \int_0^\tau \tau^{-\alpha}\|u_s'(z)\|_{\Omega_s} dz \\
&\le \frac{C}{\Gamma(1-\alpha)} B(1-\alpha, \alpha) + C\tau^{-\alpha}\tau^{\alpha} \le C,
\end{aligned}$$

where we have used Theorem 4 at above two inequalities. In addition, $B(1-\alpha, \alpha)$ is a Beta function. For fixed $n \ge 2$, one has

$$\begin{aligned}
\|K_{n,0}\|_{\Omega_s} &= \left\| \frac{1}{\Gamma(1-\alpha)} \int_0^\tau (n\tau - z)^{-\alpha} u_s'(z)dz - \int_0^\tau b_1^{(n)}\tau^{-\alpha} u_s'(z)dz \right\|_{\Omega_s} \\
&\le \frac{C}{\Gamma(1-\alpha)} (n\tau - \tau)^{-\alpha} \int_0^\tau z^{\alpha-1}dz + \frac{C\Gamma(n-\alpha)}{\Gamma(1-\alpha)\Gamma(n)} \int_0^\tau \tau^{-\alpha} z^{\alpha-1} dz \\
&\le \frac{C}{\Gamma(1-\alpha)} \left(n\tau - \frac{n}{2}\tau\right)^{-\alpha} \tau^{\alpha} + \frac{C}{\Gamma(1-\alpha)} n^{-\alpha}\big(1 + O(n^{-1})\big)\tau^{-\alpha}\frac{\tau^{\alpha}}{\alpha} \\
&\le Cn^{-\alpha} + \frac{C}{\Gamma(1-\alpha)} (n^{-\alpha} + Cn^{-\alpha-1})\tau^{-\alpha}\frac{\tau^{\alpha}}{\alpha} \\
&\le Cn^{-\alpha} + Cn^{-\alpha-1} \le Cn^{-\alpha}.
\end{aligned}$$

Therefore, we have

$$\|K_{n,0}\|_{\Omega_s} \le Cn^{-\alpha}, \quad n \ge 1.$$

Let us consider the terms $\sum_{j=1}^{n-2} K_{n,j}$, $n \ge 3$. Let $z = \xi\tau$, we have

$$\begin{aligned}
\sum_{j=1}^{n-2} K_{n,j} &= \sum_{j=1}^{n-2} \left[\int_{j\tau}^{(j+1)\tau} \frac{(n\tau - z)^{-\alpha}}{\Gamma(1-\alpha)} u_s'(z) - \frac{\Gamma(n-j-\alpha)}{\Gamma(1-\alpha)\Gamma(n-j)} \tau^{-\alpha} u_s'(z)dz \right] \\
&= \frac{1}{\Gamma(1-\alpha)} \sum_{j=1}^{n-2} \left[\int_j^{j+1} \frac{\tau^{1-\alpha}}{(n-\xi)^{\alpha}} u_s'(\xi\tau) - \frac{\Gamma(n-j-\alpha)}{\Gamma(n-j)} \tau^{1-\alpha} u_s'(\xi\tau) d\xi \right] \\
&= \frac{1}{\Gamma(1-\alpha)} \sum_{j=1}^{n-2} \int_j^{j+1} \tau^{1-\alpha} \left((n-\xi)^{-\alpha} - \frac{\Gamma(n-j-\alpha)}{\Gamma(n-j)} \right) u_s'(\xi\tau) d\xi.
\end{aligned}$$

In terms of a well-known asymptotic formula in [34]

$$z^{b-a}\frac{\Gamma(z+a)}{\Gamma(z+b)} = 1 + O(z^{-1}), \quad z \to \infty, \tag{17}$$

we have

$$\sum_{j=1}^{n-2} K_{n,j} = \frac{\tau^{1-\alpha}}{\Gamma(1-\alpha)} \sum_{j=1}^{n-2} \int_j^{j+1} \left((n-\xi)^{-\alpha} - \frac{\Gamma(n-\xi+\xi-j-\alpha)}{\Gamma(n-\xi+\xi-j)} \right) u_s'(\xi\tau) d\xi$$

$$= \frac{\tau^{1-\alpha}}{\Gamma(1-\alpha)} \sum_{j=1}^{n-2} \int_j^{j+1} \left((n-\xi)^{-\alpha} - \frac{\left(1+O\left((n-\xi)^{-1}\right)\right)}{(n-\xi)^{\alpha}} \right) u_s'(\xi\tau) d\xi$$

$$= \frac{\tau^{1-\alpha}}{\Gamma(1-\alpha)} \sum_{j=1}^{n-2} \int_j^{j+1} -(n-\xi)^{-\alpha} O\left((n-\xi)^{-1}\right) u_s'(\xi\tau) d\xi.$$

Then, there exists a constant C such that

$$\left\| \sum_{j=1}^{n-2} K_{n,j} \right\|_{\Omega_s} \leq C\tau^{1-\alpha} \sum_{j=1}^{n-2} \int_j^{j+1} (n-\xi)^{-\alpha-1} \| u_s'(\xi\tau) \|_{\Omega_s} d\xi$$

$$\leq C\tau^{1-\alpha} \sum_{j=1}^{n-2} \int_j^{j+1} (n-\xi)^{-\alpha-1} (\xi\tau)^{\alpha-1} d\xi \leq C \sum_{j=1}^{n-2} (n-j-1)^{-\alpha-1} j^{\alpha-1}$$

$$= C \left(\sum_{j=1}^{\lceil \frac{n}{2} \rceil -1} (n-j-1)^{-\alpha-1} j^{\alpha-1} + \sum_{j=\lceil \frac{n}{2} \rceil}^{n-2} (n-j-1)^{-\alpha-1} j^{\alpha-1} \right)$$

$$\leq C \left(\sum_{j=1}^{\lceil \frac{n}{2} \rceil -1} (n-\lceil \frac{n}{2} \rceil)^{-\alpha-1} j^{\alpha-1} + (\frac{n}{2})^{-2} \sum_{j=\lceil \frac{n}{2} \rceil}^{n-2} (n-j-1)^{-\alpha-1} j^{\alpha+1} \right)$$

$$\leq C \left(n^{-\alpha-1} \sum_{j=1}^{\lceil \frac{n}{2} \rceil -1} j^{\alpha-1} + n^{-2} \int_0^{n-1} (n-z-1)^{-\alpha-1} z^{\alpha+1} dz \right)$$

$$\leq C(n^{-\alpha} + n^{-1}) \leq Cn^{-\alpha}.$$

It remains to estimate $K_{n,n-1}$, $n \geq 2$.

$$\| K_{n,n-1} \|_{\Omega_s} = \left\| \int_{(n-1)\tau}^{n\tau} \frac{(n\tau - z)^{-\alpha}}{\Gamma(1-\alpha)} u_s'(z) dz - b_n^{(n)} \tau^{-\alpha} (u_s(t_n) - u_s(t_{n-1})) \right\|_{\Omega_s}$$

$$\leq \left\| \int_{(n-1)\tau}^{n\tau} \frac{(n\tau - z)^{-\alpha}}{\Gamma(1-\alpha)} u_s'(z) dz \|_{\Omega_s} + \| \int_{(n-1)\tau}^{n\tau} \tau^{-\alpha} u_s'(z) dz \right\|_{\Omega_s}$$

$$\leq (1+\Gamma(1-\alpha)) \left\| \int_{(n-1)\tau}^{n\tau} \frac{(n\tau - z)^{-\alpha}}{\Gamma(1-\alpha)} u_s'(z) dz \right\|_{\Omega_s}$$

$$\leq C(1+\Gamma(1-\alpha)) \int_{(n-1)\tau}^{n\tau} \frac{(n\tau - z)^{-\alpha}}{\Gamma(1-\alpha)} z^{\alpha-1} dz.$$

Let $z = \big(\nu + (n-1)\big)\tau$. Then it follows from Lemma 7 that

$$\begin{aligned}\|K_{n,n-1}\|_{\Omega_s} &\le C(1+\Gamma(1-\alpha))\int_0^1 \frac{(1-\nu)^{-\alpha}}{\Gamma(1-\alpha)}(\nu+n-1)^{\alpha-1}d\nu\\ &= C\frac{(1+\Gamma(1-\alpha))}{\alpha}\left({}^c\partial_t^\alpha (t+n-1)^\alpha\right)(1)\\ &= C\frac{(1+\Gamma(1-\alpha))}{\alpha}O(n-1)^{-(2-\alpha)}\\ &\le Cn^{-(2-\alpha)}.\end{aligned}$$

Taken together, these observations deduce that the local truncation error satisfying

$$\|({}^c\partial_t^\alpha u_s)(t_n) - D_{t_n}^\alpha u_s(\cdot)\|_{\Omega_s} \le C(n^{-\alpha} + n^{-(2-\alpha)}) \le Cn^{-\alpha}.$$

This proof is completed.

Remark 2 By virtue of the results in Theorem 4 deduced by resolvent family theory, we can clearly see the influence of weak regularity at the initial time on the convergence rates of the truncation errors. The most previous convergence analysis of the existing schemes requires that the problem (1) has a unique and sufficiently smooth solution. In principle, it is impossible to obtain higher global convergence rates for the discretization methods under the consideration of weak regularity near the initial time $t = 0$. We adopt the idea of layering in temporal direction to establish the truncation error estimates of the fully discrete difference schemes. Due to the limited regularity of the solution, we could interpret the result $O(\tau^\alpha t_n^{-\alpha})$ to read: under the uniform grid, the convergence rates of the local truncation errors could reach $O(\tau^\alpha)$ at a fixed distance from the initial time $t = 0$.

4 Global Error Analysis

We consider the global convergence rates of the implicit difference scheme (9) and the explicit difference scheme (10) in this section. We begin our analysis of the fully discrete difference schemes (9) and (10) with a pair of stability theorems.

Theorem 6 *Suppose condition (B) holds with $\omega = 0$. Then the implicit difference scheme (9) is stable, i.e.,*

$$\|\widetilde{u}_s^n\|_{\Omega_s\times Q_\tau} \le M\|\phi_s\|_{\Omega_s} + M\exp(\alpha)(n\tau)^\alpha \|f_s^n\|_{\Omega_s\times Q_\tau},$$

where $n\tau \in [0, T]$, M is a constant which is independent of s and τ.

Proof In terms of the expression of the solution of the implicit difference scheme in Theorem 2, Remark 1 and Lemma 1, we have

$$
\begin{aligned}
\|\widetilde{u}_s^n\|_{\Omega_s\times Q_\tau} &= \sup_{1\le j\le n} \|\widetilde{u}_s(j\tau)\|_{\Omega_s} \\
&\le \sup_{1\le j\le n} \|\sum_{m=1}^{j} c_m^{(j)} R_s^m \phi_s\|_{\Omega_s} + \sup_{1\le j\le n} \|\tau^\alpha \sum_{k=1}^{j}\sum_{m=1}^{j-k+1} d_{m,k}^{(j)} R_s^m f_s(k\tau)\|_{\Omega_s} \\
&\le M\|\phi_s\|_{\Omega_s} \sum_{m=1}^{n} c_m^{(n)} + M\tau^\alpha \sum_{k=1}^{n}\sum_{m=1}^{n-k+1} d_{m,k}^{(n)} \sup_{1\le j\le n} \|f_s(j\tau)\|_{\Omega_s} \\
&= M\|\phi_s\|_{\Omega_s} + M\tau^\alpha S_n \sup_{1\le j\le n} \|f_s(j\tau)\|_{\Omega_s} \\
&\le M\|\phi_s\|_{\Omega_s} + M\exp(\alpha)(n\tau)^\alpha \|f_s^n\|_{\Omega_s\times Q_\tau}.
\end{aligned}
$$

The proof is completed.

Then, we discuss the stability of the explicit difference schemes (10). Analyzing the stability of the explicit difference scheme in a similar way as the we analyzed the implicit difference scheme, we conclude the scheme (10) is stable in the maximum norm.

Theorem 7 *Suppose condition* (B) *holds with* $\omega = 0$. *Assume that there exists a number* $M_1 \ge 1$ *satisfying* $\|\tau^\alpha B_s\|_{\Omega_s\to\Omega_s} < \alpha/(M_1+2)$. *Then the explicit difference scheme* (10) *is stable, i.e.,*

$$
\|\hat{u}_s^n\|_{\Omega_s\times Q_\tau} \le M\|\phi_s\|_{\Omega_s} + M\exp(\alpha)(n\tau)^\alpha \|f_s^n\|_{\Omega_s\times Q_\tau},
$$

where $n\tau \in [0, T]$, M *is a constant which is independent of* s *and* τ.

Proof By means of the expression of the solution of the explicit difference scheme in Theorem 3, it follows from Lemmas 2 and 3 that

$$
\begin{aligned}
\|\hat{u}_s^n\|_{\Omega_s\times Q_\tau} &= \sup_{1\le j\le n} \|\hat{u}_s(j\tau)\|_{\Omega_s} \\
&\le \sup_{1\le j\le n} \left\| \sum_{k=0}^{j} \gamma_k^{(j)} \hat{R}_s^k \phi_s\|_{\Omega_s} + \sup_{1\le j\le n} \|\tau^\alpha \sum_{k=0}^{j-1}\sum_{m=0}^{j-k-1} \delta_{m,k}^{(j)} \hat{R}_s^m f_s(k\tau) \right\|_{\Omega_s} \\
&\le M\|\phi_s\|_{\Omega_s} \sup_{1\le j\le n} \sum_{k=0}^{j} \gamma_k^{(j)} + M\tau^\alpha \sup_{1\le j\le n} \|f_s(j\tau)\|_{\Omega_s} \sum_{k=0}^{n-1}\sum_{m=0}^{n-k-1} \delta_{m,k}^{(n)} \\
&= M\|\phi_s\|_{\Omega_s} + M\tau^\alpha \|f_s^n\|_{\Omega_s\times Q_\tau} \sum_{k=0}^{n-1}\sum_{m=1}^{n-k} d_{m,k+1}^{(n)}.
\end{aligned}
$$

Change the bounds of the index of summation in the last equality, i.e., let $j = k + 1$. By using Lemma 1, one has

$$
\begin{aligned}
\|\hat{u}_s^n\|_{\Omega_s \times Q_\tau} &\leq M\|\phi_s\|_{\Omega_s} + M\tau^\alpha \|f_s^n\|_{\Omega_s \times Q_\tau} \sum_{j=1}^{n} \sum_{m=1}^{n-j+1} d_{m,j}^{(n)} \\
&= M\|\phi_s\|_{\Omega_s} + M\tau^\alpha \|f_s^n\|_{\Omega_s \times Q_\tau} S_n \\
&\leq M\|\phi_s\|_{\Omega_s} + M \exp(\alpha)(n\tau)^\alpha \|f_s^n\|_{\Omega_s \times Q_\tau}.
\end{aligned}
$$

The proof is completed.

Next, we will make a more detailed study of error analysis for the difference schemes (9) and (10). Initially, we present a result with respect to the coefficients which might be helpful to estimate the errors.

Lemma 8 *For any $n \in \mathbb{N}^+$, we have*

$$
\sum_{j=1}^{n} \sum_{m=1}^{n-j+1} d_{m,j}^{(n)} j^{-\alpha} \leq \Gamma(1-\alpha).
$$

Proof By Lemma 2 and equality (12), we can obtain

$$
\sum_{j=1}^{n} \sum_{m=1}^{n-j+1} d_{m,j}^{(n)} b_{n-m-j+1}^{(n-m)} = \sum_{j=0}^{n-1} \sum_{m=1}^{n-j} d_{m,j+1}^{(n)} b_{n-m-j}^{(n-m)} = \sum_{j=0}^{n-1} \sum_{m=0}^{n-j-1} \delta_{m,j}^{(n)} b_{n-m-j}^{(n-m)} = 1.
$$

In terms of a completed version of the asymptotic formula (17) (also see [34]),

$$
\Gamma(z+a)/\Gamma(z+\beta) = z^{\alpha-\beta}\left[1 + 1/2z^{-1}(a-\beta)(a+\beta-1) + O\left(z^{-2}\right)\right],
$$

one has

$$
b_{n-m-j+1}^{(n-m)} = \frac{1}{\Gamma(1-\alpha)} \frac{\Gamma(j-\alpha)}{\Gamma(j)} = \frac{j^{-\alpha}}{\Gamma(1-\alpha)}\left[1 + \frac{\alpha(\alpha+1)}{2} j^{-1} + O\left(j^{-2}\right)\right],
$$

where $0 < \alpha < 1$, $j = 1, 2, \ldots$ Then, we have $j^{-\alpha} \leq \Gamma(1-\alpha) b_{n-m-j+1}^{(n-m)}$. Since $d_{m,j}^{(n)} \geq 0$, we have

$$
\sum_{j=1}^{n} \sum_{m=1}^{n-j+1} d_{m,j}^{(n)} j^{-\alpha} \leq \Gamma(1-\alpha) \sum_{j=1}^{n} \sum_{m=1}^{n-j+1} d_{m,j}^{(n)} b_{n-m-j+1}^{(n-m)} \leq \Gamma(1-\alpha).
$$

The proof is completed.

Now, we give temporal error estimates for the implicit difference scheme (9).

Theorem 8 *Let $u_s(\cdot)$ be the solution of problem (7) and $\widetilde{u}_s^n$ be the solution of implicit difference scheme (9). Assume that $\phi_s \in D(B_s^\ell)$ with $\ell\alpha > 1$, $f_s(\cdot) \in C^1([0,T];\Omega_s)$ hold. Then, there exists a positive constant C such that*

$$\|\widetilde{u}_s^n - u_s(t_n)\|_{\Omega_s} \le C\tau^\alpha,$$

where C is a constant independent of s, $t_n \in [0,T]$.

Proof Let $\delta_s(t_n) = u_s(t_n) - \widetilde{u}_s^n$, where $u_s(t_n)$ represents the value of the solution of the semidiscrete scheme (7) at mesh point $t = t_n$. Obviously, we have $\delta_s(0) = 0$ and

$$\begin{aligned} D_{t_n}^\alpha \delta_s(\cdot) - B_s\delta_s(t_n) &= D_{t_n}^\alpha\big(u_s(\cdot) - \widetilde{u}_s^n\big) - B\big(u_s(t_n) - \widetilde{u}_s^n\big) \\ &= D_{t_n}^\alpha u_s(\cdot) - B_s u_s(t_n) - f_s^n \\ &= D_{t_n}^\alpha u_s(\cdot) - {}^c\partial_t^\alpha u_s(t_n) := \widetilde{r}_s(t_n). \end{aligned}$$

According to Theorem 2, we have

$$\delta_s(t_n) = \tau^\alpha \sum_{j=1}^{n} \sum_{m=1}^{n-j+1} d_{m,j}^{(n)} R_s^m \widetilde{r}_s(t_j).$$

Then, according to Remark 1, Theorem 5 and Lemma 8, we have

$$\begin{aligned} \|\delta_s(t_n)\|_{\Omega_s} &= \left\| \tau^\alpha \sum_{j=1}^{n} \sum_{m=1}^{n-j+1} d_{m,j}^{(n)} R_s^m \widetilde{r}_s(t_j) \right\|_{\Omega_s} \\ &\le \tau^\alpha \sum_{j=1}^{n} \sum_{m=1}^{n-j+1} d_{m,j}^{(n)} \|R_s^m\|_{\Omega_s\to\Omega_s} \|\widetilde{r}_s(t_j)\|_{\Omega_s} \\ &\le C\tau^\alpha \sum_{j=1}^{n} \sum_{m=1}^{n-j+1} d_{m,j}^{(n)} j^{-\alpha} \le C\tau^\alpha. \end{aligned}$$

Therefore, the proof is completed.

By Theorem 5 and equality (11), we obtain the following convergence theorem for the fully discrete explicit difference scheme (10).

Theorem 9 *Let $u_s(\cdot)$ be the solution of problem (7) and $\hat{u}_s^n$ be the solution of explicit difference scheme (10). Assume that $\|\tau^\alpha B_s\|_{\Omega_s\to\Omega_s} < \alpha/(M_1+2)$ with some $M_1 \ge 1$ and $\phi_s \in D(B_s^\ell)$ with $\ell\alpha > 1$, $f_s(\cdot) \in C^1([0,T];\Omega_s)$ hold. Then, there exists a positive constant C such that*

$$\|u_s(t_n) - \hat{u}_s^n\|_{\Omega_s} \le C\tau^\alpha,$$

where C is independent of s and τ.

Proof To prove the required convergence rate, we proceed in a way similar to the proof of Theorem 8. Let $\zeta_s(t_n) = u_s(t_n) - \hat{u}_s^n$, where $u_s(t_n)$ represents the value of the solution of problem (7) at mesh point $t = t_n$. Then, we have

$$D_{t_n}^\alpha(u_s - \zeta_s)(\cdot) = B_s(u_s(t_{n-1}) - \zeta_s(t_{n-1})) + f_s^{n-1},$$

and $\zeta_s(0) = 0$. Thereupon, one yields

$$\begin{aligned}
D_{t_n}^\alpha \zeta_s(\cdot) &= B_s\zeta_s(t_{n-1}) + D_{t_n}^\alpha u_s(\cdot) - B_s u_s(t_{n-1}) - f_s^{n-1} \\
&= B_s\zeta_s(t_{n-1}) + D_{t_n}^\alpha u_s(\cdot) - {}^c\partial_t^\alpha u_s(t_{n-1}) \\
&= B_s\zeta_s(t_{n-1}) + D_{t_n}^\alpha u_s(\cdot) - {}^c\partial_t^\alpha u_s(t_n) + {}^c\partial_t^\alpha u_s(t_n) - {}^c\partial_t^\alpha u_s(t_{n-1}) \\
&= B_s\zeta_s(t_{n-1}) + \widetilde{r}_s(t_n) + B_s\big(u_s(t_n) - u_s(t_{n-1})\big) + f_s(t_n) - f_s(t_{n-1}) \\
&= B_s\zeta_s(t_{n-1}) + \hat{r}_s(t_{n-1}),
\end{aligned}$$

where $\hat{r}_s(t_{n-1}) = \widetilde{r}_s(t_n) + B_s\big(u_s(t_n) - u_s(t_{n-1})\big) + f_s(t_n) - f_s(t_{n-1})$. It follows from formula (11) that for all $n \ge 1$,

$$\zeta_s(t_n) = \tau^\alpha \sum_{j=0}^{n-1}\sum_{m=0}^{n-j-1} \delta_{m,j}^{(n)} \hat{R}_s^m \hat{r}_s(t_j).$$

When $n = 1$, in terms of the condition $\|\tau^\alpha B_s\|_{\Omega_s\to\Omega_s} < \alpha/(M_1+2)$, there exists $\xi \in (0,\tau)$ such that

$$\begin{aligned}
\|\hat{r}_s(t_0)\|_{\Omega_s} &= \|\widetilde{r}_s(t_1) + B_s\big(u_s(t_1) - u_s(t_0)\big) + f_s(t_1) - f_s(t_0)\|_{\Omega_s} \\
&\le C\tau^\alpha\tau^{-\alpha} + \|B_s\|_{\Omega_s\to\Omega_s}\int_0^\tau \|u_s'(z)\|_{\Omega_s} dz + \|f_s'(\xi)\|_{\Omega_s}(t_1 - t_0) \\
&\le C + \|B_s\|_{\Omega_s\to\Omega_s}\int_0^\tau z^{\alpha-1}dz + C\tau \le C + \|\tau^\alpha B_s\|_{\Omega_s\to\Omega_s} + C\tau \le C,
\end{aligned}$$

where we employ the mean value theorem and the result in Theorem 4. When $n \ge 2$, by virtue of Lemma 7, there exist $\eta, \xi_2 \in (t_{n-1}, t_n)$ such that

$$
\begin{aligned}
\|\hat{r}_s(t_{n-1})\|_{\Omega_s} &\leq \|\widetilde{r}_s(t_n)\|_{\Omega_s} + \|\tau^\alpha B_s\|_{\Omega_s\to\Omega_s} \frac{\|u_s(t_n) - u_s(t_{n-1})\|_{\Omega_s}}{\tau^\alpha} \\
&\quad + \|f_s'(\xi_2)\|_{\Omega_s}(t_n - t_{n-1}) \\
&< Cn^{-\alpha} + \frac{\alpha}{M_1+2}\left\| \int_{(n-1)\tau}^{n\tau} \tau^{-\alpha} u_s'(z)dz\|_{\Omega_s} + \tau\|f_s'(\xi_2)\right\|_{\Omega_s} \\
&\leq Cn^{-\alpha} + \frac{C\alpha\Gamma(1-\alpha)}{M_1+2}\int_{(n-1)\tau}^{n\tau} \frac{(n\tau - z)^{-\alpha}}{\Gamma(1-\alpha)} z^{\alpha-1}dz + C\tau \\
&\leq Cn^{-\alpha} + \frac{C\alpha\Gamma(1-\alpha)}{M_1+2}\int_0^1 \frac{(1-\nu)^{-\alpha}}{\Gamma(1-\alpha)}(\nu+n-1)^{\alpha-1}d\nu + Cn^{-1} \\
&\leq Cn^{-\alpha} + \frac{C\alpha\Gamma(1-\alpha)}{\alpha(M_1+2)}\left({}^c\partial_t^\alpha(t+n-1)^\alpha\right)(1) + Cn^{-1} \\
&= Cn^{-\alpha} + \frac{C\Gamma(1-\alpha)}{(M_1+2)} O(n-1)^{-(2-\alpha)} + Cn^{-1} \\
&\leq Cn^{-\alpha} + Cn^{-(2-\alpha)} + Cn^{-1} \leq Cn^{-\alpha}.
\end{aligned} \tag{18}
$$

Hence, we have $\|\hat{r}_s(t_{n-1})\|_{\Omega_s} \leq Cn^{-\alpha}$, for all $n \geq 1$. Then according to Lemmas 3 and 8, we obtain

$$
\begin{aligned}
\|\zeta_s(t_n)\|_{\Omega_s} &\leq \tau^\alpha \sum_{j=0}^{n-1}\sum_{m=0}^{n-j-1} \delta_{m,j}^{(n)} \|\hat{R}_s^m\|_{\Omega_s\to\Omega_s} \|\hat{r}_s(t_j)\|_{\Omega_s} \leq C\tau^\alpha \sum_{j=0}^{n-1}\sum_{m=0}^{n-j-1} \delta_{m,j}^{(n)} (j+1)^{-\alpha} \\
&\leq C\tau^\alpha \sum_{j=0}^{n-1}\sum_{m=1}^{n-j} d_{m,j+1}^{(n)} (j+1)^{-\alpha} \leq C\tau^\alpha \sum_{j=1}^{n}\sum_{m=1}^{n-j+1} d_{m,j}^{(n)} j^{-\alpha} \leq C\tau^\alpha.
\end{aligned} \tag{19}
$$

Therefore, the proof is completed.

Summarizing, Theorem 8 shows that the implicit difference scheme based on Ashyralyev's fractional difference derivative has a convergence rate $O(\tau^\alpha)$. On the other hand, Theorem 9 shows that the convergence rate of the explicit difference scheme can also reach α-th accuracy under some additional conditions.

5 Numerical Results

In order to demonstrate the effectiveness of our theoretical analysis, some experiments are now presented. We consider the case $d = 1$, i.e., $\Omega = (0, \pi) \subset \mathbb{R}^1$. Consider the following time fractional subdiffusion equation

$$
{}^c\partial_t^\alpha u(x,t) - \frac{\partial^2 u(x,t)}{\partial x^2} = \psi(x,t), \quad 0 < x < \pi, \quad 0 < t \leq 2, \tag{20}
$$

with the following initial and boundary value conditions:

$$u(x,0) = 2\sin(2x), \quad 0 < x < \pi,$$
$$u(0,t) = u(\pi,t) = 0, \quad 0 \le t \le 2,$$

where $\psi(x,t) = \left(2t^{2-\alpha}/\Gamma(3-\alpha) + \Gamma(1+\alpha)\right)\sin(2x) + 4(t^{\alpha} + t^2 + 2)\sin(2x)$. The analytical solution of the above problem is $u(x,t) = (t^{\alpha} + t^2 + 2)\sin(2x)$. Obviously, the solution has limited smoothness since $u \in C^1$ can not hold.

In this numerical result, we focus on the temporal convergence rates. The second-order spatial derivatives are discretised by means of standard approximations [19] with a convergence rate $O(h^2)$,

$$\frac{\partial^2 u(x_s,t_n)}{\partial x^2} \approx \Delta_x^2 u(x_s,t_n) := \frac{u_{s+1}^n - 2u_s^n + u_{s-1}^n}{h^2}.$$

Then, we have a fully discrete difference scheme,

$$\begin{cases} D_{t_n}^{\alpha}\overline{u}_s(\cdot) = \Delta_x^2 \overline{u}_s^n + \psi_s^n, \\ \overline{u}_s(0) = 2\sin(2x_s), \end{cases} \tag{21}$$

which works in the uniform mesh $Q_{h\tau} = Q_h \times Q_\tau = \{x_s = sh, s = 0, 1, \ldots, S\} \times Q_\tau$, $h = \pi/S$ is the space step size, $\tau = 2/N$ is the time step size. Next, we check the convergence of the implicit difference method (9). The maximum L^2 error, defined by

$$\varrho^{S,N} = \max_{1\le n\le N} \left\| u_s(t_n) - \overline{u}_s^n \right\|_{\Omega_s} = \max_{1\le n\le N} \sqrt{h\sum_{s=1}^{S-1}\left(\overline{u}_s^n - u(x_s,t_n)\right)^2},$$

is adopted in this example. The solution curves of Eq. (20) and the scheme (21) with different α, $\alpha = 0.1$, $\alpha = 0.4$, $\alpha = 0.9$, are shown in Fig. 1 to confirm the convergence of the numerical results. We list the maximum L^2 errors and the observed experimental orders of convergence for scheme (21) when $\alpha = 0.4$, $\alpha = 0.7$ and $\alpha = 0.9$ in Table 1, respectively.

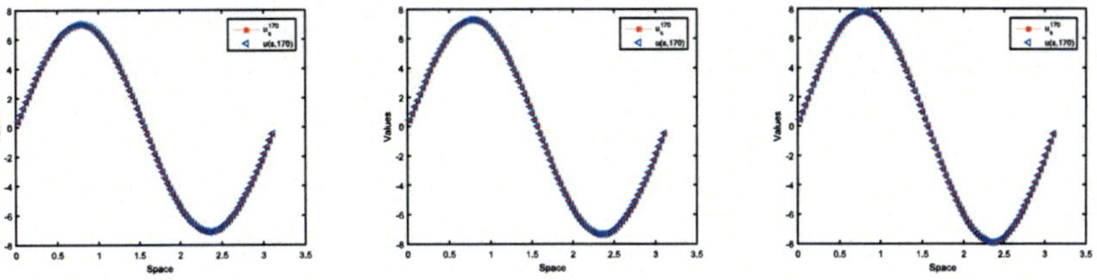

(a) α=0.1. (b) α=0.4. (c) α=0.9.

Fig. 1 The solutions curves at $T = 2$ with $N = 170$, $S = 100$ for different α

Table 1 Errors and numerical convergence orders for different α, S=1000

τ	$\alpha = 0.4$		$\alpha = 0.7$		$\alpha = 0.9$	
	Error	Order	Error	Order	Error	Order
1/100	1.367e-2		3.854e-3		6.063e-4	
1/200	1.145e-2	0.255	2.534e-3	0.605	3.657e-4	0.729
1/400	9.424e-3	0.281	1.623e-3	0.643	2.070e-4	0.821
1/800	7.637e-3	0.303	1.023e-3	0.665	1.139e-4	0.862
1/1600	6.108e-3	0.322	6.394e-4	0.679	6.182e-5	0.882
1/3200	4.831e-3	0.338	3.971e-4	0.687	3.334e-5	0.891

For completeness sake, the implementation methods are briefly described here. In the numerical examples, $\varpi^{S,N}$ denotes the observed orders of convergence for t component, which is computed using the standard formula $\varpi^{S,N} = \log_2 \left(\varrho^{S,N}/\varrho^{S,2N}\right)$. According to the experiment data in Table 1, we can find that the lower-order accuracy can be encountered if α is close to zero.

6 Conclusion

In this work, projection methods are applied to obtain the semidiscrete difference schemes. A finite difference algorithm is used to approximate the time Caputo fractional derivative. Based on the theory of resolvent family and the boundness of the derivative of the solution, we prove the convergence of the proposed implicit and explicit difference schemes for the fractional subdiffusion equation by virtue of the idea of layering in temporal direction. Under the consideration about the weak singularity of solutions, we show that the orders of the global convergence of two schemes in temporal direction could reach $O(\tau^{\alpha})$. A numerical simulation is given. The theoretical and experimental results show that the convergence rates could reach the α-th order accuracy.

Acknowledgements This work was supported by the National Natural Science Foundation of China (Grant Nos. 11571300 and 11871064) and the Russian Science Foundation (Grant No. N20-11-20085, 23-21-00005).

References

1. Adams EE, Gelhar LW (1992) Field study of dispersion in a heterogeneous aquifer: 2. Spatial moments analysis. Water Resour Res 28(12):3293–3307. https://doi.org/10.1029/92WR01757
2. Alikhanov AA (2015) A new difference scheme for the time fractional diffusion equation. J Comput Phys 280:424–438. https://doi.org/10.1016/j.jcp.2014.09.031

3. Ashyralyev A (2009) A note on fractional derivatives and fractional powers of operators. J Math Anal Appl 357(1):232–236. https://doi.org/10.1016/j.jmaa.2009.04.012
4. Ashyralyev A, Cuevas C, Piskarev S (2008) On well-posedness of difference schemes for abstract elliptic problems in $L^p([0, T]; E)$ spaces. Numer Funct Anal Optim 29(1–2):43–65. https://doi.org/10.1080/01630560701872698
5. Ashyralyev A, Emirov N, Cakir Z (2014) Well-posedness of fractional parabolic differential and difference equations with Dirichlet-Neumann conditions. Electron J Differ Equ 2014(97):1–17. https://ejde.math.txstate.edu/
6. Ashyralyev A, Hicdurmaz B (2020) Bounded solutions of second order of accuracy difference schemes for semilinear fractional schrödinger equations. Fract Calc Appl Anal 23(6):1723–1761. https://doi.org/10.1515/fca-2020-0086
7. Ashyralyev A, Piskarev S, Weis L (2002) On well-posedness of difference schemes for abstract parabolic equations in $L^p([0, T]; E)$ spaces. Numer Funct Anal Optim 23(7–8):669–693. https://doi.org/10.1081/NFA-120016264
8. Bajlekova E (2001) Fractional evolution equations in banach spaces. Eindhoven University of Technology, Dissertation, Eindhoven
9. Cao J, Li C, Chen Y (2015) Compact difference method for solving the fractional reaction-subdiffusion equation with Neumann boundary value condition. Int J Comput Math 92(1):167–180. https://doi.org/10.1080/00207160.2014.887702
10. Carvalho AN, Piskarev S (2006) A general approximation scheme for attractors of abstract parabolic problems. Numer Funct Anal Optim 27(7–8):785–829. https://doi.org/10.1080/01630560600882723
11. Cuesta E, Lubich C, Palencia C (2006) Convolution quadrature time discretization of fractional diffusion-wave equations. Math Comp 75:673–696. https://doi.org/10.1090/S0025-5718-06-01788-1
12. Chen H, Holland F, Stynes M (2019) An analysis of the Grünwald-Letnikov scheme for initial-value problems with weakly singular solutions. Appl Numer Math 139:52–61. https://doi.org/10.1016/j.apnum.2019.01.004
13. Chen H, Stynes M (2019) Error analysis of a second-order method on fitted meshes for a time-fractional diffusion problem. J Sci Comput 79:624–647. https://doi.org/10.1007/s10915-018-0863-y
14. Diethelm K (2010) The analysis of fractional differential equations, vol 2004. Springer, Berlin Heidelberg
15. Dixon J (1985) On the order of the error in discretization methods for weakly singular second kind non-smooth solutions. BIT 25(4):623–634. https://doi.org/10.1007/BF01936141
16. Gao G, Sun Z, Zhang H (2014) A new fractional numerical differentiation formula to approximate the Caputo fractional derivative and its applications. J Comput Phys 259:33–50. https://doi.org/10.1016/j.jcp.2013.11.017
17. Hatano Y, Hatano N (1998) Dispersive transport of ions in column experiments: an explanation of long-tailed profiles. Water Resour Res 34(5):1027–1033. https://doi.org/10.1029/98WR00214
18. Huang J, Zhang J, Arshad S, Tang Y (2021) A numerical method for two-dimensional multi-term time-space fractional nonlinear diffusion-wave equations. Appl Numer Math 159:159–173. https://doi.org/10.1016/j.apnum.2020.09.003
19. Huang J, Zhao Y, Arshad S, Li K, Tang Y (2018) Alternating direction implicit schemes for the two-dimensional time fractional nonlinear super-diffusion equations. J Comp Math 37(3):297–315. https://doi.org/10.4208/jcm.1802-m2017-0196
20. Jin B, Li B, Zhou Z (2017) Correction of high-order BDF convolution quadrature for fractional evolution equations. SIAM J Sci Comput 39(6):A3129–A3152. https://doi.org/10.1137/17m1118816
21. Jin B, Li B, Zhou Z (2018) Numerical analysis of nonlinear subdiffusion equations. SIAM J Numer Anal 56(1):1–23. https://doi.org/10.1137/16m1089320
22. Jin B, Li B, Zhou Z (2020) Subdiffusion with time-dependent coefficients: improved regularity and second-order time stepping. Numer Math 145(4):883–913. https://doi.org/10.1007/s00211-020-01130-2

23. Kilbas AA, Srivastava HM, Trujillo JJ (2006) Theory and applications of fractional differential equations. North-Holland mathematics studies, vol 204. Elsevier Science B.V., Amsterdam
24. Lubich Ch (1986) Discretized fractional calculus. SIAM J Math Anal 17(3):704–719. https://doi.org/10.1137/0517050
25. Li M, Chen C, Li F (2010) On fractional powers of generators of fractional resolvent families. J Funct Anal 259:2702–2726. https://doi.org/10.1016/j.jfa.2010.07.007
26. Liu L, Fan Z, Li G, Piskarev S (2019) Maximal regularity for fractional Cauchy equation in Höder space and its approximation. Comput Methods Appl Math 19(4):779–796. https://doi.org/10.1515/cmam-2018-0185
27. Liu L, Fan Z, Li G, Piskarev S (2021) Discrete almost maximal regularity and stability for fractional differential equations in $L^p([0,1],\Omega)$. Appl Math Comput 389:125574. https://doi.org/10.1016/j.amc.2020.125574
28. Li C, Li M (2018) Hölder regularity for abstract fractional Cauchy problems with order in $(0,1)$. J Appl Math Phys 6:310–319. https://doi.org/10.4236/jamp.2018.61030
29. Liu R, Li M, Piskarev S (2015) Approximation of semilinear fractional Cauchy problem. Comput Methods Appl Math 15(2):203–212. https://doi.org/10.1515/cmam-2015-0001
30. Liu R, Li M, Piskarev S (2015) Stability of difference schemes for fractional equations. J Differ Equ 51(7):904–924. https://doi.org/10.1134/S0012266115070095
31. Liu R, Li M, Piskarev S (2017) The order of convergence of difference schemes for fractional equations. Numer Funct Anal Optim 38(6):754–769. https://doi.org/10.1080/01630563.2017.1297825
32. Li Y, Sun H, Feng Z (2016) Fractional abstract Cauchy problem with order $\alpha \in (1,2)$. Dyn Partial Differ Equ 13:155–177. https://doi.org/10.4310/DPDE.2016.v13.n2.a4
33. Lin Y, Xu C (2007) Finite difference/spectral approximations for the time-fractional diffusion equation. J Comput Phys 225(2):1533–1552. https://doi.org/10.1016/j.jcp.2007.02.001
34. Macrobert TM (1955) Higher transcendental functions. Nature 175:317–317. https://doi.org/10.1038/175317a0
35. Meerschaert M, Tadjeran C (2004) Finite difference approximations for fractional advection-dispersion flow equations. J Comput Appl Math 172(1):65–77. https://doi.org/10.1016/j.cam.2004.01.033
36. Metzler R, Klafter J (2000) The random walk's guide to anomalous diffusion: a fractional dynamics approach. Phys Rep 339(1):1–77. https://doi.org/10.1016/S0370-1573(00)00070-3
37. Nigmatullin RR (1986) The realization of the generalized transfer equation in a medium with fractal geometry. Phys Status Solidi (b) 133(1):425–430. https://doi.org/10.1002/pssb.2221330150
38. Oldham KB, Spanier J (1974) The fractional calculus. Academic Press, New York
39. Podlubny I (1999) Fractional differential equations. In: Conference: mathematics in science and engineering. vol 198, Academic Press, San Diego
40. Piskarev S (1979) Approximation of holomorphic semigroup. Tartu Riikl Ül Toimetised 3–14
41. Piskarev S, Siegmund S (2019) Approximations of stable manifolds in the vicinity of hyperbolic equilibrium points for fractional differential equations. Nonlinear Dyn 95(1):685–697. https://doi.org/10.1007/s11071-018-4590-6
42. Stynes M, O'Riordan E, Gracia JL (2017) Error analysis of a finite difference method on graded meshes for a time-fractional diffusion equation. SIAM J Numer Anal 55(2):1057–1079. https://doi.org/10.1137/16M1082329
43. Sakamoto K, Yamamoto M (2011) Initial value/boundary value problems for fractional diffusion-wave equations and applications to some inverse problems. J Math Anal Appl 382(1):426–447. https://doi.org/10.1016/j.jmaa.2011.04.058
44. Vainikko G (1978) Approximative methods for nonlinear equations (two approaches to the convergence problem). Nonlinear Anal 2(6):647–687. https://doi.org/10.1016/0362-546X(78)90013-5

45. Zeng F, Li C, Liu F, Turner I (2013) The use of finite difference/element approaches for solving the time-fractional subdiffusion equation. SIAM J Sci Comput 35(6):A2976–A3000. https://doi.org/10.1137/130910865
46. Zhang J, Aleroev TS, Tang Y, Huang J (2021) Numerical schemes for time-space fractional vibration equations. Adv Appl Math Mech 13(4):806–826

Degenerate Quasilinear Equations with Dzhrbashyan—Nersesian Derivatives and Applications

Marina Plekhanova and Elizaveta Izhberdeeva

Abstract Quasilinear equations with Dzhrbashyan—Nersesyan derivatives in Banach spaces are studied. The existence of a unique classical solution for a Showalter type initial value problem is proved for equation, which contains a degenerate linear operator at the oldest derivative. This result and results for the corresponding degenerate linear equation, which were obtained by authors earlier, are applied to the consideration of initial boundary value problems for linearized and nonlinear systems of partial differential equations with the Dzhrbashyan—Nersesyan time derivative, which describes the dynamics of viscoelastic fluids.

Keywords Dzhrbashyan—Nersesyan derivative · Degenerate evolution equation · Quasilinear equation · Viscoelastic fluid

1 Introduction

One of the rapidly developing areas of the modern mathematics is the theory of differential equations of fractional order and its applications [1–4]. Among the many different definitions of a fractional derivative, the Riemann—Liouville [5] and the Gerasimov—Caputo [5–7] derivatives are the most commonly used. In this paper, we consider equations with the Dzhrbashyan—Nersesyan fractional derivative [8], generalizes the Riemann—Liouville and Gerasimov—Caputo derivatives.

Evolution equations and systems of equations, not solved with respect to the oldest time derivative, or simply degenerate evolution equations studied in this work are often encountered among nonclassical equations of mathematical physics. In this paper, within the framework of the proposed abstract problems we study initial-

M. Plekhanova (✉) · E. Izhberdeeva
Chelyabinsk State University, Chelyabinsk, Russia
e-mail: mariner79@mail.ru

E. Izhberdeeva
e-mail: elizaveta.izhberdeeva@gmail.com
URL: http://www.csu.ru/Lists/List4/sotrudnik.aspx?ID=67

V. Vasilyev (ed.), *Differential Equations, Mathematical Modeling and Computational Algorithms*, Springer Proceedings in Mathematics & Statistics 423,
https://doi.org/10.1007/978-3-031-28505-9_8

boundary value problems for such systems of equations, describing a motion of viscoelastic fluids, with a Dzhrbashyan—Nersesyan time fractional derivative.

Let $T > 0$, $z : (0, T] \to \mathcal{Z}$ for some Banach space $\mathcal{Z}$. The Riemann—Liouville fractional integral of an order $\alpha > 0$ of the function z has the form

$$J_t^{\alpha} z(t) := \int_{t_0}^{t} \frac{(t-s)^{\alpha-1}}{\Gamma(\alpha)} z(s)ds, \quad t > 0.$$

The Riemann—Liouville fractional derivative of an order $\alpha > 0$ for the function z is defined as ${}^R D_t^{\alpha} z(t) := D_t^m J_t^{m-\alpha} z(t)$, where $m - 1 < \alpha \leq m \in \mathbf{N}$, $D_t^m := \frac{d^m}{dt^m}$ is the derivative of the integer order $m \in \mathbf{N}$. Further, we will use the notations ${}^R D_t^{\alpha} := D_t^{\alpha}$, $D_t^{-\alpha} := J_t^{\alpha}$ for $\alpha > 0$. The Gerasimov—Caputo fractional derivative of an order $\alpha > 0$ is defined as

$${}^C D_t^{\alpha} z(t) := {}^R D_t^{\alpha} \left(z(t) - \sum_{k=0}^{m-1} z^{(k)}(t_0) \frac{(t-t_0)^k}{k!} \right).$$

Now let us define the Dzhrbashyan—Nersesyan fractional derivative [8]. For a sequence $\{\alpha_k\}_0^n = \{\alpha_0, \alpha_1, \ldots, \alpha_n\} \subset \mathbf{R}$, such that $0 < \alpha_k \leq 1$, $k = 0, 1, \ldots, n \in \mathbf{N}$, denote

$$\sigma_k := \sum_{j=0}^{k} \alpha_j - 1, \quad k = 0, 1, \ldots, n,$$

so $-1 < \sigma_k \leq k - 1$. Further, the condition $\sigma_n > 0$ is assumed to be satisfied. We define the following differential operations

$$D^{\sigma_0} z(t) := D_t^{\alpha_0 - 1} z(t), \tag{1}$$

$$D^{\sigma_k} z(t) := D_t^{\alpha_k - 1} D_t^{\alpha_{k-1}} D_t^{\alpha_{k-2}} \ldots D_t^{\alpha_0} z(t), \quad k = 1, 2, \ldots, n. \tag{2}$$

The Dzhrbashyan—Nersesyan fractional differentiation of an order σ_n, associated with the sequence $\{\alpha_k\}_0^n$, is defined by relations (1), (2), it includes the Riemann—Liouville fractional derivative ($\alpha_0 \in (0, 1)$, $\alpha_k = 1$, $k = 1, 2, \ldots, n$) and the Gerasimov—Caputo fractional derivative ($\alpha_k = 1$, $k = 0, 1, \ldots, n-1$, $\alpha_n \in (0, 1)$) as partial cases.

Various differential equations with Dzhrbashyan—Nersesyan derivatives were considered in the works of A.V. Pskhu. For example, in [9] a fundamental solution of the diffusion-wave equation with a time Dzhrbashyan—Nersesyan fractional derivative is obtained; in [10] issues of the solvability are studied for the case of a discrete time Dzhrbashyan—Nersesyan fractional derivative.

Here in the second section the unique solvability theorem is proved, while actively using the results obtained in the earlier paper [11] for linear equations, for a Cauchy type problem to a quasilinear equation in a Banach space with Dzhrbashyan—Nersesyan fractional derivatives, which is resolved with respect to the oldest derivative. In the third section this result is used for the study of a unique solution existence issues for a quasilinear equation in a Banach space with Dzhrbashyan—Nersesyan fractional derivatives and with a degenerate linear operator at the oldest of them. The fourth section contains the application of obtained abstract results for the quasilinear degenerate equations in Banach spaces to the study of a initial-boundary value problems for a linearized and a nonlinear models of dynamics of viscoelastic Kelvin—Voigt fluid.

2 Quasilinear Nondegenerate Equation

Let $\mathcal{Z}$ is a Banach space, $A \in \mathcal{L}(\mathcal{Z})$, i. e A is linear bounded operator from $\mathcal{Z}$ to $\mathcal{Z}$, denote by Z an open set in $\mathbf{R} \times \mathcal{Z}^n$, operator $B : Z \to \mathcal{Z}$ is nonlinear, generally speaking, $t_0 \in \mathbf{R}$, $T > t_0$. Consider the initial value problem for a nonlinear equation

$$D^{\sigma_n} z(t) = Az(t) + B(t, D^{\sigma_0} z(t), D^{\sigma_1} z(t), \dots, D^{\sigma_{n-1}} z(t)), \tag{3}$$

$$D^{\sigma_k} z(t_0) = z_k, \quad k = 0, 1, \dots, n-1. \tag{4}$$

The function $z \in C((t_0, T]; \mathcal{Z})$ is called a solution of problem (3), (4) on $(t_0, T]$, if $D_t^{\sigma_k} z \in C([t_0, T]; \mathcal{Z})$, $k = 0, 1, \dots, n-1$, $D_t^{\sigma_n} z \in C((t_0, T]; \mathcal{Z})$, for all $t \in (t_0, T]$ elements $(t, D^{\sigma_0} z(t), D^{\sigma_1} z(t), \dots, D^{\sigma_{n-1}} z(t))$ belong to the set Z, equality (3) and conditions (4) are satisfied.

Firstly recall the result on the corresponding linear equation.

Theorem 1 ([11]) *Let $A \in \mathcal{L}(\mathcal{Z})$, $0 < \alpha_k \leq 1$, $k = 0, 1, \dots, n$, $\sigma_n > 0$, $\alpha_0 + \alpha_n > 1$, $g \in C([t_0, T]; \mathcal{Z})$, $z_k \in \mathcal{Z}$, $k = 0, 1, \dots, n-1$. Then there exists a unique solution of problem (4) for a linear equation $D^{\sigma_n} z(t) = Az(t) + g(t)$ and it has the form*

$$z(t) = \sum_{k=0}^{n-1} (t - t_0)^{\sigma_k} E_{\sigma_n, \sigma_k + 1}((t - t_0)^{\sigma_n} A) z_k +$$

$$+ \int_{t_0}^{t} (t - s)^{\sigma_n - 1} E_{\sigma_n, \sigma_n}((t - s)^{\sigma_n} A) g(s) ds.$$

Lemma 1 *Let $A \in \mathcal{L}(\mathcal{Z})$, $0 < \alpha_k \leq 1$, $k = 0, 1, \dots, n$, $\sigma_n > 0$, $\alpha_0 + \alpha_n > 1$, $B \in C(Z; \mathcal{Z})$, $(t_0, z_0, \dots, z_{n-1}) \in Z$. Then a function $z \in C((t_0, t_1]; \mathcal{Z})$, such that $D^{\sigma_k} z \in C([t_0, t_1]; \mathcal{Z})$, $k = 0, 1, \dots, n-1$, is a solution of problem (3), (4) on $(t_0, t_1]$, if and only if for $t \in (t_0, t_1]$*

$$z(t) = \sum_{k=0}^{n-1}(t-t_0)^{\sigma_k} E_{\sigma_n,\sigma_k+1}((t-t_0)^{\sigma_n}A)z_k+$$

$$+\int_{t_0}^{t}(t-s)^{\sigma_n-1}E_{\sigma_n,\sigma_n}((t-s)^{\sigma_n}A)B(s, D^{\sigma_0}z(s), D^{\sigma_1}z(s), \dots, D^{\sigma_{n-1}}z(s))ds. \quad (5)$$

Proof If a function z is a solution of problem (3), (4), then the map from $[t_0, t_1]$ to $\mathcal{Z}$ of the form $t \to B(t, D^{\sigma_0}z(t), D^{\sigma_1}z(t), \dots, D^{\sigma_{n-1}}z(t))$ is continuous. Therefore, as it is proved in Theorem 1, equality (5) holds.

Let $z \in C((t_0, t_1]; \mathcal{Z})$, such that $D^{\sigma_k}z \in C([t_0, t_1]; \mathcal{Z})$, $k = 0, 1, \dots, n-1$, satisfies (5), then the mapping $t \to B(t, D^{\sigma_0}z(t), D^{\sigma_1}z(t), \dots, D^{\sigma_{n-1}}z(t))$ acts from $[t_0, t_1]$ to $\mathcal{Z}$ continuously and as in the proof of Theorem 1 one can directly check that z is a solution of the problem (3), (4).

Denote $\bar{x} := (x_1, x_2, \dots, x_n)$, $S_\delta(\bar{x}) = \{\bar{y} \in \mathcal{Z}^n : \|y_k - x_k\|_{\mathcal{Z}} \le \delta,\ k = 1, 2, \dots, n\}$.

Theorem 2 *Let $A \in \mathcal{L}(\mathcal{Z})$, $0 < \alpha_k \le 1$, $k = 0, 1 \dots, n$, $\sigma_n > 0$, $\alpha_0 + \alpha_n > 1$, Z be an open set in $\mathbf{R} \times \mathcal{Z}^n$, $B \in C^2(Z; \mathcal{Z})$. Then for each $(t_0, z_0, z_1, \dots, z_{n-1}) \in Z$ there exists a unique solution of problem* (3), (4) *on $(t_0, t_1]$ for some $t_1 > t_0$.*

Proof According to Lemma 1, it suffices to prove that for some $t_1 > t_0$ Eq. (5) has a unique solution $z \in C((t_0, t_1]; \mathcal{Z})$, such that $D^{\sigma_k}z \in C([t_0, t_1]; \mathcal{Z})$, $k = 0, 1, \dots, n-1$.

Let $y(t) := D^{\sigma_0}z(t) = D_t^{\alpha_0-1}z(t)$, then $D^{\sigma_k}z(t) = D_t^{\alpha_k-1}D_t^{\alpha_{k-1}}\dots D_t^1 y(t)$. For the set $\{\alpha_0, \alpha_1, \dots, \alpha_n\}$ define a new set $\{\beta_0 = 1, \beta_1 = \alpha_1, \dots, \beta_n = \alpha_n\}$, numbers $\rho_0 := 0$, $\rho_k := \sum\limits_{l=0}^{k}\beta_l - 1 = \sum\limits_{l=1}^{k}\beta_l > 0$ and the corresponding Dzhrbashyan—Nersesyan fractional derivatives $D^{\rho_0}y(t) = y(t) = D^{\sigma_0}z(t)$, $D^{\rho_k}y(t) = D_t^{\beta_k-1}D_t^{\beta_{k-1}}D_t^{\beta_{k-2}}\dots D^{\beta_0}y(t) = D^{\sigma_k}z(t)$ for $k = 1, 2, \dots, n$. It follows from (4) that $y(t_0) = z_0$, $D^{\rho_k}y(t_0) = z_k$, $k = 1, 2, \dots, n-1$.

Denote $B^y(s) := B(s, y(s), D^{\rho_1}y(s), \dots, D^{\rho_{n-1}}y(s))$. Choose $\tau > 0, \delta > 0$, such that $V = [t_0, t_0 + \tau] \times S_\delta(\overline{z}) \subset Z$, where $\overline{z} = (z_0, z_1, \dots, z_{n-1})$ is the vector of initial data from (4). Consider the Banach space $C^{\{\beta_k\}_0^{n-1}}([t_0, t_0 + \tau]; \mathcal{Z})$ with the norm

$$\|v\|_{C^{\{\beta_k\}_0^{n-1}}([t_0,t_0+\tau];\mathcal{Z})} := \|v\|_{C([t_0,t_0+\tau];\mathcal{Z})} + \max_{t\in[t_0,t_0+\tau]}\|(t-t_0)^{1-\beta_1}v'(t)\|_{\mathcal{Z}}+$$

$$+\sum_{k=1}^{n-1}\left(\|D^{\rho_k}v\|_{C([t_0,t_0+\tau];\mathcal{Z})} + \max_{t\in[t_0,t_0+\tau]}\|(t-t_0)^{1-\beta_{k+1}}D^{\rho_k+1}v(t)\|_{\mathcal{Z}}\right).$$

Denote by S_τ the set of functions y from $C^{\{\beta_k\}_0^{n-1}}([t_0, t_0 + \tau]; \mathcal{Z})$, such that

$$\|D^{\rho_k} y(t) - z_k\|_{\mathcal{Z}} \le \delta, \ \|(t-t_0)^{1-\beta_n} D_t^1 D^{\rho_{n-1}} y(t)\|_{\mathcal{Z}} \le \|B(t_0, z_0, \dots, z_{n-1})\|_{\mathcal{Z}} + \delta,$$

$$\left\|(t-t_0)^{1-\beta_{k+1}} D_t^1 D^{\rho_k} y(t) - \frac{z_{k+1}}{\Gamma(\rho_{k+1}-\rho_k)}\right\|_{\mathcal{Z}} \le \delta, \ k = 0, 1, \dots, n-2,$$

for $t_0 \le t \le t_0 + \tau$, $k = 0, 1, \dots, n-1$. In S_τ we define the metric $d(y, v) = \|y - v\|_{C^{\{\beta_k\}_0^{n-1}}([t_0, t_0+\tau]; \mathcal{Z})}$. Note that for

$$\tilde{y} := z_0 + \sum_{k=1}^{n-1} \frac{(t-t_0)^{\rho_k} z_k}{\Gamma(\rho_k+1)} \quad D^{\rho_k}|_{t=t_0} \tilde{y}(t) = z_k, \quad k = 0, 1, \dots, n-1,$$

$$(t-t_0)^{1-\beta_{k+1}} D_t^1 D^{\rho_k} \tilde{y}(t) = \frac{z_{k+1}}{\Gamma(\rho_{k+1}-\rho_k)}, \quad k = 0, 1, \dots, n-2,$$

$(t-t_0)^{1-\beta_n} D_t^1 D^{\rho_n} \tilde{y}(t) = 0$, therefore, $\tilde{y} \in S_\tau$. Define

$$G(y)(t) := \sum_{k=0}^{n-1} (t-t_0)^{\rho_k} E_{\rho_n, \rho_k+1}((t-t_0)^{\rho_n} A) z_k +$$

$$+ \int_{t_0}^{t} (t-s)^{\rho_n - 1} E_{\rho_n, \rho_n}((t-s)^{\rho_n} A) B^y(s) ds$$

and note that (5) has the form $y(t) = G(y)(t)$. Denote for $k = 0, 1, \dots, n-1$

$$B_t^y(s) := \frac{\partial B}{\partial t}(t, y_0, y_1, \dots, y_{n-1}), \quad B_k^y(s) := \frac{\partial B}{\partial y_k}(t, y_0, y_1, \dots, y_{n-1}).$$

For $y \in C^{\{\beta_k\}_0^{n-1}}([t_0, t_0+\tau]; \mathcal{Z})$ by virtue of Theorem 2 [11] $D^{\rho_k} G(y) \in C([t_0, t_0+\tau]; \mathcal{Z})$, $k = 0, 1, \dots, n-1$. Moreover,

$$(t-t_0)^{1-\beta_{k+1}} D_t^1 D^{\rho_k} G(y)(t) = \sum_{l=k+1}^{n-1} (t-t_0)^{\rho_l - \rho_k - \beta_{k+1}} E_{\rho_n, \rho_l - \rho_k}((t-t_0)^{\rho_n} A) z_l +$$

$$+ (t-t_0)^{\rho_n - \rho_k - \beta_{k+1}} E_{\rho_n, \rho_n - \rho_k}((t-t_0)^{\rho_n} A) B^y(t_0) +$$

$$+ (t-t_0)^{1-\beta_{k+1}} \int_{t_0}^{t} (t-s)^{\rho_n - \rho_k - 1} E_{\rho_n, \rho_n - \rho_k}((t-s)^{\rho_n} A) D_s^1 B^y(s) ds,$$

$$\left\|(t-t_0)^{1-\beta_{k+1}}\int_{t_0}^{t}(t-s)^{\rho_n-\rho_k-1}E_{\rho_n,\rho_n-\rho_k}((t-s)^{\rho_n}A)D_s^1B^y(s)ds\right\|_{\mathcal{Z}} \le$$

$$\le (t-t_0)^{\rho_n-\rho_k-\beta_{k+1}+1}E_{\rho_n,\rho_n-\rho_k}((t-t_0)^{\rho_n}\|A\|_{\mathcal{L}(\mathcal{Z})})\|B^y(t_0)\|_{\mathcal{Z}}+$$

$$+(\rho_n-\rho_k-\beta_{k+1})\left\|\int_{t_0}^{t}(t-s)^{\rho_n-\rho_k-\beta_{k+1}-1}E_{\rho_n,\rho_n-\rho_k}((t-s)^{\rho_n}A)B^y(s)ds\right\|_{\mathcal{Z}} +$$

$$+(t-t_0)^{\rho_n-\rho_k-\beta_{k+1}+1}E_{\rho_n,\rho_n-\rho_k}((t-t_0)^{\rho_n}\|A\|_{\mathcal{L}(\mathcal{Z})})\max_{t\in[t_0,t_0+\tau]}\|B_t^y(t)\|_{\mathcal{Z}}+$$

$$+(t-t_0)^{\rho_n-\rho_k}E_{\rho_n,\rho_n-\rho_k}((t-t_0)^{\rho_n}\|A\|_{\mathcal{L}(\mathcal{Z})})\times$$

$$\times\left(\sum_{k=0}^{n-1}\max_{t\in[t_0,t_0+\tau]}\|B_k^y(t)\|_{\mathcal{L}(\mathcal{Z})}\max_{t\in[t_0,t_0+\tau]}[(t-t_0)^{1-\beta_{k+1}}\|D_t^1D^{\rho_k}y(t)\|_{\mathcal{Z}}]\right)ds\to 0$$

for $t\to t_0+$, because $\rho_l-\rho_k-\beta_{k+1}=\rho_l-\rho_{k+1}>0$ for $l=k+2,k+3,\dots,n$. Here we also use the inequality $(t-t_0)^{\beta}\le 2^{\beta-1}[(t-s)^{\beta}+(s-t_0)^{\beta}]$ for $s\in[t_0,t]$. Therefore, for $k=0,1,\dots,n-2$

$$\lim_{t\to t_0+}(t-t_0)^{1-\beta_{k+1}}D_t^1D^{\rho_k}G(y)(t)=\frac{z_{k+1}}{\Gamma(\rho_{k+1}-\rho_k)},$$

$$\lim_{t\to t_0+}(t-t_0)^{1-\beta_n}D_t^1D^{\rho_{n-1}}G(y)(t)=B^y(t_0)=B(t_0,z_0,z_1,\dots,z_{n-1}),$$

$G(y)\in C^{\{\beta_k\}_0^{n-1}}([t_0,t_0+\tau];\mathcal{Z})$, and for small enough $\tau>0$ we have $G(y)\in S_\tau$.

If $\|D^{\rho_k}y(t)-z_k\|_{\mathcal{Z}}\le\delta$ for $t_0\le t\le t_0+\tau$, $k=0,1,\dots,n-1$, then, decreasing $\tau>0$, if necessary, we obtain

$$\|D^{\rho_k}G(y)(t)-z_k\|_{\mathcal{Z}}\le\frac{\delta}{2}+\tau^{\rho_n-\rho_k}E_{\rho_n,\rho_n-\rho_k}(\tau^{\rho_n}\|A\|_{\mathcal{L}(\mathcal{Z})})(2c\delta n+K)\le\delta$$

for $t_0\le t\le t_0+\tau$, $k=0,1,\dots,n-1$, where $K=\max_{s\in[t_0,t_0+\tau]}\|B^{\tilde{y}}(s)\|_{\mathcal{Z}}$. Here we use for $h\in C([t_0,t_0+\tau];\mathcal{Z})$ the equalities

$$D^{\rho_k}\int_{t_0}^{t}(t-s)^{\rho_n-1}E_{\rho_n,\rho_n}((t-s)^{\rho_n}A)h(s)ds=\int_{t_0}^{t}\sum_{j=0}^{\infty}\frac{(t-s)^{\rho_nj+\rho_n-\rho_k-1}A^j}{\Gamma(\rho_nj+\rho_n-\rho_k)}h(s)ds$$

$$=\int_{t_0}^{t}(t-s)^{\rho_n-\rho_k-1}E_{\rho_n,\rho_n-\rho_k}((t-s)^{\rho_n}A)h(s)ds.$$

Further,

$$\left\|(t-t_0)^{1-\beta_{k+1}} D_t^1 \int_{t_0}^{t} (t-s)^{\rho_n-\rho_k-1} E_{\rho_n,\rho_n-\rho_k}((t-s)^{\rho_n} A)(B^y(s)-B^v(s))ds\right\|_{\mathcal{Z}}$$

$$\leq \left[2\tau^{\rho_n-\rho_{k+1}+1} + (\rho_n-\rho_{k+1})\tau^{\rho_n-\rho_{k+1}} + \left(\tau^{1-\beta_{k+1}} + \|B^y(t_0)\|_{\mathcal{Z}} + \delta\right)\tau^{\rho_n-\rho_k}\right] \times$$

$$\times E_{\rho_n,\rho_n-\rho_k}(\tau^{\rho_n}\|A\|_{\mathcal{L}(\mathcal{Z})}) c \|y-v\|_{C^{\{\beta_k\}_0^{n-1}}([t_0,t_0+\tau];\mathcal{Z})}. \tag{6}$$

For each $t \in [t_0, t_0+\tau]$, $r = 0, 1, \ldots, m-1$, $y, v \in S_\tau$, due to (6) we get

$$\|G(y)(t) - G(v)(t)\|_{C^{\{\beta_k\}_0^{n-1}}([t_0,t_0+\tau];\mathcal{Z})} \leq$$

$$\leq c_2\tau^{\alpha_n}\|y-v\|_{C^{\{\beta_k\}_0^{n-1}}([t_0,t_0+\tau];\mathcal{Z})} \leq \frac{d(y,v)}{2}$$

for sufficiently small $\tau > 0$. Therefore, the operator G has a unique fixed point y in S_τ. Hence,

$$z(t) = D_t^1 \int_{t_0}^{t} \frac{(t-s)^{\alpha_0-1}}{\Gamma(\alpha_0)} y(s)ds = \frac{(t-t_0)^{\alpha_0-1}}{\Gamma(\alpha_0)} z_0 - \int_{t_0}^{t} \frac{(t-s)^{\alpha_0-1}}{\Gamma(\alpha_0)} y'(s)ds.$$

Since $y \in C^{\{\beta_k\}_0^{n-1}}([t_0, t_0+\tau]; \mathcal{Z})$, we have

$$\left\|\int_{t_0}^{t} \frac{(t-s)^{\alpha_0-1}}{\Gamma(\alpha_0)} y'(s)ds\right\|_{\mathcal{Z}} \leq \frac{\Gamma(\alpha_1)(t-t_0)^{\alpha_0+\alpha_1-1}}{\Gamma(\alpha_0+\alpha_1)}\|y\|_{C^{\{\beta_k\}_0^{n-1}}([t_0,t_0+\tau];\mathcal{Z})}.$$

Therefore, $z = D_t^{1-\alpha_0} y \in C((t_0, T]; \mathcal{Z})$, $D^{\sigma_k} z = D_t^{\rho_k} y \in C([t_0, T]; \mathcal{Z})$, $k = 0, 1, \ldots, n-1$, z is a solution of the problem (3), (4) on $(t_0, t_0+\tau]$. The uniqueness of a solution follows from the uniqueness of a fixed point of G.

3 Quasilinear Degenerate Equation

When solving a degenerate evolution equation, some theory of such equations will be required. Let $L \in \mathcal{L}(\mathcal{X}; \mathcal{Y})$, i. e L is linear bounded operator from $\mathcal{X}$ to $\mathcal{Y}$, $\ker L \neq \{0\}$, $M \in \mathcal{C}l(\mathcal{X}; \mathcal{Y})$, i. e it is linear closed operator from $\mathcal{X}$ to $\mathcal{Y}$ with a dense domain D_M in $\mathcal{X}$. Define L-resolvent set of an operator M as $\rho^L(M) := \{\mu \in \mathbb{C} : (\mu L - M)^{-1} \in \mathcal{L}(\mathcal{Y}; \mathcal{X})\}$ and denote $R_\mu^L(M) := (\mu L - M)^{-1}L$, $L_\mu^L(M) := L(\mu L - M)^{-1}$.

An operator M is called (L, σ)-bounded, if

$$\exists a > 0 \quad \forall \mu \in \mathbf{C} \quad (|\mu| > a) \Rightarrow (\mu \in \rho^L(M)).$$

Lemma 2 ([12, pp. 89, 90]) *Let an operator M be (L, σ)-bounded, $\gamma = \{\mu \in \mathbf{C} : |\mu| = r > a\}$. Then the operators*

$$P = \frac{1}{2\pi i} \int_\gamma R^L_\mu(M)\, d\mu \in \mathcal{L}(\mathcal{X}), \quad Q = \frac{1}{2\pi i} \int_\gamma L^L_\mu(M)\, d\mu \in \mathcal{L}(\mathcal{Y})$$

are projectors.

Put $\mathcal{X}^0 := \ker P$, $\mathcal{X}^1 := \operatorname{im} P$, $\mathcal{Y}^0 := \ker Q$, $\mathcal{Y}^1 := \operatorname{im} Q$. Denote by L_k (M_k) the restriction of the operator L (M) on $\mathcal{X}^k$ $(D_{M_k} = D_M \cap \mathcal{X}^k)$, $k = 0, 1$.

Theorem 3 ([12, p. 90, 91]) *Let an operator M be (L, σ)-bounded. Then*
(i) $M_1 \in \mathcal{L}(\mathcal{X}^1; \mathcal{Y}^1)$, $M_0 \in \mathcal{C}l(\mathcal{X}^0; \mathcal{Y}^0)$, $L_k \in \mathcal{L}(\mathcal{X}^k; \mathcal{Y}^k)$, $k = 0, 1$;
(ii) *there exist operators* $M_0^{-1} \in \mathcal{L}(\mathcal{Y}^0; \mathcal{X}^0)$, $L_1^{-1} \in \mathcal{L}(\mathcal{Y}^1; \mathcal{X}^1)$.

Denote $G := M_0^{-1} L_0$. For $p \in \mathbf{N}_0 := \mathbf{N} \cup \{0\}$ an operator M is called (L, p)-bounded, if it is (L, σ)-bounded, $G^p \neq 0$, $G^{p+1} = 0$. Thus, for $(L, 0)$-bounded operator M we have $L_0 = 0$.

Consider the Showalter type [12] initial value problem

$$D^{\sigma_k} P x(t_0) = x_k, \quad k = 0, 1, \ldots, n - 1, \tag{7}$$

for a linear inhomogeneous fractional order equation

$$D^{\sigma_n} L x(t) = M x(t) + g(t), \tag{8}$$

where D^{σ_n} is the Dzhrbashyan—Nersesyan fractional derivative, which corresponds to the set of numbers $\{\alpha_0, \alpha_1, \ldots, \alpha_n\}$, $0 < \alpha_k \leq 1$, $k = 0, 1, \ldots, n$, $g \in C([0, T]; \mathcal{Y})$.

A function $x : (t_0, T] \to D_M$ is called a solution of problem (7), (8), if $Mx \in C((t_0, T]; \mathcal{Y})$, $D^{\sigma_k} P x \in C([t_0, T]; \mathcal{X})$, $k = 0, 1, \ldots, n - 1$, $D^{\sigma_n} L x \in C((t_0, T]; \mathcal{Y})$, equality (8) holds for all $t \in (t_0, T]$ and conditions (7) hold.

Theorem 4 ([11, p. 10]) *Let an operator M be (L, p)-bounded, $0 < \alpha_k \leq 1$, $k = 0, 1 \ldots, n$, $\sigma_n > 0$, $\alpha_0 + \alpha_n > 1$, $g \in C([t_0, T]; \mathcal{Y})$, $(D^{\sigma_n} G)^l M_0^{-1}(I - Q) g \in C((t_0, T]; \mathcal{X})$, $l = 0, 1, \ldots, p$, $x_k \in \mathcal{X}^1$, $k = 0, 1, \ldots, n - 1$. Then there exists a unique solution of problem* (7), (8) *and it has the form*

$$x(t) = \sum_{k=0}^{n-1}(t-t_0)^{\sigma_k} E_{\sigma_n,\sigma_k+1}((t-t_0)^{\sigma_n} L_1^{-1}M)x_k +$$

$$+\int_{t_0}^{t}(t-s)^{\sigma_n-1}E_{\sigma_n,\sigma_n}((t-s)^{\sigma_n}L_1^{-1}M)L_1^{-1}Qg(s)ds - \sum_{l=0}^{p}(D^{\sigma_n}G)^l M_0^{-1}(I-Q)g(t).$$

Let X be an open set in $\mathbf{R}\times\mathcal{X}^n$, an operator $N : X \to \mathcal{Y}$ is nonlinear. Consider the Showalter type initial value problem for the nonlinear equation

$$D^{\sigma_n}Lx(t) = Mx(t) + N(t, D^{\sigma_0}x(t), D^{\sigma_1}x(t), \ldots, D^{\sigma_{n-1}}x(t)), \qquad (9)$$

$$D^{\sigma_k}Px(t_0) = x_k, \quad k = 0, 1, \ldots, n-1. \qquad (10)$$

Denote $V := X \cap (\mathbf{R}\times(\mathcal{X}^1)^n)$. In the following theorem we will use the assumption that for all $(t_0, z_0, \ldots, z_{n-1}) \in X$ equality $N(t_0, z_0, \ldots, z_{n-1}) = N_1(t_0, Pz_0, \ldots, Pz_{n-1})$ is satisfied with some $N_1 : V \to \mathcal{Y}$. A function $x \in C((t_0, t_1]; D_M)$ is called a solution of problem (9), (10) on $(t_0, t_1]$, if for all $k = 0, 1, \ldots, n-1$ $D^{\sigma_k}Px \in C([t_0, t_1]; \mathcal{X})$, $D^{\sigma_n}Lx \in C((t_0, t_1]; \mathcal{X})$, for all $t \in (t_0, t_1]$ $(t, D^{\sigma_0}Px(t), D^{\sigma_1}Px(t), \ldots, D^{\sigma_{n-1}}Px(t)) \in V$, equality (9) is satisfied and conditions (10) are valid.

Theorem 5 *Let* $0 < \alpha_k \le 1$, $k = 0, 1 \ldots, n$, $\sigma_n > 0$, $\alpha_0 + \alpha_n > 1$, *an operator* M *be* $(L, 0)$*-bounded,* X *be an open set in the space* $\mathbf{R}\times\mathcal{X}^n$, $N : X \to \mathcal{Y}$, *for all* $(t_0, z_0, \ldots, z_{n-1}) \in X$ *equality* $N(t_0, z_0, \ldots, z_{n-1}) = N_1(t_0, Pz_0, \ldots, Pz_{n-1})$ *is true with some* $N_1 \in C^2(V; \mathcal{Y})$. *Then for arbitrary* $(t_0, x_0, x_1, \ldots, x_{n-1}) \in V$ *there exists* $t_1 > t_0$, *such that problem* (9), (10) *has a unique solution on* $(t_0, t_1]$.

Proof Let us introduce the notations $v(t) := Px(t)$, $w(t) := (I-P)x(t)$, $S_1 := L_1^{-1}M_1$. We act on Eq. (9) by the operator $M_0^{-1}(I-Q)$, by the operator $L_1^{-1}Q$ and obtain a problem for the system of equations on mutually complementary subspaces $\mathcal{X}^1$ and $\mathcal{X}^0$

$$D^{\sigma_n}v(t) = S_1 v(t) + L_1^{-1}QN_1(t, D^{\sigma_0}v(t), D^{\sigma_1}v(t), \ldots, D^{\sigma_{n-1}}v(t)), \qquad (11)$$

$$D^{\sigma_k}v(t_0) = x_k, \quad k = 0, 1, \ldots, n-1, \qquad (12)$$

$$0 = w(t) + M_0^{-1}(I-Q)N_1(t, D^{\sigma_0}v(t), D^{\sigma_1}v(t), \ldots, D^{\sigma_{n-1}}v(t)).$$

Here we use the equality $L_0 = 0$, which is valid due to $(L, 0)$-boundedness of the operator M. Since V is an open set in the space $\mathbf{R}\times(\mathcal{X}^1)^n$, $L_1^{-1}QN_1 \in C^2(V; \mathcal{X})$, then problem (11), (12) has a unique solution v on $(t_0, t_1]$ with some $t_1 > t_0$ by Theorem 2. Hence, $w(t) = -M_0^{-1}(I-Q)N_1(t, D^{\sigma_0}v(t), D^{\sigma_1}v(t), \ldots, D^{\sigma_{n-1}}v(t))$, where v is a solution of problem (11), (12). Thus, there is a unique solution $x(t) = v(t) + w(t)$ to problem (9), (10).

4 Fractional Models of Viscoelastic Fluid Dynamics

Consider an initial-boundary value problem for a system of equations

$$(1-\chi\Delta)D^{\sigma_n,t}v = \nu\Delta v - \nabla p + f, \quad (x,t) \in \Omega \times (t_0,T), \tag{13}$$

$$\nabla \cdot v = 0, \quad (x,t) \in \Omega \times (t_0,T), \tag{14}$$

$$v = 0, \quad (x,t) \in \partial\Omega \times (t_0,T), \tag{15}$$

$$D^{\sigma_k,t}v(x,t_0) = w_k(x), \quad x \in \Omega, \quad k = 0,1,\ldots,n-1. \tag{16}$$

Here $d \in \mathbf{N}$, $\Omega \subset \mathbf{R}^d$ is a bounded domain with a smooth boundary $\partial\Omega$, $\chi, \nu \in \mathbf{R}$. The vector function of velocity $v = (v_1, v_2, \ldots, v_d)$ and pressure gradient $\nabla p = r = (r_1, r_2, \ldots, r_d)$ are unknown, function $f : \Omega \times [t_0, T) \to \mathbf{R}^d$ is given, $0 < \alpha_k \le 1$, $D^{\sigma_k,t}$ is the Dzhrbashyan—Nersesyan fractional derivative with respect to time t, $k = 0, 1 \ldots, n$. System (13), (14) presents the linearized at zero solution Kelvin—Voigt model of viscoelastic fluid dynamics [13].

In order to reduce the initial boundary value problem (13)–(16) to abstract problem (7), (8), we introduce the Lebesgue space $\mathbf{L}_2 := (L_2(\Omega))^d$ and Sobolev spaces $\mathbf{H}^1 := (W_2^1(\Omega))^d$, $\mathbf{H}^2 := (W_2^2(\Omega))^d$. The closure of the subspace $\mathcal{L} := \{v \in (C_0^\infty(\Omega))^d : \nabla \cdot v = 0\}$ in $\mathbf{L}_2$ is denoted by $\mathbf{H}_\sigma$; $\mathbf{H}_\sigma^1$ is the closure of $\mathcal{L}$ in $\mathbf{H}^1$. We will use also the notation $\mathbf{H}_\sigma^2 := \mathbf{H}_\sigma^1 \cap \mathbf{H}^2$. The orthogonal complement of $\mathbf{H}_\sigma$ in $\mathbf{L}_2$ is denoted by $\mathbf{H}_\pi$, the corresponding orthoprojections are $\Sigma : \mathbf{L}_2 \to \mathbf{H}_\sigma$, $\Pi = I - \Sigma$.

Consider an operator $A = \Sigma\Delta$ in $\mathcal{L}$. This operator, extended to a closed operator in $\mathbf{H}_\sigma$ with domain $\mathbf{H}_\sigma^2$, has a real negative discrete spectrum of finite multiplicity condensing only to $-\infty$ [14]. The eigenvalues of $\{\lambda_k\}$ are numbered in nonincreasing order, taking into account their multiplicity. An orthonormal system $\{\varphi_k\}$ of corresponding eigenfunctions forms a basis in $\mathbf{H}_\sigma$.

We define spaces and operators as follows:

$$\mathcal{X} = \mathbf{H}_\sigma^2 \times \mathbf{H}_\pi, \quad \mathcal{Y} = \mathbf{L}_2 = \mathbf{H}_\sigma \times \mathbf{H}_\pi, \tag{17}$$

$$L = \begin{pmatrix} I-\chi A & \mathbf{O} \\ -\chi\Pi\Delta & \mathbf{O} \end{pmatrix} \in \mathcal{L}(\mathcal{X};\mathcal{Y}), \quad M = \begin{pmatrix} \nu A & \mathbf{O} \\ \nu\Pi\Delta & -I \end{pmatrix} \in \mathcal{L}(\mathcal{X};\mathcal{Y}). \tag{18}$$

Lemma 3 *Let the spaces $\mathcal{X}$ and $\mathcal{Y}$ be defined in (17) and the operators L and M be defined in (18), $\chi, \nu \in \mathbf{R}$, $\chi \neq 0$, $\chi^{-1} \notin \sigma(A)$. Then the operator M is $(L,0)$-bounded,*

$$P = \begin{pmatrix} I & \mathbf{O} \\ \nu(I-\chi A)^{-1}\Pi\Delta & \mathbf{O} \end{pmatrix}, \quad Q = \begin{pmatrix} I & \mathbf{O} \\ -\chi\Pi\Delta(I-\chi A)^{-1} & \mathbf{O} \end{pmatrix}. \tag{19}$$

Proof In [15, Theorem 7] it was shown that the operator M is $(L,0)$-bounded and

$$(\mu L - M)^{-1} = \begin{pmatrix} (\mu I - (\mu\chi+\nu)A)^{-1} & \mathbf{O} \\ (\mu\chi+\nu)\Pi\Delta(\mu I - (\mu\chi+\nu)A)^{-1} & I \end{pmatrix}.$$

This implies the equalities

$$R^L_\mu(M) = \begin{pmatrix} (\mu I - (\mu\chi+\nu)A)^{-1}(I-\chi A) & \mathbf{O} \\ (\mu\chi+\nu)\Pi\Delta(\mu I - (\mu\chi+\nu)A)^{-1}(I-\chi A) - \chi\Pi\Delta & \mathbf{O} \end{pmatrix} =$$

$$= \begin{pmatrix} (\mu I - \nu(I-\chi A)^{-1}A)^{-1} & \mathbf{O} \\ (\chi\nu(I-\chi A)^{-1}A + \nu I)\Pi\Delta(\mu I - \nu(I-\chi A)^{-1}A)^{-1} & \mathbf{O} \end{pmatrix} =$$

$$= \begin{pmatrix} (\mu I - \nu(I-\chi A)^{-1}A)^{-1} & \mathbf{O} \\ \nu(I-\chi A)^{-1}\Pi\Delta(\mu I - \nu(I-\chi A)^{-1}A)^{-1} & \mathbf{O} \end{pmatrix},$$

$$L^L_\mu(M) = \begin{pmatrix} (I-\chi A)(\mu I - (\mu\chi+\nu)A)^{-1} & \mathbf{O} \\ -\chi\Pi\Delta(\mu I - (\mu\chi+\nu)A)^{-1} & \mathbf{O} \end{pmatrix} =$$

$$= \begin{pmatrix} (\mu I - \nu A(I-\chi A)^{-1})^{-1} & \mathbf{O} \\ -\chi\Pi\Delta(I-\chi A)^{-1}(\mu I - \nu A(I-\chi A)^{-1})^{-1} & \mathbf{O} \end{pmatrix}.$$

Using the expansion into the Neumann series, e.g.,

$$(\mu I - \nu A(I-\chi A)^{-1})^{-1} = \sum_{n=0}^{\infty} \mu^{-n-1}[A(I-\chi A)^{-1}]^n,$$

the form of projections P, Q and the residue theorem, we obtain equalities (19).

Theorem 6 *Let* $\chi,\nu \in \mathbf{R}$, $\chi \neq 0$, $\chi^{-1} \notin \sigma(A)$, $0 < \alpha_k \leq 1$, $k = 0,1,\dots,n$, $\sigma_n > 0$, $\alpha_0 + \alpha_n > 1$, $f \in C([t_0,T];\mathbf{L}_2)$, $w_k \in \mathbf{H}^2_\sigma$, $k = 0,1,\dots n-1$. *Then there exists a unique solution of problem* (13)–(16).

Proof If we use spaces (17), then due to Lemma 3 operators (18) and the function $g(t) = f(\cdot,t)$ satisfies the conditions of Theorem 4. Note also that due to form (19) of the projection P setting conditions (16) with $w_k \in \mathbf{H}^2_\sigma$ is equivalent to problem (7) with $x_k \in \mathcal{X}^1$, $k = 0,1,\dots,n-1$.

For $n = 1$ consider the nonlinear Kelvin—Voigt model of viscoelastic dynamics [13]

$$(1-\chi\Delta)D^{\sigma_1,t}v = \nu\Delta v - (v\cdot\nabla)v - \nabla p + f, \quad (x,t) \in \Omega \times (t_0,t_1). \tag{20}$$

$$\nabla \cdot v = 0, \quad (x,t) \in \Omega \times (t_0,t_1), \tag{21}$$

$$v = 0, \quad (x,t) \in \partial\Omega \times (t_0, t_1), \tag{22}$$

$$v(x, t_0) = w_0(x), \quad x \in \Omega. \tag{23}$$

Let $\alpha_0 = 1$, $\alpha_1 \in (0, 1]$, then $D^{\sigma_0,t}v = v$, for $\alpha_1 \in (0, 1)$ $D^{\sigma_1,t}v = J_t^{1-\alpha_1} D_t^1 v = {}^C D_t^{\alpha_1} v$ is the Gerasimov—Caputo derivative. For $\alpha_1 = 1$ $D^{\sigma_1,t}v = v_t$ is usual partial derivative with respect to time t.

Theorem 7 *Let* $\chi, \nu \in \mathbf{R}$, $\chi \neq 0$, $\chi^{-1} \notin \sigma(A)$, $\alpha_0 = 1$, $0 < \alpha_1 \le 1$, $f \in C^2([t_0, T]; \mathbf{L}_2)$, $w_0 \in \mathbf{H}_\sigma^2$. *Then for some* $t_1 \in (t_0, T]$ *there exists a unique solution of the problem* (20)–(23) *on* $\Omega \times (t_0, t_1)$.

Proof Under the conditions of this theorem $\sigma_1 = \alpha_1 > 0$, $\alpha_0 + \alpha_n = \alpha_0 + \alpha_1 = 1 + \alpha_1 > 1$.

It follows from the form of P that condition (23) with $w_0 \in \mathbf{H}_\sigma^2$ is equivalent to (7) with $n = 1$, $x_0 \in \mathcal{X}^1$, and the nonlinear operator $N(v, r) = -(v \cdot \nabla)v + f$ depends only on the projection of the vector $(v, r) \in \mathcal{X}$ on the subspace $\mathcal{X}^1 = \mathbf{H}_\sigma^2 \times \nu(I - \chi A)^{-1}\Pi\Delta[\mathbf{H}_\sigma^2]$. Indeed, $N(v, r) = N(P_1 P(v, r), r)$, where due to (19) $P(v, r) = (v, \nu(I - \chi A)^{-1}\Pi\Delta v)$, $P_1 P(v, r) = P_1(v, \nu(I - \chi A)^{-1}\Pi\Delta v) = v$ is the projection on the first component. So,

$$N_1(P(v, r)) := N(P_1 P(v, r), r) = -(P_1 P(v, r) \cdot \nabla)P_1 P(v, r) + f = -(v \cdot \nabla)v + f.$$

By the Sobolev theorem embedding $H^2(\Omega) \subset L_{q'}(\Omega)$ is valid for $d \le 4$ or at $q' \le 2d/(d-4)$ for $d > 4$, $H^2(\Omega) \subset H_q^1(\Omega)$ is true for $d \le 2$ or at $q \le 2d/(d-2)$ for $d > 2$. Take any $q > 1$ for $d \le 2$, or $q = 2d/(d-2)$ and $q' = q/(q-1) = 2d/(d+2) > 1$ for $d > 2$, then $\|N(v, r)\|_{\mathbf{L}_2}^2 \le 2\|v\|_{\mathbf{L}_{q'}}^{1/q'} \|v\|_{\mathbf{H}_q^1}^{1/q} + 2\|f\|_{\mathbf{L}_2}^2$ and $N : \mathcal{X} \to \mathcal{Y}$.

The first and second order Frechet derivatives have the form for $h \in \mathcal{X}$

$$N_1'(v, r)h = -(P_1 Ph \cdot \nabla)v - (v \cdot \nabla)P_1 Ph, \quad \langle N_1''(v, r)h, h\rangle = -2(P_1 Ph \cdot \nabla)P_1 Ph.$$

Therefore, the second order Frechet derivative of the operator N is constant with respect to v and $N_1 \in C^2(\mathcal{X}; \mathcal{Y})$. Therefore, all the conditions on N_1 of Theorem 5 for $X = \mathbf{R} \times \mathcal{X}$ are fulfilled.

The work is funded by the grant of the Russian Scientific Foundation, project 22-21-20095.

References

1. Samko SG, Kilbas AA, Marichev OI (1993) Fractional integrals and derivatives. Theory and applications. Gordon and Breach Science Publishers, Philadelphia, USA
2. Kiryakova V (1994) Generalized fractional calculus and applications. Longman Scientific & Technical, Harlow, Great Britain
3. Podlubny I (1999) Fractional differential equations. Academic Press, Boston, USA
4. Tarasov VE (2011) Fractional dynamics: applications of fractional calculus to dynamics of particles, fields and media. Springer, New York, USA
5. Kilbas AA, Srivastava HM, Trujillo JJ (2006) Theory and applications of fractional differential equations. Elsevier Science Publishing, Amsterdam, Netherlands
6. Gerasimov AN (1948) Generalization of linear laws of deformation and their application to problems of internal friction. Appl Math Mech 12(3):251–260 (In Russian)
7. Caputo M (1967) Linear model of dissipation whose Q is almost frequency independent: II. Geophys J R Astron Soc 13:529–539
8. Dzhrbashyan MM, Nersesyan AB (1968) Fractional derivatives and the Cauchy problem for differential equations of fractional order. Izvestiya Akademii Nauk Armyanskoy SSR Matematika 3(1):3–28 (In Russian)
9. Pskhu AV (2009) The fundamental solution of a diffusion-wave equation of fractional order. Izvestiya: Math 73(2):351–392. https://doi.org/10.4213/im2429
10. Pskhu AV (2016) Fractional diffusion equation with a discretely distributed differentiation operator. Sib Elektron Math Rep 13:1078–1098. https://doi.org/10.17377/semi.2016.13.086
11. Fedorov VE, Plekhanova MV, Izhberdeeva EM (2021) Initial value problems of linear equations with the Dzhrbashyan—Nersesyan derivative in banach spaces. Symmetry 13(6):1058. https://doi.org/10.3390/sym13061058
12. Sviridyuk GA, Fedorov VE (2003) Linear Sobolev type equations and degenerate semigroups of operators. VSP, Utrecht, Netherlands
13. Mainardi F, Spada G (2011) Creep, relaxation and viscosity properties for basic fractional models in rheology. Eur Phys J Spec Top 193:133–160
14. Ladyzhenskaya OA (1969) The mathematical theory of viscous incompressible flow. Gordon and Breach, Science Publishers, New York, London, Paris
15. Fedorov VE, Davydov PN (2013) Semilinear degenerate evolution equations and nonlinear systems of hydrodynamic type. Trudy Inst Mat i Mekh UrO RAN 19(4):267–278. (In Russian)

On a K-Homogeneous Metric

Marina V. Polovinkina and Igor P. Polovinkin

Abstract We consider a Riemannian metric which generates the Beltrami-Laplace operator coinciding with the B-elliptic operator up to a factor.

Keywords B-elliptic operator · Riemannian metric · Laplace Beltrami operator · Isometry group · Killing conditions · Lobachevsky geometry

1 K-Homogeneous Metric

Let $\gamma = (\gamma_1, \ldots, \gamma_n)$ be a vector with fixed numbers γ_i, $i = 1 \ldots, n$, which are not equal to zero at the same time. We denote by R^n_+ the set of points $x = (x_1, \ldots, x_n) \in R^n$ such that $x_i \in R$, when $\gamma_i = 0$, $x_i \in (0, +\infty)$, when $\gamma_i \neq 0$.

If $\gamma_i \neq 0$, the variable x_i is called singular. As usual, we will use the notation

$$(x)^\gamma = \prod_{i=1}^{n} x_i^{\gamma_i}, \; x = (x_1, \ldots, x_n) \in R^n_+.$$

Let the function $u(x)$ be twice continuously differentiable in R^n_+.

We define the operator Δ_{B_γ} by the formula

$$\Delta_{B_\gamma} u = \sum_{i=1}^{n} \frac{\partial^2 u}{\partial x_i^2} + \sum_{i=1}^{n} \frac{\gamma_i}{x_i} \frac{\partial u}{\partial x_i}. \tag{1}$$

M. V. Polovinkina
Voronezh State University of Engineering Technologies, Voronezh, Russian Federation
e-mail: polovinkina-marina@yandex.ru

I. P. Polovinkin (✉)
Voronezh State University, Voronezh, Russian Federation
e-mail: polovinkin@yandex.ru

Belgorod State National Research University (BelGU), Belgorod, Russian Federation

V. Vasilyev (ed.), *Differential Equations, Mathematical Modeling and Computational Algorithms*, Springer Proceedings in Mathematics & Statistics 423,
https://doi.org/10.1007/978-3-031-28505-9_9

Operators of the form (1) have been studied by I. A. Kipriyanov and his disciples (see [1–5]).

The aim of this section is to find a positively defined in R^n_+ symmetric quadratic form (metric)

$$ds^2 = \sum_{i=1}^{n}\sum_{j=1}^{n} g_{ij}\, dx_i\, dx_j\,,$$

such that the Beltrami–Laplace operator (see [6])

$$\Delta_\omega = \frac{1}{\sqrt{|g|}}\sum_{i=1}^{n}\frac{\partial}{\partial x_i}\sum_{k=1}^{n} g^{ik}\sqrt{|g|}\frac{\partial}{\partial x_k} \tag{2}$$

would coincide with the operator Δ_{B_γ} up to a multiplier. Here the functions g^{ij}, $i, j = 1, \dots, n$, are the entries of the matrix $\|g^{ij}\|$, which is the inverse of the matrix $\|g_{ij}\|$ (covariant metric tensor),

$$g = \det \|g_{ij}\|.$$

The study of the properties of elliptic differential operators using Riemannian metrics has a long history (see, for example, [7, 8]).

Theorem 1 *If $n \geq 3$, the entries of the matrix $\|g_{ij}\|$ are defined by formulas*

$$g_{ij} = \delta_{ij}\prod_{i=1}^{n} x_i^{K_i} = \delta_{ij}x^K,\ i, j = 1, \dots, n,\ K = (K_1, \dots, K_n), \tag{3}$$

where

$$K_i = 2\gamma_i/(n-2), \tag{4}$$

δ_{ij} *is the Kronecker symbol.*

Proof Indeed, since $g_{ij} = 0$ for $i \neq j$, substituting (3) into (2), we get:

$$\Delta_\omega u = \frac{1}{\sqrt{|g|}}\sum_{k=1}^{n}\frac{\partial}{\partial x_k}\left(g^{kk}\sqrt{|g|}\frac{\partial u}{\partial x_k}\right), \tag{5}$$

where

$$|g| = g = x^{nK} = \prod_{i=1}^{n}\prod_{i=1}^{N} n\, x_i^{K_i} = \prod_{i=1}^{n} x_i^{2n\gamma_i/(n-2)}, \tag{6}$$

$$g^{kk} = x^{-K} = \prod_{i=1}^{n} x_i^{-2\gamma_i/(n-2)}. \tag{7}$$

Taking into account (6) and (7), it is possible to rewrite (5) in the following form:

$$\Delta_\omega u = \frac{1}{x^{nK/2}} \sum_{j=1}^{n} \frac{\partial}{\partial x_j} \left(x^{-K} x^{Kn/2} \frac{\partial u}{\partial x_j} \right) =$$

$$= x^{-K} \sum_{j=1}^{n} \frac{\partial^2 u}{\partial x_j^2} + x^{-Kn/2} \sum_{j=1}^{n} \frac{\partial u}{\partial x_j} \left(\prod_{l=1}^{n} x_l^{K_l(n-2)/2} \right) \frac{\partial u}{\partial x_j} =$$

$$= x^{-K} \sum_{j=1}^{n} \frac{\partial^2 u}{\partial x_j^2} + x^{-Kn/2} \sum_{j=1}^{n} \prod_{l=1}^{n} x_l^{K_l(n-2)/2} \frac{K_j(n-2)}{2} x_j^{-1} \frac{\partial u}{\partial x_j} =$$

$$= x^{-K} \sum_{j=1}^{n} \frac{\partial^2 u}{\partial x_j^2} + x^{-K} \sum_{j=1}^{n} \frac{K_j(n-2)}{2x_j} \frac{\partial u}{\partial x_j} = x^{-K} \Delta_{B_\gamma} u,$$

so

$$\Delta_\omega u = x^{-K} \Delta_{B_\gamma} u, \tag{8}$$

which was required to be proved.

We will consider the set R_+^n equipped with a Riemannian metric

$$ds^2 = x^K \sum_{i=1}^{n} dx_i^2, \quad K \in R, \tag{9}$$

as a Riemannian space; we will denote it by KI_n, and we will call metric (9) the K-homogeneous Kipriyanov metric.

Theorem 2 *If $n = 2$, the problem of finding a metric satisfying equality (8) has no solution.*

Proof Let

$$g_{11} = E, \; g_{12} = g_{21} = F, \; g_{22} = G.$$

Then

$$g = \det \|g_{ij}\| = EG - F^2, g^{ij} = (-1)^{i+j} \frac{g_{ij}}{EG - F^2}.$$

Hence

$$\Delta_\omega u = G/|g| \frac{\partial^2 u}{\partial x_1^2} + E/|g| \frac{\partial^2 u}{\partial x_2^2} -$$

$$- 2F/\,|g| \frac{\partial^2 u}{\partial x_1 \partial x_2} + \Phi \left(\frac{\partial u}{\partial x_1}, \frac{\partial u}{\partial x_2} \right), \tag{10}$$

where Φ denotes a summand that depends only on the first-order derivatives of the function u. In order for expression (10) to coincide up to a multiplier with (1), it is necessary that $F \equiv 0$. This will entail equalities

$$g = EG,\ g^{11} = 1/E,\ g^{22} = 1/G,\ g_{12} = g_{21} = g^{12} = g^{21} = 0.$$

Therefore,

$$\Delta_\omega u = \frac{1}{\sqrt{|EG|}}\left(\frac{\partial}{\partial x_1}\left(\sqrt{\left|\frac{G}{E}\right|}\frac{\partial u}{\partial x_1}\right) + \frac{\partial}{\partial x_2}\left(\sqrt{\left|\frac{E}{G}\right|}\frac{\partial u}{\partial x_2}\right)\right) =$$

$$= \frac{1}{E}\frac{\partial^2 u}{\partial x_1^2} + \frac{1}{G}\frac{\partial^2 u}{\partial x_2^2} + \frac{\partial}{\partial x_1}\sqrt{\left|\frac{G}{E}\right|}\,\frac{\partial u}{\partial x_1} + \frac{\partial}{\partial x_2}\sqrt{\left|\frac{E}{G}\right|}\,\frac{\partial u}{\partial x_2}.$$

The first two terms must have the same coefficients, from where $E = G$. Then the last two terms are equal to zero, which means that it is impossible to find a metric satisfying equality (8) for $n = 2$.

2 Investigation of Isometric Transformations for the K-Homogeneous Kipriyanov Metric

The fulfillment of the Killing requirements

$$\sum_{s=1}^{n}\left(\xi_s\frac{\partial g_{ij}}{\partial x_s} + g_{is}\frac{\partial \xi_s}{\partial x_j} + g_{js}\frac{\partial \xi_s}{\partial x_i}\right) = 0,\ i, j = 1, \ldots, n.$$

is a necessary and sufficient condition for a one-parameter group G with an infinitesimal operator

$$X = \sum_{i=1}^{n}\xi_i(x)\frac{\partial}{\partial x_i}$$

to be an isometry group.

Obviously,

$$\frac{\partial g_{ij}}{\partial x_s} = \delta_{ij}\frac{K_s x^K}{x_s}.$$

Therefore, the Killing equations will take the form

$$\sum_{s=1}^{n}\left(\delta_{ij}\xi_s K_s x^{K-1} + x^K\left(\frac{\partial \xi_i}{\partial x_j} + \frac{\partial \xi_j}{\partial x_i}\right)\right) = 0,\ i, j = 1, \ldots, n.$$

By summing and reducing by x^K, we get

$$\delta_{ij}\sum_{s=1}^{N}\frac{\xi_s K_s}{x_s}+\frac{\partial\xi_i}{\partial x_j}+\frac{\partial\xi_j}{\partial x_i}=0,\ i,j=1,\dots,n. \tag{11}$$

For $i\neq j$, Eq. (11) can be written as

$$\frac{\partial\xi_i}{\partial x_j}+\frac{\partial\xi_j}{\partial x_i}=0,\ i,j=1,\dots,n,\ i\neq j. \tag{12}$$

For $i=j$, Eq. (11) can be written in the form

$$2\frac{\partial\xi_j}{\partial x_j}+\sum_{s=1}^{n}\frac{K_s\xi_s}{x_s}=0,\ i=1,\dots,n. \tag{13}$$

The vector

$$\xi=(\xi_1,\dots,\xi_n),\ \ \xi_j=C\,x^p x_j, \tag{14}$$

where

$$p=(p_1,\dots,p_n),\ \ p_1=p_2=\dots=p_n=\beta=-\sum_{l=1}^{n}K_l/2-1, \tag{15}$$

is a solution to system (13), which can be checked by direct verification. Substituting (14) into (12), taking into account (15), we obtain

$$0\equiv\frac{\partial\xi_i}{\partial x_j}+\frac{\partial\xi_j}{\partial x_i}=\beta x^p\left(\frac{x_i}{x_j}+\frac{x_j}{x_i}\right),\ i,j=1,\dots,n,\ i\neq j.$$

Hence we get

$$p_1=p_2=\dots=p_n=\beta=-\sum_{l=1}^{n}K_l/2-1=0, \tag{16}$$

or, what is the same,

$$\sum_{l=1}^{n}K_l=-2, \tag{17}$$

and considering (4),

$$\sum_{i=1}^{n}\gamma_i=2-N. \tag{18}$$

3 Characteristics of the K-Homogeneous Kipriyanov Metric in the Case of a Single Singular Variable

There is a well-known case of fulfillment of condition (16), or, what is the same, (18). When $\gamma_1 = \gamma_2 = \cdots = \gamma_{n-1} = 0$, $\gamma_n = 2 - n$, $K = -2$, the space KI_n is the Poincare model of the $n-$dimensional Lobachevsky geometry. Next, we will consider the case of $\gamma_1 = \gamma_2 = \cdots = \gamma_{n-1} = 0$, $\gamma_n = 0$. Metric (3) will now take the form

$$g_{ij} = \delta_{ij} x_n^K, \ i, j = 1, \ldots, n, \tag{19}$$

where

$$K = 2\gamma/(n-2), \tag{20}$$

δ_{ij} is the Kronecker symbol.

The following facts are established by direct calculation.

Theorem 3 *The Christoffel symbols of the first kind, corresponding to metric (19), have the form*

$$\Gamma_{ij,k} = \frac{K x_n^{K-1}}{2} (\delta_{ik}\delta_{jn} + \delta_{jk}\delta_{in} - \delta_{ij}\delta_{kn}).$$

Proof From the definition of the Christoffel symbols of the first kind, taking into account (19)–(20), we obtain:

$$\Gamma_{ij,k} = \frac{1}{2} \left(\frac{\partial g_{ik}}{\partial x_j} + \frac{\partial g_{jk}}{\partial x_i} - \frac{\partial g_{ij}}{\partial x_k} \right) =$$

$$= \frac{1}{2} (\delta_{ik}\delta_{jn} K x_n^{K-1} + \delta_{jk}\delta_{in} K x_n^{K-1} - \delta_{ij}\delta_{kn} K x_n^{K-1}),$$

which was required to be proved.

Theorem 4 *The Christoffel symbols of the second kind, corresponding to metric (9), have the form*

$$\Gamma_{ij}^k = \frac{K}{2x_n} (\delta_{ik}\delta_{jn} + \delta_{jk}\delta_{in} - \delta_{ij}\delta_{kn}).$$

Proof From the definition of the Christoffel symbols of the second kind and the previous theorem, we obtain:

$$\Gamma_{ij}^k = \sum_{h=1}^{n} g^{kh} \Gamma_{ij,h} = \frac{K}{2} \sum_{h=1}^{n} \delta_{kh} x_n^{-K} x_n^{K-1} (\delta_{ih}\delta_{jn} + \delta_{jh}\delta_{in} - \delta_{ij}\delta_{hn}) =$$

$$\frac{K}{2x_n} (\delta_{ki}\delta_{jn} + \delta_{kj}\delta_{in} - \delta_{ij}\delta_{kn}).$$

The theorem is proved.

Theorem 5 *The components of the Riemann tensor, corresponding to metric (9), have the form*

$$R^l_{ijk} = (\frac{K^2}{4x_n^2} - \frac{K}{2x_n^2})(\delta_{li}\delta_{in}\delta_{kn} + \delta_{ik}\delta_{jn}\delta_{ln} - \delta_{ij}\delta_{kn}\delta_{ln} - \delta_{lk}\delta_{in}\delta_{jn}) +$$
$$+\frac{K^2}{4x_n^2}(\delta_{ij}\delta_{lk} - \delta_{ik}\delta_{lj}).$$

Proof In accordance to definition, the components of the Riemann tensor are calculated by the formulas

$$R^l_{ijk} = \frac{\partial \Gamma^l_{ik}}{\partial x_j} - \frac{\partial \Gamma^l_{ij}}{\partial x_k} + \sum_{m=1}^{n} (\Gamma^m_{ik}\Gamma^l_{mj} - \Gamma^m_{ij}\Gamma^l_{mk}).$$

We will calculate the partial derivatives included in these formulas. We have

$$\frac{\partial \Gamma^k_{ij}}{\partial x_s} = -\frac{K}{2x_n^2}\delta_{sn}(\delta_{ki}\delta_{jn} + \delta_{kj}\delta_{in} - \delta_{ij}\delta_{kn}),$$

from where, we obtain

$$\frac{\partial \Gamma^l_{ik}}{\partial x_j} = -\frac{K}{2x_n^2}\delta_{jn}(\delta_{li}\delta_{kn} + \delta_{lk}\delta_{in} - \delta_{ik}\delta_{ln}),$$

$$\frac{\partial \Gamma^l_{ij}}{\partial x_k} = -\frac{K}{2x_n^2}\delta_{kn}(\delta_{li}\delta_{jn} + \delta_{lj}\delta_{in} - \delta_{ij}\delta_{ln}).$$

Therefore,

$$\frac{\partial \Gamma^l_{i}}{\partial x_j} - \frac{\partial \Gamma^l_{ij}}{\partial x_s} = -\frac{K}{2x_n^2}(\delta_{jn}\delta_{lk}\delta_{in} - \delta_{jn}\delta_{ik}\delta_{ln} - \delta_{kn}\delta_{lj}\delta_{in} + \delta_{kn}\delta_{ij}\delta_{ln}).$$

Now we will calculate the last term in the definition. Taking into account Theorem 4, we find:

$$\sum_{m=1}^{n} (\Gamma^m_{ik}\Gamma^l_{mj} - \Gamma^m_{ij}\Gamma^l_{mk}) =$$

$$= \frac{K^2}{4x_n^2}\sum_{m=1}^{n} (\delta_{mi}\delta_{kn}\delta_{lm}\delta_{jn} + \delta_{mi}\delta_{kn}\delta_{lj}\delta_{mn} -$$

$$-\delta_{mi}\delta_{kn}\delta_{mj}\delta_{ln} + \delta_{mk}\delta_{in}\delta_{lm}\delta_{jn} +$$

$$+\delta_{mk}\delta_{in}\delta_{lj}\delta_{mn} - \delta_{mk}\delta_{in}\delta_{mj}\delta_{ln} - \delta_{ik}\delta_{mn}\delta_{lm}\delta_{jn} - \delta_{ik}\delta_{mn}\delta_{lj}\delta_{mn} +$$

$$\delta_{ik}\delta_{mn}\delta_{mj}\delta_{ln} - \delta_{mi}\delta_{jn}\delta_{lm}\delta_{kn} - \delta_{mi}\delta_{jn}\delta_{lk}\delta_{mn} + \delta_{mi}\delta_{jn}\delta_{mk}\delta_{ln} -$$

$$-\delta_{mj}\delta_{in}\delta_{lm}\delta_{kn} - \delta_{mj}\delta_{in}\delta_{lk}\delta_{mn} + \delta_{mj}\delta_{in}\delta_{mk}\delta_{ln} + \delta_{ij}\delta_{mn}\delta_{lm}\delta_{kn} +$$

$$+\delta_{ij}\delta_{mn}\delta_{lk}\delta_{mn} - \delta_{ij}\delta_{mn}\delta_{mk}\delta_{ln}).$$

Hence, taking into account the properties of the Kronecker symbol, in particular, the formulas

$$\delta_{il} = \delta_{li},$$

$$\sum_{m=1}^{n} \delta_{mi}\delta_{lm} = \delta_{il},$$

after identical transformations, we obtain a statement of the theorem.

Theorem 6 *The components of the Ricci tensor, corresponding to metric (9), have the form*

$$R_{ij} = \frac{K}{4x_n^2}\left((K-2)(2-n)\delta_{in}\delta_{jn} + (K(n-2)+2)\delta_{ij}\right).$$

Proof Directly from the definition of the components of the Ricci tensor

$$R_{ij} = \sum_{k=1}^{n} R_{ijk}^{k},$$

after identical transformations, we come to the validity of the theorem.

Theorem 7 *The curvature of the space KI_n is calculated by the formula*

$$R = \frac{Kn(n-2)}{x_n^{K+2}} = \frac{2\gamma n}{x_n^{(2\gamma+2n-4)/(n-2)}}.$$

Proof From the definition of curvature

$$R = \sum_{i=1}^{n}\sum_{j=1}^{n} g^{ij} R_{ij},$$

we come to the statement of the theorem by performing summation and identical transformations.

4 Investigation of Geodesic Lines for a K-Homogeneous Kipriyanov Metric

Theorem 8 *The system of equations for geodesic lines in the space KI_n can be reduced to a system of the first order*

$$\frac{dx_k}{ds} = \frac{C_k}{x_n^K}\,, \quad k = 1, \ldots, n-1, \tag{21}$$

$$\left(\frac{dx_n}{ds}\right)^2 = \frac{C_n}{x_n^K} - \frac{B^2}{x_n^{2K}}\,, \tag{22}$$

where

$$B = \sqrt{\sum_{k=1}^{n-1} C_k^2}. \tag{23}$$

Proof The system of equations for geodesic lines in a given metric $\|g_{ij}\|$ has the form

$$\frac{d^2x_k}{ds^2} + \sum_{i=1}^{n}\sum_{j=1}^{n} \Gamma_{ij}^k \frac{dx_i}{ds}\frac{dx_j}{ds} = 0, \; k = 1, 2, \ldots, n,$$

where s is the natural parameter (arc length). In our case, using the calculated Christoffel symbols, we can write this system as

$$\frac{d^2x_k}{ds^2} + \frac{K}{x_n}\frac{dx_n}{ds}\frac{dx_k}{ds} = 0, \; k = 1, \ldots, n-1, \tag{24}$$

$$\frac{d^2x_n}{ds^2} - \frac{K}{2x_n}\sum_{i=1}^{n}\left(\frac{dx_i}{ds}\right)^2 + \frac{K}{2x_n}\left(\frac{dx_n}{ds}\right)^2 = 0. \tag{25}$$

Equation (24) can be written as

$$x_n^{-K}\frac{d}{ds}\left(x_n^K\frac{dx_k}{ds}\right) = 0, \; k = 1, \ldots, n-1. \tag{26}$$

Multiplying (26) by x_n^K, integrating and dividing by x_n^K, we get

$$\frac{dx_k}{ds} = \frac{C_k}{x_n^K}, \; k = 1, \ldots, n-1. \tag{27}$$

Substituting (27) into (25), we get

$$\frac{d^2x_n}{ds^2} - \frac{KB^2}{2x_n^{2K+1}} + \frac{K}{2x_n}\left(\frac{dx_n}{ds}\right)^2 = 0. \tag{28}$$

Equation (28) admits a reduction of the order in a standard way. Suppose

$$p = p(x_n) = \frac{dx_n}{ds}, \ v = p^2.$$

Then

$$\frac{d^2x_n}{ds^2} = p'p = \frac{1}{2}v'.$$

After that, Eq. (28) will be reduced to the form

$$v' + \frac{K}{x_n}v = \frac{B^2K}{x_n^{2K+1}},$$

which is equivalent to the equation

$$\frac{d}{dx_n}(x_n^K v) = \frac{B^2K}{x_n^{K+1}}.$$

Integrating and dividing by x_n^K, we get

$$v = p^2 = \left(\frac{dx_n}{ds}\right)^2 = \frac{C_n}{x_n^K} - \frac{B^2}{x_n^{2K}}.$$

It is known [9], that geodesic lines have the property

$$\sum_{i=1}^{n}\sum_{j=1}^{n} g^{ij}\frac{dx_i}{ds}\frac{dx_j}{ds} = const.$$

In the case under consideration, this will lead to equality

$$\sum_{i=1}^{n} x_n^K\left(\frac{dx_i}{ds}\right)^2 = const. \tag{29}$$

From (21), it is easily deduced that the constant in equality (29) coincides with C_n.

References

1. Kipriyanov IA (1967) FourierBessel transforms and imbedding theorems for weighted classes. Proc Steklov Inst Math 89:130–213
2. Kipriyanov IA (1997) Singular elliptic boundary-value problems. NaukaPhysmatlit, Moskow (in Russian)
3. Katrakhov VV, Sitnik SM (2018) The transmutation method and boundary-value problems for singular elliptic equations. Contemporary mathematics. Fundam Dir 64(2):211–426. https://doi.org/10.22363/2413-3639-2018-64-2-211-426
4. Sitnik SM, Shishkina EL (2019) Method of transmutations for differential equations with Bessel operators (in Russian) Fizmatlit, Moscow
5. Lyahov LN (2007) B-hypersingular integrals and their applications to the description of the Kupriyanov functional classes and to integral equations with B-potential nuclei LSPU, Lipetsk (in Russian)
6. Helgason S (1964) Differential geometry, lie groups, and symmetric spaces. Graduate Studies in Mathematics, 34, American Mathematical Society Providence, Rhode Island
7. Il'in VA (1969) Fourier series in fundamental systems of functions of the Beltrami operator. Differ Uravn 5(11):1940–1978 (in Russian)
8. Feller W (1930) Über die Lösungen der linearen partiellen Differentialgleichungen zweiter Ordnung vom elliptischen pus. Math Ann 102:633–649
9. Eisenhart LP (1949) Riemannian geometry. Princeton

Biooscillators in Models of Genetic Networks

Felix Sadyrbaev, Inna Samuilik, and Valentin Sengileyev

Abstract We study periodic attractors in a system of ordinary differential equations, which is used to model genetic regulatory networks. The systems of order two and four are considered, which posess the periodic attractors. The systems of order three and six are considered also.

Keywords Dynamical systems · Mathematical models · Genetic networks · Periodic attractors

1 Introduction

Genetic regulatory networks exist in any cell of any living organism. They are responsible for many important functions, including the morphogenesis, reactions to non-favorable influences and more. The understanding of principles of their functioning is necessary for the purposes of managing and control of them. The experimental data, collecting by the experts in the field and their teams, usually are of huge volume and require simplifications and systematizations. As in other natural sciences, the mathematical models can be of great help.

There are different kinds of mathematical models for genetic regulatory networks (GRN in short). These models use Boolean algebras, the Graph theory and more. The interested reader can gain the necessary information from the sources [1–5]. One of the more effective tools for the purposes of modelling the behavior of GRN and

F. Sadyrbaev (✉)
Institute of Mathematics and Computer Science, University of Latvia, 29 Rainis boul., Riga LV-1459, Latvia
e-mail: felix@latnet.lv

F. Sadyrbaev · V. Sengileyev
Daugavpils University, 13 Vienibas Street Daugavpils, Daugavpils LV-5401, Latvia

I. Samuilik
Riga Technical University, 1 Kalku Street, Riga LV-1658, Latvia

V. Vasilyev (ed.), *Differential Equations, Mathematical Modeling and Computational Algorithms*, Springer Proceedings in Mathematics & Statistics 423,
https://doi.org/10.1007/978-3-031-28505-9_10

tracking the evolution of GRN in time is the theory of differential equations. This model consists of a vectorial system [8]

$$X' = F(WX - \Theta) - X, \tag{1}$$

where X is the n-dimensional vector of the current state of a network, W stands for the so called regulatory matrix, and Θ is the parameter vector, which defines the individual properties of any gene. Any element of a network can be imagined as a separate element (more precisely, any x_i is the rate of expression of proteins, sending to other elements of a network to form a collective response to current threats), called a gene. The interaction of genes for short (relatively) periods of time is described by the regulatory matrix W. For instance, for a two-element network with the state vector $X(t)(x_1(t), x_2(t))$, the regulatory matrix in a general form is

$$W = \begin{pmatrix} w_{11} & w_{12} \\ w_{21} & w_{22} \end{pmatrix}. \tag{2}$$

The element w_{ij} means the influence of x_j on x_i. The positive value corresponds to the activation, the negative to the inhibition (repression), the zero entry means no interactions. It is to be mentioned here that the regulatory matrices, obtained for the real networks, mostly are sparse, with great zero fields [6, 7]. Consider three types of interactions, namely,

$$W_a = \begin{pmatrix} 1 & 1 \\ 1 & 1 \end{pmatrix}, \quad W_i = \begin{pmatrix} 0 & -2 \\ -2 & 0 \end{pmatrix}, \quad W_m = \begin{pmatrix} 2 & 1 \\ -1 & 2 \end{pmatrix}. \tag{3}$$

The first one corresponds to the activation, the second one can be classified as the inhibition, and the last matrix corresponds to the mixed activation-inhibition case. The respective system of differential equations is of the form

$$\begin{cases} x_1' = \frac{1}{1+e^{-\mu_1(w_{11}x_1+w_{12}x_2-\theta_1)}} - x_1, \\ x_2' = \frac{1}{1+e^{-\mu_2(w_{21}x_1+w_{22}x_2-\theta_2)}} - x_2, \end{cases} \tag{4}$$

Notice, that in the absence of interrelation between genes, the system turns to the linear one

$$\begin{cases} x_1' = -x_1, \\ x_2' = x_2, \end{cases} \tag{5}$$

representing the natural exponential decay. In all three cases there are attracting sets in a (x_1, x_2)-phase plane. Let us look at the pictures. In these pictures, $\mu_1 = \mu_2 = 4$, $\theta_1 = (w_{11} + w_{12})/2$, $\theta_2 = (w_{21} + w_{22})/2$. The curves in blue and red are two nullclines (Figs. 1, 2 and 3).

The role of attractors in systems of the form (5) cannot be overestimated. Future states $X(t)$ depend on the attractors, their locations, and their properties.

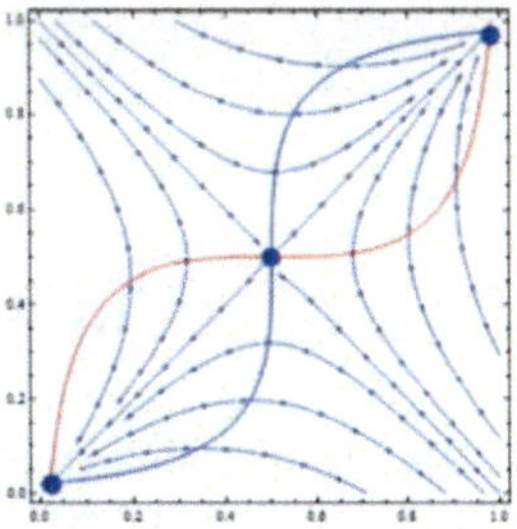

Fig. 1 The vector field and the nullclines for the system (2) with the matrix W_a

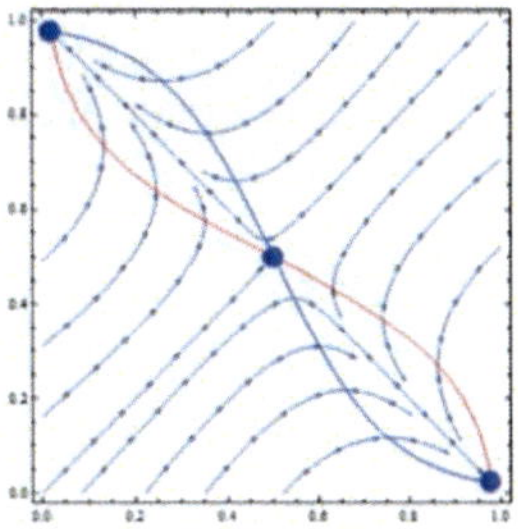

Fig. 2 The vector field and the nullclines for the system (2) with the matrix W_i

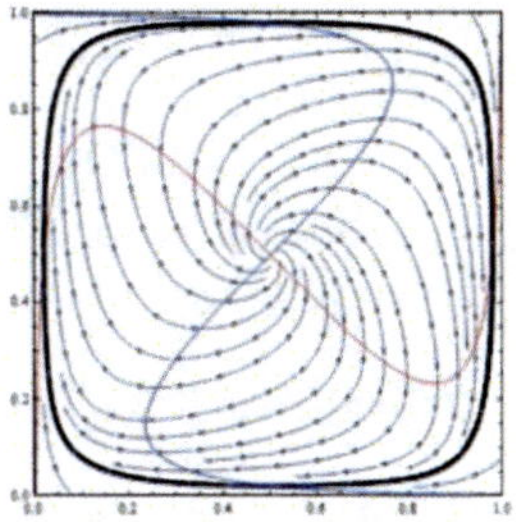

Fig. 3 The vector field and the nullclines for the system (2) with the matrix W_m

2 General Models

The general system, modelling the n-dimensional GRN network, is

$$\begin{cases} \dfrac{dx_1}{dt} = \dfrac{1}{1+e^{-\mu_1(w_{11}x_1+w_{12}x_2+\cdots+w_{1n}x_n-\theta_1)}} - x_1, \\ \dfrac{dx_2}{dt} = \dfrac{1}{1+e^{-\mu_2(w_{21}x_1+w_{22}x_2+\cdots+w_{2n}x_n-\theta_2)}} - x_2, \\ \dots \\ \dfrac{dx_n}{dt} = \dfrac{1}{1+e^{-\mu_n(w_{n1}x_1+w_{n2}x_2+\cdots+w_{nn}x_n-\theta_n)}} - x_n. \end{cases} \tag{6}$$

It involves the regulatory matrix

$$W = \begin{pmatrix} w_{11} & w_{12} & \dots & w_{1n} \\ w_{21} & w_{22} & \dots & w_{2n} \\ \dots \\ w_{n1} & w_{n2} & \dots & w_{nn} \end{pmatrix}. \quad (7)$$

where the interrelation of elements of a network is encrypted.

The right hand side of the system (6) consists of a linear and nonlinear part. The nonlinearity is represented by a sigmoidal function. Sigmoidal functions are those, which are continuous and smooth, and monotonically increase from zero to unity. They have exactly one inflection point. In the system (6) the sigmoidal function $f(z) = \frac{1}{1+e^{-\mu z}}$ is used, as in many other sources.

The nullclines of the system (6) are given by the equations in the system

$$\begin{cases} x_1 = \dfrac{1}{1 + e^{-\mu_1(w_{11}x_1 + w_{12}x_2 + \dots + w_{1n}x_n - \theta_1)}}, \\ x_2 = \dfrac{1}{1 + e^{-\mu_2(w_{21}x_1 + w_{22}x_2 + \dots + w_{2n}x_n - \theta_2)}}, \\ \dots \\ x_n = \dfrac{1}{1 + e^{-\mu_n(w_{n1}x_1 + w_{n2}x_2 + \dots + w_{nn}x_n - \theta_n)}}. \end{cases} \quad (8)$$

The critical points of the system (6), also called *the equilibria*, are solutions x_1^* to x_n^* of the system (8).

The standard local analysis of the critical points can be made by considering the linearized at a critical point system.

Proposition 1 *The unit cube in a phase space is invariant.*

Proof Consider the faces $x_1 = 0$ and $x_1 = 1$. The first component of the vector field, defined by the system (6), is

$$\frac{1}{1 + e^{-\mu_1(w_{11}x_1 + w_{12}x_2 + \dots + w_{1n}x_n - \theta_1)}} - x_1.$$

This values at $x_1 = 0$ are strictly positive, since the sigmoidal function is positive. At $x_1 = 1$ the vector field is strictly negative, since the values of the sigmoidal function satisfy the inequality $\frac{1}{1+e^{-\mu_1(w_{11}x_1+w_{12}x_2+\dots+w_{1n}x_n-\theta_1)}} < 1$ for any choice of the variables and parameters.

Proposition 2 *All critical points of the system (6) locate in the unit cube of the previous proposition.*

Proof Consider the system (8), which defines the nullclines of (6). As was mentioned, all critical points are solutions of (8). Consider the first equation in (8). The values of the variable x_1 are in the open interval $(0, 1)$. This is true because the right side contains the sigmoidal function $\frac{1}{1+e^{-\mu_1(w_{11}x_1+w_{12}x_2+\dots+w_{1n}x_n-\theta_1)}}$ which takes values in the same interval $(0, 1)$. Hence the x_1 coordinate of any critical point must be in

(0, 1). Considering the remaining equations in the system (8), the conclusion can be made that for any critical point $(x_1^*, \dots, x_n^*)$ the coordinates satisfy $0 < x_i^* < 1$ for any $i = 1, \dots, n$.

Remark 1 More information on mathematical modelling of GRN by systems of ordinary differential equations can be found in [6, 7, 9, 11–15].

3 Four-Dimensional Systems

Suppose the model of four-element GRN is studied. The respective dynamical system is

$$\begin{cases} \dfrac{dx_1}{dt} = \dfrac{1}{1 + e^{-\mu_1(w_{11}x_1 + w_{12}x_2 + w_{13}x_3 + w_{14}x_4 - \theta_1)}} - x_1, \\ \dfrac{dx_2}{dt} = \dfrac{1}{1 + e^{-\mu_2(w_{21}x_1 + w_{22}x_2 + w_{23}x_3 + w_{24}x_4 - \theta_2)}} - x_2, \\ \dfrac{dx_3}{dt} = \dfrac{1}{1 + e^{-\mu_2(w_{31}x_1 + w_{32}x_3 + w_{33}x_3 + w_{34}x_4 - \theta_3)}} - x_3, \\ \dfrac{dx_4}{dt} = \dfrac{1}{1 + e^{-\mu_n(w_{41}x_1 + w_{42}x_2 + w_{43}x_3 + w_{44}x_4 - \theta_4)}} - x_4. \end{cases} \tag{9}$$

with the regulatory matrix

$$W = \begin{pmatrix} w_{11} & w_{12} & w_{13} & w_{14} \\ w_{21} & w_{22} & w_{23} & w_{24} \\ w_{31} & w_{32} & w_{33} & w_{34} \\ w_{41} & w_{42} & w_{43} & w_{44} \end{pmatrix}. \tag{10}$$

Theorem 1 *Attractors of various kinds are possible in system (9).*

These attractors were constructed and studied in [15–17].

Theorem 2 *Periodic attractors are possible in system (9).*

These attractors were constructed in [12, 15–17].

Example 1 Consider system of the form (9) with the regulatory matrix

$$W = \begin{pmatrix} 2 & -1 & 0 & 0 \\ 1 & 2 & 0 & 0 \\ 0 & 0 & 2 & -1 \\ 0 & 0 & 1 & 2 \end{pmatrix}. \tag{11}$$

This system possesses the four-dimensional attractor which is composed of two identical periodic solutions of the two-dimensional system, considered in the introduction (for the matrix W_m). This periodic attractor cannot be seen, but the projections

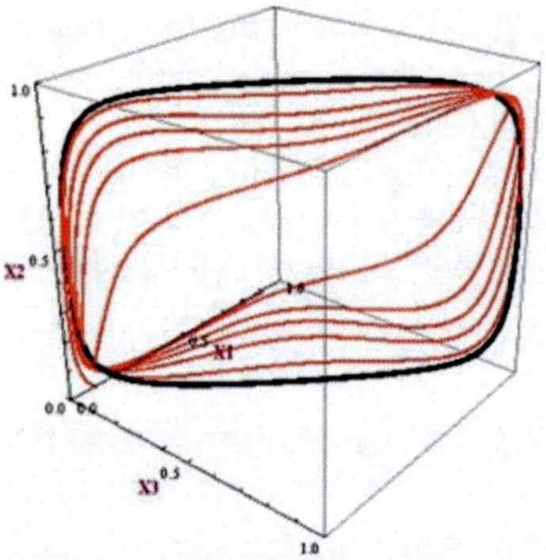

Fig. 4 Projections on (x_1, x_2, x_3)

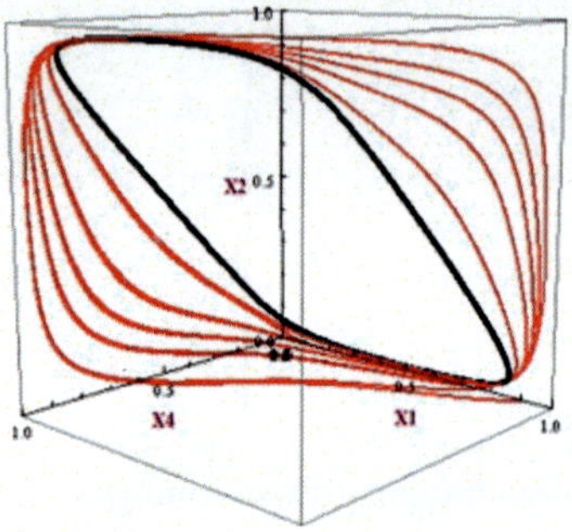

Fig. 5 Projections on (x_1, x_2, x_4)

onto two-dimensional and three-dimensional subspaces are possible to visualize. In Figs. 4 and 5 projections of the periodic attractor and several trajectories, which start at $x_1(0) = 0.1 + 0.2i$, $i = 0, 1, 2, 4$. The periodic attractor is in black.

Suppose that the regulatory matrix W for the system (9) is the block matrix with two dimensional blocks on the main diagonal. For the convenient reference call these blocks B_1 and B_2. Other spaces are left filled with zeros. Let the two dimensional systems, corresponding to these blocks, be

$$\begin{cases} x_1' = \frac{1}{1+e^{-\mu_1(w_{11}x_1+w_{12}x_2-\theta_1)}} - x_1, \\ x_2' = \frac{1}{1+e^{-\mu_2(w_{21}x_1+w_{22}x_2-\theta_2)}} - x_2, \end{cases} \tag{12}$$

and

$$\begin{cases} x_3' = \frac{1}{1+e^{-\mu_3(w_{33}x_3+w_{34}x_4-\theta_3)}} - x_3, \\ x_4' = \frac{1}{1+e^{-\mu_4(w_{43}x_3+w_{44}x_4-\theta_4)}} - x_4. \end{cases} \tag{13}$$

Let both systems have the stable limit cycles with the periods T_1 and T_2.

Theorem 3 *Suppose that the four-dimensional attractor is obtained from the two two-dimensional limit cycles. If their periods relate as $mT_2 = nT_1$, where m and n are positive integers, then this attractor is a four-dimensional closed curve.*

Proof Let $L_1 = \{(x_1(t), x_2(t)) : t \in T_1\}$ and $L_2 = \{(x_3(t), x_4(t)) : t \in T_2\}$ be the above mentioned limit cycles. Without loss of generality, one may say, that both limit cycles start at $t = 0$. This is possible due to the autonomy of both two-dimensional systems. At the time $t_1 = nT_1$ both solutions (x_1, x_2) and (x_3, x_4) are the same as at the point $t = 0$. This follows from the condition $mT_2 = nT_1$.

The proof immediately follows.

Example 2 Consider system of the form (9) with the regulatory matrix

$$W = \begin{pmatrix} 1.1 & -1 & 0 & 0 \\ 1 & 1.3 & 0 & 0 \\ 0 & 0 & 2 & -1 \\ 0 & 0 & 1 & 2 \end{pmatrix}. \tag{14}$$

This system possesses the four-dimensional attractor which is composed of two identical periodic solutions of the two-dimensional system, considered in the introduction (for the matrix W_m). This periodic attractor cannot be seen, but the projections onto two-dimensional and three-dimensional subspaces are possible to visualize. In Figs. 6 and 7 projections of the attractor and several trajectories are depicted. The periods of both two-dimensional limit cycle relate as $T_2 : T_1 = 2 : 1$. Therefore the periodic attractor is the four-dimensional closed curve.

Remark 2 More examples of this kind can be found in [16].

Fig. 6 Projections on (x_1, x_2, x_3)

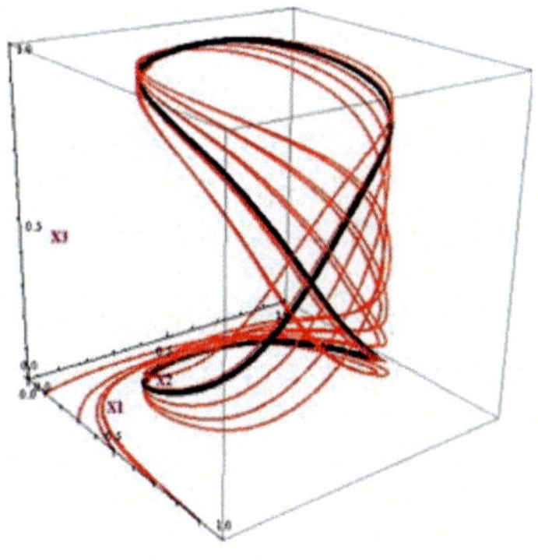

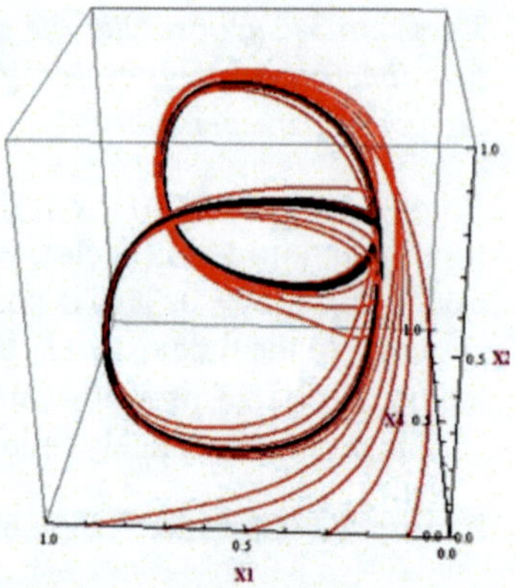

Fig. 7 Projections on (x_1, x_2, x_4)

4 Three-Dimensional Systems and Six-Dimensional Systems

We pass to the three-dimensional systems (3D for brevity) in this section. It is an easy matter to construct a 3D system with the stable periodic solution, if examples of 2D systems are known. Consider the regulatory matrix

$$W_3 = \begin{pmatrix} 2 & 1 & 0 \\ -1 & 2 & 0 \\ 0 & 0 & w_{33} \end{pmatrix},$$

where the third x_3 nullcline is a single plane or a union of three planes depending on the number of roots of the equation (with respect to x_3)

$$x_3 = \frac{1}{1 + e^{-\mu_3(w_{33}x_3 - \theta_3)}}.$$

Then the 2D limit cycle (recall Fig. 3), corresponding to the 2×2 block in the left upper corner, appears as the 3D limit cycle in the 3D system with the above regulatory matrix. That case was studied in details in the conference paper [10].

Example 3 We will consider less trivial example of a 3D limit cycle, obtained numerically. Consider 3D system of the form (6) with the regulatory matrix

$$W = \begin{pmatrix} k & 0 & -1 \\ -1 & k & 0 \\ 0 & -1 & k \end{pmatrix}. \tag{15}$$

Other parameters are chosen as $\mu_i = 5, \theta_i = (k-1)/2, i = 1, 2, 3$. This system was shown [17] to have a limit cycle for $k \in [0.36, 2.34]$. The visualization of this limit cycle will appear in the current text later, as a 3D projection of some 6D-attractor.

Set $k = 0.36$. The respective periodic solution (the limit cycle) has the period $T_1 = 6.23$. This is, of course, the approximate value. We wish to find the value of k, for which the periodic solution (also the limit cycle), has the period $T_2 = 2T_1$ (also approximately). Such a value was found, it is $k = 1.165$.

Our intent is to construct the 6D-system of the form (6), which is composed of two 3D-systems, corresponding to $k_1 = 0.36$ and $k_2 = 1.165$.

Consider the regulatory matrix

$$\begin{pmatrix} k_1 & 0 & -1 & 0 & 0 & 0 \\ -1 & k_1 & 0 & 0 & 0 & 0 \\ 0 & -1 & k_1 & 0 & 0 & 0 \\ 0 & 0 & 0 & k_2 & 0 & -1 \\ 0 & 0 & 0 & -1 & k_2 & 0 \\ 0 & 0 & 0 & 0 & -1 & k_2 \end{pmatrix}, \tag{16}$$

and the corresponding system of the form (6). Other parameters are $\mu_i = 5$, $\theta_i = (k-1)/2$, $i = 1, \ldots, 6$, where $k = k_1$ for $i = 1, 2, 3$ and $k = k_2$ for $i = 4, 5, 6$.

This system possesses the 6D attractor which is composed of two 3D periodic solutions (limit cycles). The three dimensional projections can be visualized.

In Figs. 8 and 9 the projections of the periodic attractor for the 6D system with the matrix (16) are shown. In fact, they are images of the limit cycles in two 3D systems, corresponding to 3×3 blocks of the matrix (16). As was said above their periods are in the relation $T_1 : T_2 = 1 : 2$.

In Figs. 10 and 11 two more projections of the periodic attractor for the 6D system are depicted.

The theorem, similar to Theorem 1, is valid for the 6D systems of the form (6).

Theorem 4 *Suppose that the six-dimensional attractor is obtained from the two three-dimensional limit cycles. If their periods relate as $mT_2 = nT_1$, where m and n are positive integers, then this attractor is a six-dimensional closed curve.*

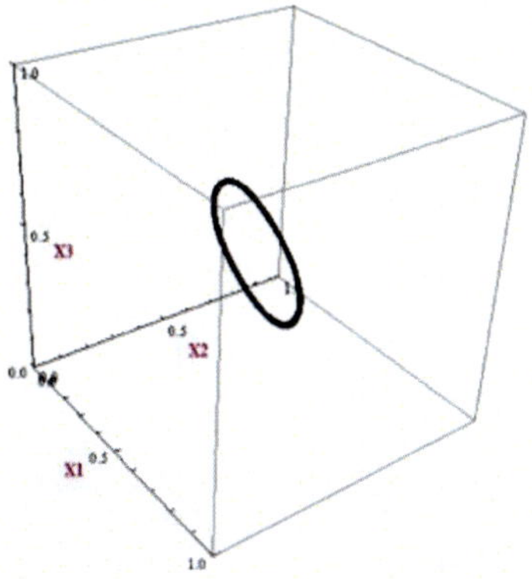

Fig. 8 Projections on (x_1, x_2, x_3)

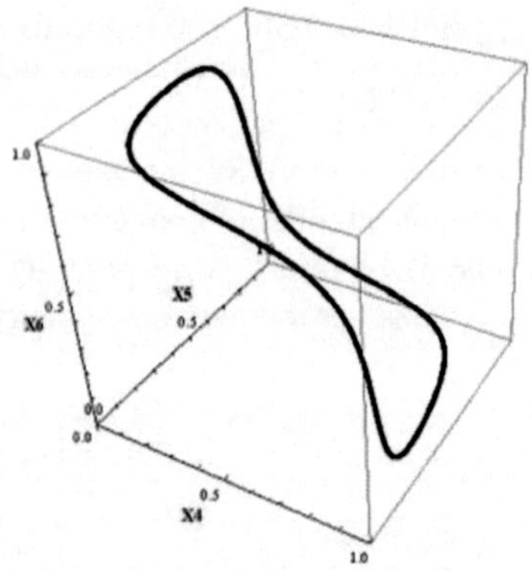

Fig. 9 Projections on (x_4, x_5, x_6)

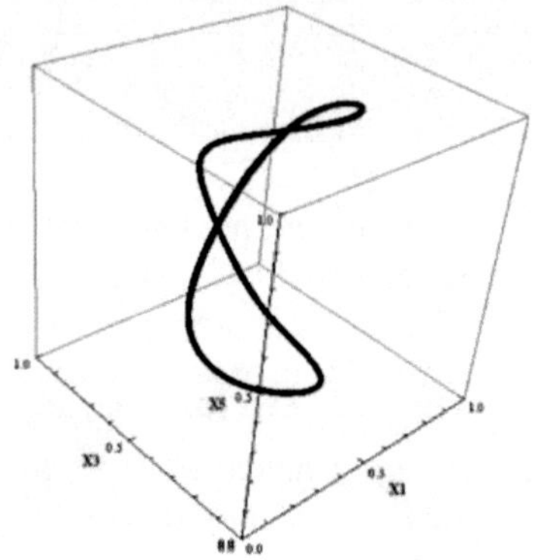

Fig. 10 Projections on (x_1, x_3, x_5)

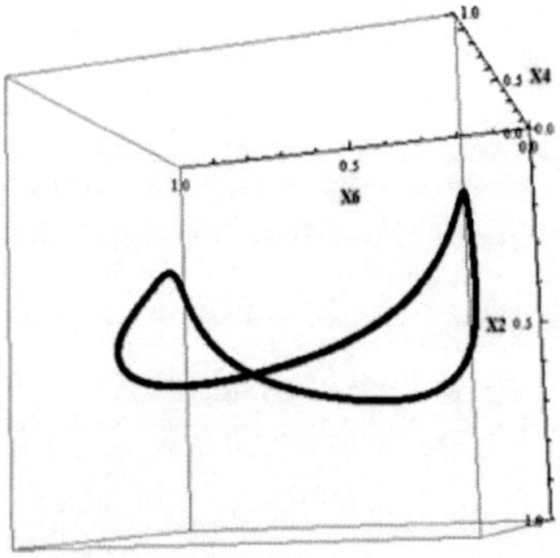

Fig. 11 Projections on (x_2, x_4, x_6)

5 Notes and Comments

Generally the following result is valid. Imagine the n dimensional GRN-type system (6) with the regulatory $n \times n$ matrix W. Let W be block matrix with the square matrices of orders n_i, placed on the main diagonal of W, $\sum_{i=1}^{k} n_i = n$. In terms of the systems of the form (6), it consists of k independent GRN-type systems of order n_i. We denote each system S_i.

Theorem 5 *Suppose each system S_i has an n_i-dimensional limit cycle L_i with the period T_i. Let there exist positive integers m_i such that*

$$m_1 T_1 = m_2 T_2 = \ldots = m_k T_k.$$

Then there exists an attractor in the n-dimensional phase space, which is the n-dimensional closed curve.

Description of biological processes, described by such attractors, is a challenge for collaborating biologists and mathematicians.

References

1. Acary V, Jong HD, Brogliato B (2014) Numerical simulation of piecewise-linear models of gene regulatory networks using complementarity system. Phys D: Nonlinear Phenom 269:103–119
2. Jong HD (2002) Modeling and simulation of genetic regulatory systems: a literature review. J Comput Biol 9(1):67–103. https://doi.org/10.1089/10665270252833208
3. Koizumi Y et al (2010) Adaptive virtual network topology control based on attractor selection. J Lightwave Technol (ISSN : 0733-8724) 28(11):1720–1731. https://doi.org/10.1109/JLT.2010.2048412
4. Schlitt T (2013) Approaches to modeling gene regulatory networks: a gentle introduction. In: Silico systems biology. Methods in molecular biology (Methods and protocols), vol 1021. Humana Press, pp 13–35. https://doi.org/10.1007/978-1-62703-450-0
5. Vijesh N, Chakrabarti SK, Sreekumar J (2013) Modeling of gene regulatory networks: a review. J Biomed Sci Eng 6:223–231
6. Wang L-Z, Su R-Q, Huang Z-G, Wang X, Wang W-X, Celso G, Lai Y-C (2016) A geometrical approach to control and controllability of nonlinear dynamical networks. Nat Commun 7, Article number: 11323. https://doi.org/10.1038/ncomms11323
7. Cornelius SP, Kath WL, Motter AE (2013) Realistic control of network dynamic. Nat Commun 4:1–9, Article number: 1942
8. Wilson HR, Cowan JD (1972) Excitatory and inhibitory interactions in localized populations of model neurons. Biophys J 12(1):1–24
9. Atslega S, Finaskins D, Sadyrbaev F (2016) On a planar dynamical system arising in the network control theory. Math Model Anal 21(3):385–398
10. Atslega S, Sadyrbaev F, Samuilik I (2021) On modelling of complex networks. Eng Rural Develop 20:10091014
11. Brokan E, Sadyrbaev F (2018) Attraction in n-dimensional differential systems from network regulation theory. Math Methods Appl Sci 41(17):7498–7509. https://doi.org/10.1002/mma.5086

12. Edwards R, Ironi L (2014) Periodic solutions of gene networks with steep sigmoidal regulatory functions. Phys D 282:1–15. https://doi.org/10.1016/j.physd.2014.04.013
13. Kozlovska O, Sadyrbaev F (2022) Models of genetic networks with given properties. WSEAS Trans Comput Res 10:43–49. https://doi.org/10.37394/232018.2022.10.6
14. Ogorelova D, Sadyrbaev F, Sengileyev V (2020) Control in inhibitory genetic regulatory network models. Contemp Math (ISSN 2705-1056) 1(5):393–400. https://doi.org/10.37256/cm.152020538
15. Sadyrbaev F, Sengileyev V (2022) Remarks on inhibition. WSEAS Trans Equ 2:16–22. https://doi.org/10.37394/232021.2022.2.4
16. Sadyrbaev F, Samuilik I, Sengileyev V (2021) On modelling of genetic regulatory networks. WSEAS Trans Electron 12:7380. https://doi.org/10.37394/232017.2021.12.10
17. Samuilik I, Sadyrbaev F, Sengileyev V (2022) Examples of periodic biological oscillators: transition to a six-dimensional system. WSEAS Trans Comput Res 10:49–54. https://doi.org/10.37394/232018.2022.10.7

Numerical Method for Problem of Scattering by a Small Thickness Dielectric Layer on a Perfectly Conductive Substrate

Alexey Setukha and Stanislav Stavtsev

Abstract In this work we consider the problems of scattering of a monochromatic wave by a dielectric body in the form of a thin layer placed on a perfectly conducting base. For this case we formulate the boundary value problem for Maxwell's equations with an impedance boundary condition and reduce it to a system of two boundary integral equations with weakly and strongly singular integrals on a perfectly conducting surface. Finally, we construct a numerical method for the considered problem which based on solution of these integral equations.

Keywords Computational electrodynamics · Integral equations · Maxwell's equations · Impedance boundary conditions

Introduction

The method of integral equations is an efficient method for solving problems of electromagnetic scattering in the monochromatic case. The problem of scattering by an ideally conducting body or screen can be reduced to solving a boundary integral equation for a tangential vector field (surface current) placed on the radiated surface. In this case, the dimension of the problem is actually reduced – instead of the original three-dimensional boundary value problem, the two-dimensional integral equation has to be solved [1, 2].

The problems of scattering of a monochromatic wave by a dielectric body can be reduced to a volume integral equation written in the domain occupied by the dielectric [1, 3, 4]. In this case, the problem becomes three-dimensional, but the advantage of this approach is that the grid is constructed only for a dielectric body.

An important class of problems corresponds to the case when a dielectric body is considered in the form of a thin layer placed on a perfectly conducting base. In this case, it is possible to write the system from the volume integral equation in

A. Setukha (✉)
Lomonosov Moscow State University, 119234 Moscow, Russia
e-mail: Setuhaav@rambler.ru

S. Stavtsev
Marchuk Institute of Numerical Mathematics of the RAS, 119333 Moscow, Russia
e-mail: sstass2000@mail.ru

V. Vasilyev (ed.), *Differential Equations, Mathematical Modeling and Computational Algorithms*, Springer Proceedings in Mathematics & Statistics 423,
https://doi.org/10.1007/978-3-031-28505-9_11

this layer and the equation on a perfectly conducting surface. If the dielectric is homogeneous, it is possible to formulate a system of boundary integral equations on a perfectly conducting surface and on the boundary of the dielectric [5–7]. Both of these approaches are general and do not exploit the small thickness of the dielectric layer.

However, an approach based on the approximate allowance for the dielectric layer by the boundary condition of the impedance type is much more economical from the computational point of view. In this work, we develop a numerical method for solving the target problem basing on such an approach.

In this paper, we formulate, the boundary value problem for Maxwell's equations with an impedance boundary condition. We reduce the considered problem to a system of two boundary integral equations with weakly and strongly singular integrals on a perfectly conducting surface. Next, we construct a numerical scheme for these equations basing on the methods of piecewise constant approximations and collocations. We use quadrature formulas, developed in [8, 9] for approximation of the integral operators.

1 Problem Statement

Let us consider the problem of electromagnetic field scattering by a thin dielectric layer placed on a perfectly conducting base. An ideally conducting base can be a system of ideally conducting bodies, each of which is limited by a closed surface, or a screens. The dielectric layer is located above the entire surface of a perfectly conducting body, or above a part of this surface, and above the entire surface of a perfectly conducting screen or above a part of the surface of this screen, on one side (see Fig. 1). One has to find the strengths of the electric and magnetic fields of the form

$$\mathbf{E}(x)e^{-i\omega t}, \quad \mathbf{H}(x)e^{-i\omega t}.$$

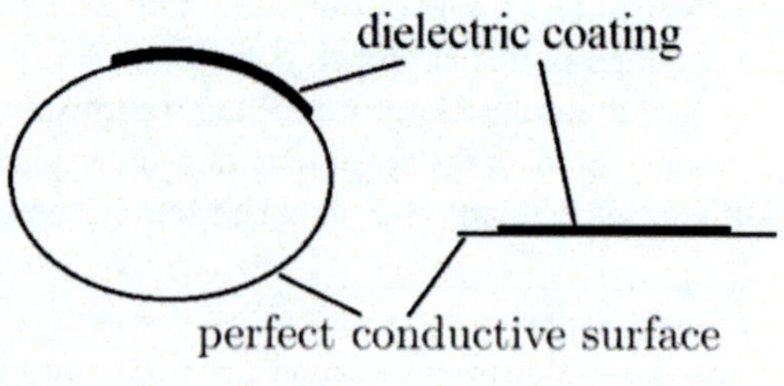

Fig. 1 Coating coverage scheme

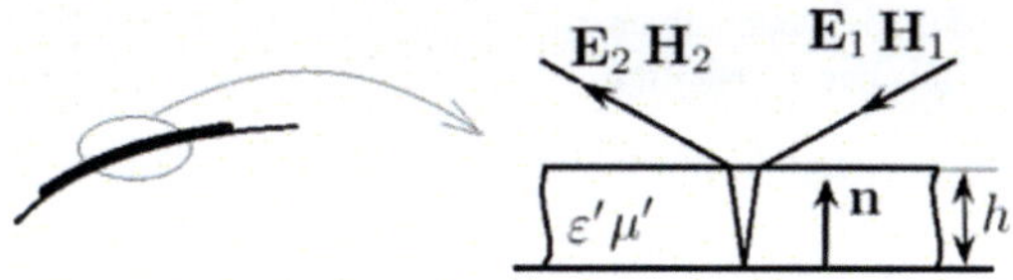

Fig. 2 Wave refraction in a dielectric layer

The spatial components of these fields must satisfy Maxwell's equations [10]:

$$rot\,\mathbf{E} = i\omega\mu\mu_0\mathbf{H}, \tag{1}$$

$$rot\,\mathbf{H} = -i\omega\varepsilon\varepsilon_0\mathbf{E}, \tag{2}$$

where $\mu_0 = 4\pi \times 10^{-7}$ H/m – vacuum permeability, $\varepsilon_0 = 1/(\mu_0 c_0^2)$ – vacuum permittivity, $c_0 = 299792458$ m/s – speed of light in vacuum, ε' – relative permittivity and μ' – relative permeability of the medium. We assume that in the outer environment we have $\varepsilon' = \mu' = 1$.

Inside of the dielectric layer, Maxwell's equations (1)–(2) also operate, but with the values ε' and μ' corresponding to the characteristics of the dielectric. The thickness of the coating h can be variable. The values ε' and μ' are assumed to be constant along the normal vector to the surface, but may change if one moves along the surface. We assume that the quantities ε' and μ' are complex in general. Note that from a physical point of view, the representation of the permittivity in the form $\varepsilon' = \varepsilon_1' + i\varepsilon_2'$ corresponds to a medium with conductivity, where there $\varepsilon_2 = \varepsilon_2'\varepsilon_0$ – a conductivity of the medium.

The main idea of the utilized model is following: if the thickness of the layer is small, then the layer can have a significant influence on the scattering of the incident wave only under the condition $|\varepsilon'\mu'| >> 1$. Let us assume that the product of the real parts of the quantities ε' and μ' is much greater than 1. Let an external field $\mathbf{E}_1, \mathbf{H}_1$ fall on some section of a perfectly conducting surface covered with a dielectric layer.

In the local consideration, we consider this field as a plane wave (Fig. 2). Under the chosen assumptions, this field is refracted inside of the dielectric layer, and this field falls on the surface of an ideal conductor in a direction close to the direction of the normal vector. Due to interaction with the surface of an ideal conductor, a reflected wave $\mathbf{E}_2, \mathbf{H}_2$ arises. This reflected wave also moves inside of the dielectric in a direction close to the direction of the normal vector.

On the boundary between the dielectric and the external media, the boundary conditions for the total field $\mathbf{E}_{tot}, \mathbf{H}_{tot}$ must be satisfied

$$\mathbf{E}_{tot} = \mathbf{E}_1 + \mathbf{E}_2, \;\; \mathbf{H}_{tot} = \mathbf{H}_1 + \mathbf{H}_2 \tag{3}$$

$$\mathbf{n} \times \mathbf{E}_{tot}^{+} = \mathbf{n} \times \mathbf{E}_{tot}^{-}, \;\; \mathbf{n} \times \mathbf{H}_{tot}^{+} = \mathbf{n} \times \mathbf{H}_{tot}^{-} \tag{4}$$

Thus, it is possible to obtain a relation for the tangential components of the electric and magnetic fields on the boundary between the dielectric and the external medium

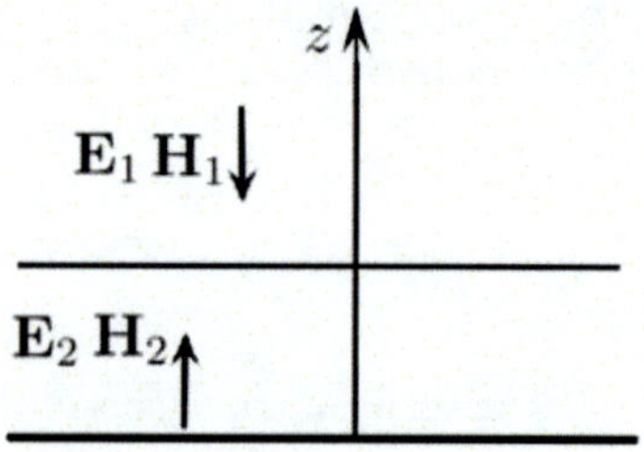

Fig. 3 Obtaining an impedance condition

by considering the reflection problem of a plane wave incident along the normal vector on a perfectly conducting plane, above which is located a dielectric with parameters ε' and μ'.

Let us consider a perfectly conducting plane. Let us introduce such Cartesian coordinates $Oxyz$ that the considered ideally conducting plane is determined by the equation $z = 0$ and assume that the half-space $z > 0$ is filled with a dielectric with parameters ε' and μ' (see Fig. 3).

Suppose that a plane wave falls on a plane, inside which the electric field is represented as:

$$\mathbf{E}_1 = \mathbf{E}_1^0 e^{-ik'z},$$

where $\mathbf{E}_1^0$ is some constant vector parallel to a perfectly conducting plane, $k' = \omega\sqrt{\varepsilon\mu}$, $\varepsilon = \varepsilon'\varepsilon_0$, $\mu = \mu'\mu_0$.

The total field is sought in the form (3), where $\mathbf{E}_2$ – is a plane wave of the form

$$\mathbf{E}_2 = \mathbf{E}_2^0 e^{ik'z}$$

moving from the plane. On a perfectly conducting plane, the next boundary condition must be satisfied

$$\mathbf{n} \times \mathbf{E}_{tot} = 0.$$

From the last condition we conclude that the total field has the form:

$$\mathbf{E}_{tot} = \mathbf{E}_1^0 \left(e^{-ik'z} - e^{ik'z}\right).$$

Then

$$\mathbf{H}_{tot} = -\frac{i}{\omega\mu} rot \left[\mathbf{E}_1^0 \left(e^{ik'z} - e^{-ik'z}\right)\right] = -\frac{k'}{\omega\mu} \left[n \times \mathbf{E}_1^0\right] \left(e^{ik'z} + e^{-ik'z}\right).$$

From the last relations, we can write for $z = h$:

$$\mathbf{n} \times [\mathbf{n} \times \mathbf{E}_{tot}] = \mathbf{n} \times \left[\mathbf{n} \times \mathbf{E}_1^0\right] \left(e^{-ik'h} - e^{ik'h}\right),$$

$$\mathbf{n} \times \mathbf{H}_{tot} = -\frac{k'}{\omega\mu}\mathbf{n} \times \left[\mathbf{n} \times \mathbf{E}_1^0\right]\left(e^{ik'h} + e^{-ik'h}\right).$$

Hence, we have:

$$\mathbf{n} \times [\mathbf{n} \times \mathbf{E}_{tot}] = Z\,[\mathbf{n} \times \mathbf{H}_{tot}]\,, \quad Z = \frac{\omega\mu}{k'}\frac{e^{ik'h} - e^{-ik'h}}{e^{ik'h} + e^{-ik'h}} = \sqrt{\frac{\mu}{\varepsilon}}\frac{e^{ik'h} - e^{-ik'h}}{e^{ik'h} + e^{-ik'h}}. \quad (5)$$

In a view of the relations (4), the following condition must be satisfied on the outer side of the dielectric layer:

$$\mathbf{n} \times \left[\mathbf{n} \times \mathbf{E}_{tot}^{+}\right] = Z\left[\mathbf{n} \times \mathbf{H}_{tot}^{+}\right]. \quad (6)$$

Now, let us return to the original complete problem. We neglect the thickness of the coatings. We assume that the entire domain Ω outside the surfaces of ideally conducting objects is occupied by the external environment and in this domain the electric and magnetic fields satisfy Maxwell's equations (1)–(2). The total electric and magnetic fields are sought in the form

$$\mathbf{E}_{tot} = \mathbf{E}_{inc} + \mathbf{E}, \quad \mathbf{H}_{tot} = \mathbf{H}_{inc} + \mathbf{H}, \quad (7)$$

where $\mathbf{E}_{inc}$, $\mathbf{H}_{inc}$ is a given incident field, $\mathbf{E} = \mathbf{E}(x)$, $\mathbf{H} = \mathbf{H}(x)$ is an unknown secondary (reflected) field. The reflected field must satisfy the radiation conditions at infinity. On closed ideally conducting surfaces and on thin screens, we set the boundary condition (6) (in areas where there is no coating, we suppose $z = 0$). In addition, on thin screens, we set the boundary condition

$$\mathbf{n} \times \mathbf{E}_{tot}^{-} = 0 \quad (8)$$

Thus, the boundary value problem is solved for Eqs. (1)–(2) with boundary conditions (6), (8) and radiation conditions at infinity.

The primary field can be, for example, the field induced by a plane wave:

$$\mathbf{E}_{inc}(x) = \mathbf{E}_0 e^{i\,\mathbf{k}\mathbf{r}}, \quad (9)$$

where $\mathbf{k}$ is an arbitrary vector that satisfies the condition $|\mathbf{k}| = k$, $\mathbf{r}$ is a radius vector of point x, $\mathbf{E}_0$ is an arbitrary vector satisfying the condition $\mathbf{E}_0\,\mathbf{k} = 0$, k is a wave number, determined by the relation

$$k = \frac{\omega}{c_0}, \quad c_0 = \frac{1}{\sqrt{\varepsilon_0\mu_0}}. \quad (10)$$

c_0 is the speed of light in vacuum, as before.

2 Reduction of the Problem to Integral Equations

We seek to find the electric field in the form [1, 2]:

$$\mathbf{E} = \frac{i}{\omega\varepsilon}\mathbf{K}[\Sigma, \mathbf{j}_E] - \mathbf{R}\left[\Sigma, \mathbf{j}_M\right], \tag{11}$$

where **K** and **R** are operators defined by formulas:

$$\mathbf{K}[\Sigma, \mathbf{j}](x) = \int\limits_{\Sigma} \{grad_x div_x[\mathbf{j}(y)\Phi(x-y)] + k^2\mathbf{j}(y)\Phi(x-y)\}d\sigma_y \tag{12}$$

$$\mathbf{R}[\Sigma, \mathbf{j}](x) = rot\int\limits_{\Sigma} \mathbf{j}(y)\Phi(x-y)d\sigma_y = \int\limits_{\Sigma} grad_x\Phi(x-y) \times \mathbf{j}(y)d\sigma_y \tag{13}$$

$$\mathbf{j}(x)\,\mathbf{n}(x) = 0,\ \ x \in \Sigma,\ \ \Phi(x-y) = \frac{1}{4\pi}\frac{e^{ikr}}{r},\ \ r = |x-y|,$$

$\mathbf{j}_E$ and $\mathbf{j}_M$ are unknown tangent vector fields on the surface Σ – electric and magnetic currents, respectively. Taking into account Eq. (1), the magnetic field has the form:

$$\mathbf{H} = \mathbf{R}[\Sigma, \mathbf{j}_E] + \frac{i}{\omega\mu}\mathbf{K}\left[\Sigma, \mathbf{j}_M\right]. \tag{14}$$

Consider the properties of the boundary values of the vector fields generated by the operators **K** and **R**.

If Σ is a smooth surface of class C^3, and **j** is a tangent vector field of class C^2, then, as shown in [8], the field $\mathbf{E} = \mathbf{K}[\Sigma, \mathbf{j}]$, which defined outside the surface Σ, has boundary values in each point $x \in S$, which is not an edge point, for which the following relation is true:

$$\mathbf{n} \times \mathbf{E}^{+} = \mathbf{n} \times \mathbf{E}^{+} = \mathbf{n} \times \mathbf{E}, \tag{15}$$

where $\mathbf{E}(x) = \mathbf{K}[\Sigma, \mathbf{j}](x)$ is the direct value of the integral operator defined by the formula (12) for $x \in \Sigma$, if the integral is understood in the sense of the Hadamard finite part.

If **j** is a tangent vector field of class $C^1[S]$ on surface S of class C^3 and $\mathbf{E} = \mathbf{R}[S, \mathbf{j}]$, then at each point $x \in S$, which is not an edge point, there are boundary values of the field **E** for which the following formula is true [11]:

$$\mathbf{n}(x) \times \mathbf{E}^{\pm}(x) = \mathbf{n}(x) \times \mathbf{E}(x) \pm \frac{1}{2}\mathbf{j}(x),\ \ x \in S, \tag{16}$$

in this formula $\mathbf{E}(x) = \mathbf{R}[S, \mathbf{j}](x)$ is a direct value of the operator $\mathbf{R}$ obtained from the formula (13) for considered x and the integral in this expression exists in the sense of the main value.

Note that expressions (11) and (14) define vector fields $\mathbf{E}$ and $\mathbf{H}$ not only in the outer domain of ideally conducting bodies and screens, but also inside ideally conducting bodies. We require that the total electromagnetic field vanishes in the domains inside ideally conducting bodies. So, we seek a solution – electric and magnetic fields $\mathbf{E}_{tot}$ and $\mathbf{H}_{tot}$ of the form (7) with $\mathbf{E}$ and $\mathbf{H}$ in the form (14), defined everywhere outside the surface and satisfying the boundary conditions (6) and (8) on the entire surface.

Then, using relations (15) and (16) for the boundary values of the operators $\mathbf{K}$ and $\mathbf{R}$, we obtain the following equations:

$$\frac{i}{\omega\varepsilon}\mathbf{n} \times [\mathbf{n} \times \mathbf{K}[\Sigma, \mathbf{j}_E]] - \mathbf{n} \times \left[\mathbf{n} \times \mathbf{R}\left[\Sigma, \mathbf{j}_M\right]\right] - \frac{1}{2}\mathbf{n} \times \mathbf{j}_M + \mathbf{n} \times [\mathbf{n} \times \mathbf{E}_0] =$$

$$= Z\,\mathbf{n} \times \mathbf{R}\left[\Sigma, \mathbf{j}_E\right] + \frac{iZ}{\omega\mu}\mathbf{n} \times \mathbf{K}[\Sigma, \mathbf{j}_M] + \frac{1}{2}z\,\mathbf{j}_E + Z\,\mathbf{n} \times \mathbf{H}_0,$$

$$\frac{i}{\omega\varepsilon}\mathbf{n} \times \mathbf{K}[\Sigma, \mathbf{j}_E] - \mathbf{n} \times \mathbf{R}\left[\Sigma, \mathbf{j}_M\right] + \frac{1}{2}\mathbf{j}_M + \mathbf{n} \times \mathbf{E}_0 = 0.$$

We can rewrite these equations as:

$$\frac{i}{\omega\varepsilon}\mathbf{K}[\Sigma, \mathbf{j}_E]_\tau - \mathbf{R}\left[\Sigma, \mathbf{j}_M\right]_\tau - \frac{1}{2}\mathbf{n} \times \mathbf{j}_M = -\mathbf{E}_{0\tau},$$

$$\mathbf{j}_M + \frac{iz}{\omega\mu}\mathbf{K}[\Sigma, \mathbf{j}_M]_\tau - \frac{1}{2}z\,\mathbf{n} \times \mathbf{j}_E + Z\mathbf{R}\left[\Sigma, \mathbf{j}_E\right]_\tau = -z\,\mathbf{H}_{0\tau}, \tag{17}$$

the index τ means the tangent component of the vector.

Thus, the problem has been reduced to a system of integral equations (17) for unknown currents $\mathbf{j}_E$ and $\mathbf{j}_M$ on the surface Σ.

3 Numerical Scheme

A numerical scheme for solving the problem arises when Eq. (17) are discretized. We apply the method of piecewise constant approximations and collocations.

The surface Σ is approximated by a system of tetragonal cells σ_i, $i = 1, \ldots, n$. On each cell we choose a collocation point x^i and let $\mathbf{n}_i = \mathbf{n}(x^i)$ be the unit normal vector to the cell σ_i in point x_i. Triangular cells can also occur, whereby a triangular cell is considered as a tetragonal cell with two identical vertices. We assume that the vertices of the cell define a contour of four segments - the edge of the cell. The point

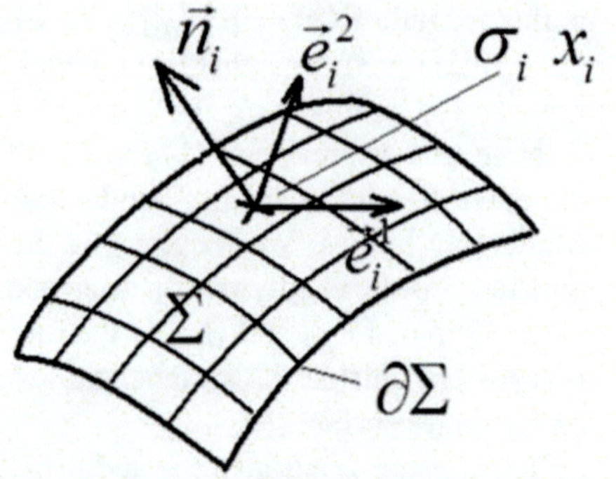

Fig. 4 Surface approximation

x^i is constructed as the intersection point of the segments connecting the midpoints of the opposite sides of the cell, and the normal vector as the normal to these segments.

Next, we build on each cell σ_i the local orthonormal basis $\mathbf{e}_i^1$, $\mathbf{e}_i^2 = \mathbf{n}_i \times \mathbf{e}_i^1$ in a plane orthogonal to the vector $\mathbf{n}_i$ (the choice of a vector $\mathbf{e}_i^1$ is arbitrary provided that it is normalized and orthogonal to the vector $\mathbf{n}_i$). In this case, the vectors $\mathbf{e}_i^1$, $\mathbf{e}_i^2$, $\mathbf{n}_i$ form a right basis (see Fig. 4).

We approximate the vector fields $\mathbf{j}_E(x)$ and $\mathbf{j}_M(x)$ by sets of values $\mathbf{j}_{E\,i} \approx \mathbf{j}_E(x^i)$, $\mathbf{j}_{M\,i} \approx \mathbf{j}_M(x^i)$ and assume that the following relation holds $(\mathbf{j}_{E\,i}, \mathbf{n}_i) = (\mathbf{j}_{M\,i}, \mathbf{n}_i) = 0$, , $i = 1, \ldots, n$. Therefore, we seek to obtain the vectors $\mathbf{j}_{E\,i}$ $\mathbf{j}_{M\,i}$ in the form:

$$\mathbf{j}_{E\,i} = j_{E\,i}^1\,\mathbf{e}_i^1 \;+\; j_{E\,i}^2\,\mathbf{e}_i^2,\;\; \mathbf{j}_{M\,i} = j_{M\,i}^1\,\mathbf{e}_i^1 \;+\; j_{M\,i}^2\,\mathbf{e}_i^2. \tag{18}$$

We approximate the operators $\mathbf{K}[\Sigma_0, \mathbf{j}]$ and $\mathbf{K}[\Sigma_0, \mathbf{j}]$ by the following expressions:

$$\mathbf{K}[\Sigma_0, \mathbf{j}] \approx \sum_{k=1}^{N} \tilde{\mathbf{K}}[\Sigma_k, \mathbf{j}_k],\;\; \mathbf{R}[\Sigma_0, \mathbf{j}] \approx \sum_{k=1}^{n} \tilde{\mathbf{R}}[\sigma_k, \mathbf{j}_k], \tag{19}$$

where $\tilde{\mathbf{K}}[\sigma_k, \mathbf{j}_k]$ and $\tilde{\mathbf{R}}[\sigma_k, \mathbf{j}_k]$ – approximations of the respective integrals for the area of surface Σ_0, approximated by cell σ_k.

The value $\tilde{\mathbf{K}}[\sigma_k, \mathbf{j}_k]$ is calculated according to the formulas, based on the extraction of the leading parts of the kernel of the integral operators, that were proposed by Ryzhakov in article [8]. Function $\mathbf{K}(\mathbf{j}, x, y)$, that is integrand in expression (12), is presented in the following form:

$$\mathbf{K}(\mathbf{j}, x, y) = \mathbf{K}^0(\mathbf{j}, x, y) + \mathbf{K}^1(\mathbf{j}, x, y), \tag{20}$$

$$\mathbf{K}^0(\mathbf{j}, x, y) = \frac{-\mathbf{j} + 3\mathbf{r}(\mathbf{r}, \mathbf{j})}{4\pi R^3},$$

$$\mathbf{K}^1(\mathbf{j}, x, y) = (\mathbf{j} - 3\mathbf{r}(\mathbf{r}, \mathbf{j}))\frac{1 - e^{ikR} + ikR\,e^{ikR}}{4\pi R^3} + (\mathbf{j} - \mathbf{r}(\mathbf{r}, \mathbf{j}))\frac{k^2 e^{ikR}}{4\pi R},$$

$R = |x - y|$, $\mathbf{r} = (x - y)/R$. At that

$$\left|\mathbf{K}^0(\mathbf{j}, x, y)\right| \leq O(|x - y|^{-3}), \quad \left|\mathbf{K}^1(\mathbf{j}, x, y)\right| \leq O(|x - y|^{-1}).$$

The approximation of $\tilde{\mathbf{K}}[\sigma_i, \mathbf{j}_k]$ is constructed according to formula:

$$\tilde{\mathbf{K}}[\sigma_k, \mathbf{j}_k] = \tilde{\mathbf{K}}^0[\sigma_k, \mathbf{j}_k] + \tilde{\mathbf{K}}^1[\sigma_k, \mathbf{j}_k], \tag{21}$$

where

$$\tilde{\mathbf{K}}^0[\sigma_k, \mathbf{j}](x) = \int_{\sigma_k} \mathbf{K}^0(\mathbf{j}_k^*(y), x, y) d\sigma_y,$$

$\mathbf{j}_k^*(y)$ – tangent field on cell σ_k, obtained at projecting vector $\mathbf{j}_k$ according to formula $\mathbf{j}_k^*(y) = (\mathbf{n}_k \times \mathbf{j}_k) \times \mathbf{n}(y)$. Integral in expression for $\tilde{\mathbf{K}}^0[\sigma_k, \mathbf{j}](x)$ is reduced to the integral along the cells boundary [8]:

$$\tilde{\mathbf{K}}^0[\sigma_k, \mathbf{j}](x) = grad \oint_{\partial\sigma_k} \frac{(\mathbf{j}_k^*(y) \times \mathbf{n}(y), \boldsymbol{\tau}(y))}{4\pi\, |x - y|} ds_y, \tag{22}$$

where $\mathbf{n}(y)$ is the vector of normal to the surface of cell σ_k, $\boldsymbol{\tau}(y)$ - tangent vector on the contour $\partial\sigma_k$ at point $y \in \partial\sigma_k$. At that, if the boundary of cell σ_k is a polygonal line, the following equation is true on each segment L that is the element of this line

$$grad \int_L \frac{(\mathbf{j}_k^*(y) \times \mathbf{n}(y), \boldsymbol{\tau}(y))}{|x - y|} ds_y = (\mathbf{j}_k \times \mathbf{n}_i, \boldsymbol{\tau}) grad \int_L \frac{1}{|x - y|} ds_y,$$

here $\boldsymbol{\tau} = (b - a)/|b - a|$, a, b – beginning and end of segment L. The latter integral is calculated analytically

$$grad \int_L \frac{1}{|x - y|} ds_y = \left(\frac{a - x}{|a - x|} + \frac{b - x}{|b - x|}\right) \frac{|b - a|}{(b - x)(a - x) + |b - x|\,|a - x|}.$$

The piecewise-constant approximation of current with the value of $\mathbf{j}_k$ on the whole cell σ_k is used for the approximation of integrals $\tilde{\mathbf{K}}^1[\sigma_k, \mathbf{j}_k]$ and $\tilde{\mathbf{R}}[\sigma_k, \mathbf{j}_k]$ over this cell:

$$\tilde{\mathbf{K}}^1[\sigma_k, \mathbf{j}](x) \approx \int_{\sigma_k} \mathbf{K}^1(\mathbf{j}_k, x, y) d\sigma_y, \quad \tilde{\mathbf{R}}[\sigma_k, \mathbf{j}](x) \approx \int_{\sigma_k} \mathbf{R}(\mathbf{j}_k, x, y) d\sigma_y.$$

These integrals over the cells are calculated using the rectangle formula with the additional partition of cells σ_k into smaller cells of the second level and smoothing of singularity in the integrands according to the scheme described in article [9]. Let's additionally partition up each cell σ_k into the cells of second level σ_k^p, $p = 1, \ldots, P_k$, and choose the collocation point $y_k^p \in \sigma_k^p$ on each of such cell. Let h' — be the maximum diameter of cells σ_k^p, $p = 1, \ldots, P_k$, $k = 1, \ldots, N$. Let:

$$\tilde{\mathbf{K}}^1[\sigma_k, \mathbf{j}_k] = \sum_{p=1}^{P_k} \mathbf{K}_1(\mathbf{j}_k, x, y_k^p)\, \theta\left(\frac{|x - y_k^p|}{\varepsilon}\right) s_k^p,$$

$$\tilde{\mathbf{R}}[\sigma_i, \mathbf{j}_k] = \sum_{p=1}^{P_k} \mathbf{R}(\mathbf{j}_k, x, y_k^p)\, \theta\left(\frac{|x - y_k^p|}{\varepsilon}\right) s_k^p, \tag{23}$$

where s_k^p — area of cell σ_k^p, $\theta(r)$ — smoothing function chosen so that $\theta(r) \in C^1[0, \infty)$, $\theta(r) = 1$ at $r \geq 1$, $0 \leq \theta(r) \leq 1$ at $0 \leq r \leq 1$, $\theta(r) = o(r)$ as $r \to 0$, ε — small parameter. In the calculations, given further, we assumed that $\theta(r) = 3r^2 - 2r^3$, $\varepsilon = 2h'$.

Then system (17) reduces to a system of linear algebraic equations:

$$\sum_{\substack{j=1,\ldots,N \\ l=1,2}} a_{k\,j}^{m\,l} j_{E\,j}^l + \sum_{\substack{j=1,\ldots,N \\ l=1,2}} b_{k\,j}^{m\,l} j_{M\,j}^l = f_k^m,$$

$$\sum_{\substack{j=1,\ldots,N \\ l=1,2}} c_{k\,j}^{m\,l} j_{E\,j}^l + \sum_{\substack{j=1,\ldots,N \\ l=1,2}} d_{k\,j}^{m\,l} j_{M\,j}^l = g_k^m, \tag{24}$$

$$k = 1, \ldots, N, \quad m = 1, 2,$$

$$a_{k\,j}^{m\,l} = \frac{i}{\omega\varepsilon}(\tilde{\mathbf{K}}[\sigma_j, \mathbf{e}_j^l](x_k), \mathbf{e}_k^m),$$

$$b_{k\,j}^{m\,l} = -(\mathbf{n}_i \times \tilde{\mathbf{R}}[\sigma_j, \mathbf{e}_j^l](x_k), \mathbf{e}_k^m) + \frac{1}{2}\delta_k^j(\mathbf{n}_i \times \mathbf{e}_k^l, \mathbf{e}_k^m),$$

$$c_{k\,j}^{m\,l} = \frac{i}{\omega\varepsilon}(\tilde{\mathbf{K}}[\sigma_j, \mathbf{e}_j^l](x_k), \mathbf{e}_k^m) + z(\mathbf{n} \times \tilde{\mathbf{R}}[\sigma_j, \mathbf{e}_j^l](x_k), \mathbf{e}_k^m) + \frac{1}{2}z\delta_k^j\delta_m^l,$$

$$d_{k\,j}^{m\,l} = -(\tilde{\mathbf{R}}[\sigma_j, \mathbf{e}_j^l](x_k), \mathbf{e}_k^m) + \frac{iz}{\omega\mu}(\mathbf{n} \times \tilde{\mathbf{K}}[\sigma_j, \mathbf{e}_j^l](x_k), \mathbf{e}_k^m) + \frac{1}{2}z\delta_k^j(\mathbf{n} \times \mathbf{e}_j^l, \mathbf{e}_k^m),$$

$$f_k^m = -(\mathbf{E}_{inc}(x^k), \mathbf{e}_k^m),$$

$$g_k^m = -(\mathbf{E}_{inc}(x_k), \mathbf{e}_k^m) - z(\mathbf{n}_i \times \mathbf{H}_{inc}(x_k), \mathbf{e}_k^m),$$

$$k, j = 1, \ldots, N, \quad m, l = 1, 2,$$

δ_i^k – kronecker symbol.

After solving the system of linear equations (24), the currents $\mathbf{j}_{E\,i}$ and $\mathbf{j}_{M\,i}$ are determined from relations (18). After that, the electric and magnetic fields are approximated by the formulas

$$\mathbf{E}(x) = \frac{i}{\omega\varepsilon}\sum_{i=1}^{N}\tilde{\mathbf{K}}[\sigma_i, \mathbf{j}_{E\,i}] - \sum_{i=1}^{N}\tilde{\mathbf{R}}[\sigma_i, \mathbf{j}_{M\,i}],$$

$$\mathbf{H}(x) = \sum_{i=1}^{N}\tilde{\mathbf{R}}[\sigma_i, \mathbf{j}_{E\,i}] + \frac{i}{\omega\mu}\sum_{i=1}^{N}\tilde{\mathbf{K}}[\sigma_i, \mathbf{j}_{M\,i}],$$

operators $\tilde{\mathbf{K}}$ and $\tilde{\mathbf{R}}$ calculated by formulas (21)–(23).

The main characteristic of the scattered electromagnetic field in the far zone is the effective scattering surface in the direction of the unit vector $\boldsymbol{\tau}$:

$$\sigma = \lim_{R\to\infty} 4\pi R^2 \frac{|\mathbf{E}(R\boldsymbol{\tau})|^2}{|\mathbf{E}_{inc}|^2}. \tag{25}$$

For evaluation of the effective scattering surface we use the following formula

$$\sigma(\boldsymbol{\tau}) = \frac{4\pi}{|\mathbf{E}_{inc}|^2}\left|\int\limits_{\Sigma} e^{-ik(\boldsymbol{\tau},y)}\left[\frac{i}{\omega\varepsilon}k^2\left(\mathbf{j}_E - \boldsymbol{\tau}\left(\mathbf{j}_E, \boldsymbol{\tau}\right)\right) - ik\left[\boldsymbol{\tau}\times\mathbf{j}_M(y)\right]\right]d\sigma_y\right|^2.$$

In the numerical solution, we use the approximate formula:

$$\sigma(\boldsymbol{\tau}) = \frac{4\pi}{|\mathbf{E}_{inc}|^2}\left|\sum_{j=1}^{N} e^{-ik(\boldsymbol{\tau},x^j)}\left[\frac{i}{\omega\varepsilon}k^2\left(\mathbf{j}_{E,j} - \boldsymbol{\tau}\left(\mathbf{j}_{E,j}, \boldsymbol{\tau}\right)\right) - ik\left[\boldsymbol{\tau}\times\mathbf{j}_{M,j}\right]\right]\sigma_j\right|^2. \tag{26}$$

4 Calculation Examples and Discussion

In order to validate the obtained results of simulations we have constructed scattering diagrams for some ideally conducting surfaces with dielectric coating.

In Fig. 5 we present scattering diagrams for ideally conducting sphere whose radius equals a with a coating having thickness $0.02a$ and without coating when one provides its radiation by plane wave (9) with a wave number $k = 10/a$. The coating corresponds a dielectric with parameters $\varepsilon' = 50$, $\mu' = 1$. In a diagram we show the dependence of value

$$\tilde{\sigma}(\alpha) = 10log\frac{\sigma(\boldsymbol{\tau}(\alpha))}{\pi a^2},$$

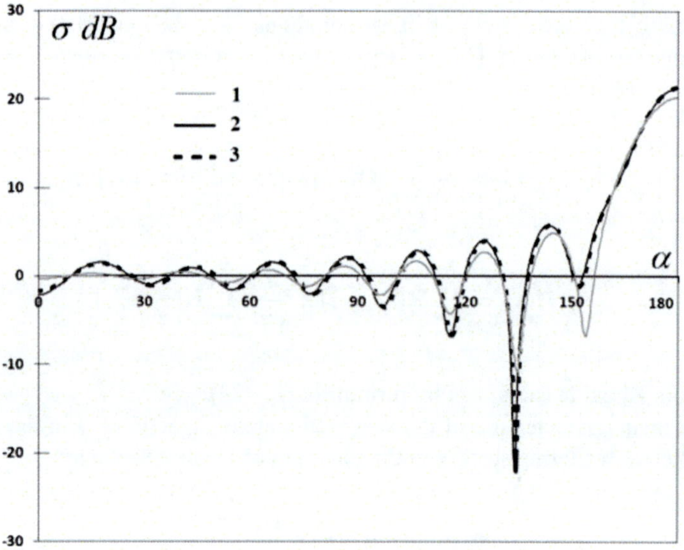

Fig. 5 Scattering diagram for sphere with coating and without coating

where α is an angle between vectors $\boldsymbol{\tau}$ and $-\mathbf{k}$, $\mathbf{E}_0$ is a polarization vector which lies within a plane basing on vectors $\boldsymbol{\tau}$ and $\mathbf{k}$.

The Curve 1 corresponds to theoretical solution for a sphere without coating and Curve 2 represents the numerical solution for sphere with coating which is constructed with use of the proposed method with impedance boundary condition. Curve 3 shows the theoretical solution for the sphere with the coating. Theoretical solutions have been constructed using a series basing on special functions [12]. These curves allow us to demonstrate an influence of the coating on the final scattering diagram as well as to show a good agreement of numerical results with analytical.

In Fig. 6 we show the reflection diagrams of the plane wave by rectangular plate of the size 1 m. We considered a plate without coating and also with coating having a thickness 0.005 m with $\varepsilon' = 50 + 50i$, $\mu' = 1$. The reflection is characterized as $\sigma(\boldsymbol{\tau}(\alpha))$, where α is the angle between the wave vector $-\mathbf{k}$ and a normal vector for a plate $\mathbf{n}$, and $\boldsymbol{\tau}$ is a vector, directed according to the law of optical reflection of the vector $\mathbf{k}$(see Fig. 6, from above). In this case we considered situations when polarization vector $\mathbf{E}_0$ (see Eq. (1.13)) lies in a plane of vectors $\mathbf{k}$ and $\mathbf{n}$ (horizontal polarization) and also orthogonal to them (vertical polarization).

Here the curves 1 correspond to calculation for a plate without coating, Curves 2 correspond to calculations with elaborated model with impedance boundary condition for a plate with coating. For comparison we demonstrate the same dependencies (Curves 3) for a plate with coating which we obtain numerically with use of the solution scattering problem in an exact statement by the method of the boundary integral equations from [6]. In the last case for the problem of scattering on a partially screened dielectric body, boundary integral equations were solved, written on

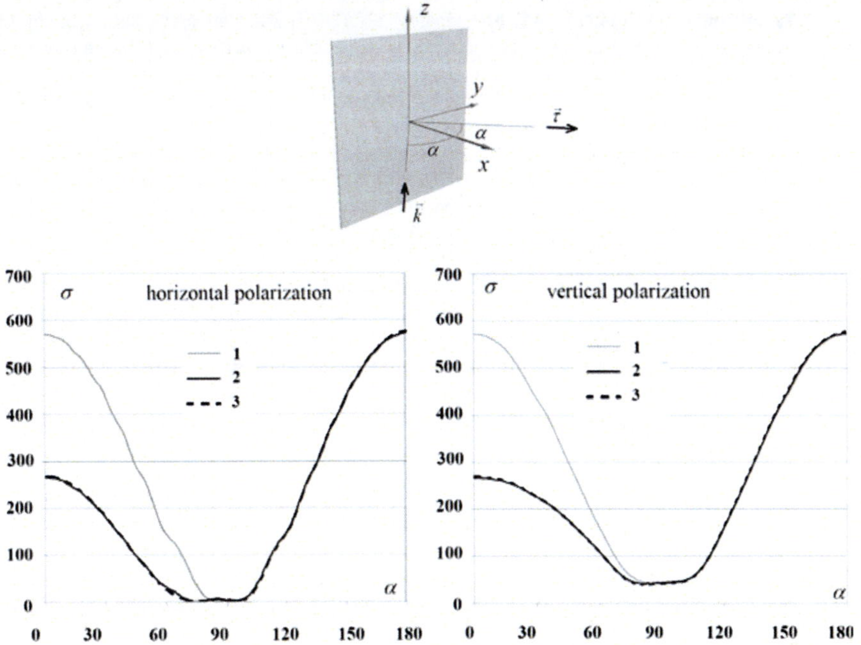

Fig. 6 Scattering diagram for plate with coating and without coating

the exact boundary of the dielectric. Curves 2 and 3 in the figure are indistinguishable and indicate the closeness of the numerical results obtained by the two indicated models. It can also be seen that when irradiated from the side of the coating, its influence is significant. When irradiated from the opposite side, the coating practically does not affect the result.

The given examples of test calculations indicate good agreement between the characteristics of the electromagnetic field in the far zone, obtained from the constructed model with the impedance boundary condition, theoretical data, and numerical results obtained by another method for the problem in the exact formulation. This indicates the performance of the model.

The research was supported by RSF, grant 19-11-00338, https://rscf.ru/project/19-11-00338/ and by the National Center for Physics and Mathematics, project No. 2 "Mathematical modeling on super-computers with exa- and zetta performance".

References

1. Volakis JL, Sertel, K (2012) Integral equation methods for electromagnetics. SciTech Publishing
2. Gibson W (2008) The method of moments in electromagnetics. Chapman and Hall/CRC, Boca Raton

3. Samokhin AB, Samokhina AS, Shestopalov YV (2017) Analysis and solution method for problems of electromagnetic wave scattering on dielectric and perfectly conducting structures. Diff Equat 53:1165–1173. https://doi.org/10.1134/S0012266117090075
4. Smirnov YG, Tsupak AA (2015) Method of integral equations in a scalar diffraction problem on a partially screened inhomogeneous body. Diff Equat 51:1225–1235. https://doi.org/10.1134/S0012266115090128
5. Davydov AG, Zakharov EV, Pimenov YV (2006) Hypersingular integral equations for the diffraction of electromagnetic waves on homogeneous magneto-dielectric bodies. Comput Math Model 17:97–104. https://doi.org/10.1007/s10598-006-0001-9
6. Setukha AV, Bezobrazova EN (2017) The method of hypersingular integral equations in the problem of electromagnetic wave diffraction by a dielectric body with a partial perfectly conducting coating. Russ J Numer Anal Math Model 32:371–380. https://doi.org/10.1515/rnam-2017-0035
7. Zakharov EV, Setukha AV (2020) Method of boundary integral equations in the problem of diffraction of a monochromatic electromagnetic wave by a system of perfectly conducting and piecewise homogeneous dielectric objects. Diff Equat 56:1153–1166. https://doi.org/10.1134/S0012266120090062
8. Zakharov EV, Ryzhakov GV, Setukha AV (2014) Numerical solution of 3D problems of electromagnetic wave diffraction on a system of ideally conducting surfaces by the method of hypersingular integral equations. Diff Equat 50:1240–1251. https://doi.org/10.1134/S0012266114090110
9. Setukha A, Fetisov S (2018) The method of relocation of boundary condition for the problem of electromagnetic wave scattering by perfectly conducting thin objects. J Comput Phys 373:631–647. https://doi.org/10.1016/j.jcp.2018.07.013
10. Hoenl H, Maue AW, Westpfahl K (1961) Theorie der Beugung. Springer, Berlin, Heidelberg
11. Colton D, Kress R (2013) Integral equation methods in scattering theory. SIAM
12. Bohren CF, Huffman DR (1998) Absorption and scattering of light by small particles. WILEY -VCH

Invariants of Dynamical Systems with Dissipation on Tangent Bundles of Low-Dimensional Manifolds

Maxim V. Shamolin

Abstract Tensor invariants (differential forms) for homogeneous dynamical systems on tangent bundles to smooth two-dimensional manifolds are presented in this paper. The connection between the presence of these invariants and the full set of the first integrals necessary for the integration of geodesic, potential and dissipative systems is shown. At the same time, the introduced force fields make the considered systems dissipative with dissipation of different signs and generalize the previously considered ones. We also represent the typical examples from rigid body dynamics.

Keywords Dynamic equations · Nonconservative force field · Integrability · Transcendental tensor invariant

1 Introduction

It is well known [1–3] that a system of differential equations can be completely integrated when it has a sufficient number of not only first integrals (scalar invariants) but also tensor invariants. For example, the order of the considered system can be reduced if there is an invariant form of the phase volume. For conservative systems, this fact is natural. However, for systems having attracting or repelling limit sets, not only some of the first integrals, but also the coefficients of the invariant differential forms involved have to consist of, generally speaking, transcendental (in the sense of complex analysis) functions [4–6].

For example, the problem of a spatial pendulum on a spherical hinge placed in material flow leads to a system on the tangent bundle of the two-dimensional sphere with a special metric on it induced by an additional symmetry group [7]. Dynamical systems describing the motion of such a pendulum have signchanging dissipation, and the complete list of first integrals consists of transcendental functions expressed

M. V. Shamolin (✉)
Lomonosov Moscow State University, Moscow 119234, Russian Federation
e-mail: shamolin.maxim@yandex.ru; shamolin@imec.msu.ru
URL: http://shamolin2.imec.msu.ru

V. Vasilyev (ed.), *Differential Equations, Mathematical Modeling and Computational Algorithms*, Springer Proceedings in Mathematics & Statistics 423,
https://doi.org/10.1007/978-3-031-28505-9_12

in terms of a finite combination of elementary functions. There are also problems concerning the motion of a point over two-dimensional surfaces of revolution, the Lobachevsky plane, etc. The results obtained are especially important in the context of a nonconservative force field present in the system [5, 6].

Below, we present tensor invariants (differential forms) for homogeneous dynamical systems on tangent bundles of smooth two-dimensional manifolds. The relation between the existence of these invariants and the existence of a complete set of first integrals necessary for the integration of geodesic, potential, and dissipative systems is shown. The force fields introduced into the considered systems make them dissipative with dissipation of different signs and generalize previously considered force fields.

2 Example: Plane Pendulum in a Jet Flow

We describe in brief some problem on a physical pendulum on a cylindrical hinge in the flow of the incoming medium. The space of positions of such a pendulum is one-dimensional circle $\mathbf{S}^1\{\theta \bmod 2\pi\}$, and the phase space is the tangent bundle $T\mathbf{S}^1\{\dot{\theta}; \theta \bmod 2\pi\}$, i.e. two-dimensional cylinder.

Under the considered model assumptions, the equation of motion of such a pendulum is written out. statement [8] is proved that the dynamical system describing the behavior of such a pendulum is trajectorically topologically equivalent to the following differential equation on a two-dimensional cylinder (an angle θ is measured 'by the flow'):

$$\ddot{\theta} + h\dot{\theta}\cos\theta + \sin\theta\cos\theta = 0, \ h > 0. \tag{1}$$

Equation (1) can be rewritten as a system on a phase cylinder $\mathbf{R}^1\{\omega\} \times \{\alpha \bmod 2\pi\}$ $(\alpha = \theta + \pi)$:

$$\dot{\alpha} = -\omega + h\sin\alpha, \ \dot{\omega} = \sin\alpha\cos\alpha, \tag{2}$$

the phase portrait of which is shown in [7].

For $h = 0$, the conservative system (2) has a smooth first integral of energy:

$$\frac{\omega^2}{2} + \frac{\sin^2\alpha}{2} = C_0 = \text{const}, \tag{3}$$

at the same time, its phase flow preserves the area on the plane $\mathbf{R}^2\{\alpha, \omega\}$, i.e. the differential 2-form is preserved

$$d\alpha \wedge d\omega. \tag{4}$$

When integrating the system, either the first integral of energy (3) or the fact of phase area conservation (4) can be used.

In the case of $h \neq 0$ is more complicated. Since the system (2) has attractive or repulsive (asymptotic) limit sets, the first integral of the system is a transcendental (in the sense of complex analysis) function, which has the form

$$\Phi_0(\alpha, \omega) = \sin\alpha \exp \Psi_0(t) = C_1 = \text{const}, \ \Psi_0(t) = \int \frac{(t-h)dt}{t^2 - ht + 1}, \ t = \frac{\omega}{\sin\alpha}, \tag{5}$$

in this case, the asymptotic limit sets are found from the system of algebraic equalities $\sin\alpha = 0, \ \omega = 0$ (see also [9]).

Since the system (2) has asymptotic limit sets there is not even an absolutely continuous function that is the density of the measure of the phase plane (cf. with [3, 7, 8]). But it is possible (along with the first integral) to present an invariant differential 2-form with coefficients that are transcendental functions, which has the form

$$T_1(\alpha, \omega) = \exp\{-h\Psi_1(t)\}\, d\alpha \wedge d\omega, \ \Psi_1(t) = \int \frac{dt}{t^2 - ht + 1}, \ t = \frac{\omega}{\sin\alpha}. \tag{6}$$

3 Example of More General System with One Degree of Freedom

We consider the smooth dynamical system on the plane $\mathbf{R}^2\{\alpha, \omega\}$ with one degree of freedom α of the following form:

$$\dot{\alpha} = -\omega + b\delta(\alpha), \ \dot{\omega} = F(\alpha); \tag{7}$$

we can rewrite this system in the form of the equation

$$\ddot{\alpha} - b\tilde{\delta}(\alpha)\dot{\alpha} + F(\alpha) = 0, \ \tilde{\delta}(\alpha) = \frac{d\delta(\alpha)}{d\alpha}. \tag{8}$$

A pair of smooth functions $(F(\alpha), \delta(\alpha))$ defines the force field in the system: the function $F(\alpha)$ describes the conservative component of the field, and the function $\delta(\alpha)$ describes possible scattering or pumping of energy in the system. For $b = 0$, the conservative system (7) has a smooth integral of energy:

$$\frac{\omega^2}{2} + 2\int_{\alpha_0}^{\alpha} F(\xi)d\xi = C_0 = \text{const}, \tag{9}$$

at the same time, its phase flow preserves the area on the plane $\mathbf{R}^2\{\alpha, \omega\}$, i.e. the differential 2-form is preserved

$$d\alpha \wedge d\omega. \tag{10}$$

When integrating the system, either the first integral of energy (9) or the fact of phase area conservation (10) can be used.

The situation is different in the case of $b \neq 0$. Since the system (7) has, generally speaking, attractive or repulsive (asymptotic) limit sets, the first integral of the system is a transcendental (in the sense of complex analysis) [10] function. Let 's give it for the next important case:

$$F(\alpha) = \lambda\delta(\alpha)\tilde{\delta}(\alpha), \ \lambda \in \mathbf{R}. \tag{11}$$

Indeed, the first integral has the form

$$\Phi(\alpha, \omega) = \delta(\alpha) \exp \Psi(t) = C_1 = \text{const}, \ \Psi(t) = \int \frac{(t-b)dt}{t^2 - bt + \lambda}, \ t = \frac{\omega}{\delta(\alpha)}, \tag{12}$$

in this case, the asymptotic limit sets are found from the system of algebraic equalities $\delta(\alpha) = 0, \ \omega = 0$ (see also [9]).

Since asymptotic limit sets appear, there is not even an absolutely continuous function that is the density of the measure of the phase plane (cf. with [7, 8]). But it is possible (along with the first integral) to present an invariant differential 2-form with coefficients that are transcendental functions.

Indeed, the desired 2-form has the form

$$T(\alpha, \omega) = \exp\{-b\Theta(t)\} d\alpha \wedge d\omega, \ \Theta(t) = \int \frac{dt}{t^2 - bt + \lambda}, \ t = \frac{\omega}{\delta(\alpha)}. \tag{13}$$

4 Invariants of Systems of Geodesic Equations

Consider a smooth two-dimensional Riemannian manifold $M^2\{\alpha, \beta\}$ with affine connectivity $\Gamma^i_{jk}(\alpha, \beta)$ and study the structure of the equations of geodesic lines on the tangent bundle $TM^2\{\dot{\alpha}, \dot{\beta}; \alpha, \beta\}$ (cf. with [11, 12]). To do this , we will further study a fairly general case of setting kinematic relations in the following form:

$$\dot{\alpha} = z_2 f_2(\alpha), \ \dot{\beta} = z_1 f_1(\alpha), \tag{14}$$

where $f_1(\alpha)$ and $f_2(\alpha)$ are sufficiently smooth functions that are not identically zero. Such coordinates z_1, z_2 in tangent space are introduced when geodesic equations are considered, for example, with three nonzero connectivity coefficients (in particular, on surfaces of rotation, Lobachevsky plane, etc.):

$$\ddot{\alpha} + \Gamma^\alpha_{\alpha\alpha}(\alpha, \beta)\dot{\alpha}^2 + \Gamma^\alpha_{\beta\beta}(\alpha, \beta)\dot{\beta}^2 = 0, \ \ddot{\beta} + 2\Gamma^\beta_{\alpha\beta}(\alpha, \beta)\dot{\alpha}\dot{\beta} = 0, \tag{15}$$

that is, the equalities are met

$$\Gamma^{\alpha}_{\alpha\beta}(\alpha, \beta) \equiv \Gamma^{\beta}_{\alpha\alpha}(\alpha, \beta) \equiv \Gamma^{\beta}_{\beta\beta}(\alpha, \beta) \equiv 0. \tag{16}$$

In the case of (14) the relations on the tangent bundle $TM^2\{z_2, z_1; \alpha, \beta\}$ will take the form

$$\begin{aligned} \dot{z}_1 = -\tfrac{f_2^2(\alpha)}{f_1(\alpha)}\Gamma^{\beta}_{\alpha\alpha}(\alpha, \beta)z_2^2 - f_2(\alpha)\left[2\Gamma^{\beta}_{\alpha\beta}(\alpha, \beta) + \tfrac{d\ln|f_1(\alpha)|}{d\alpha}\right]z_1z_2 - \\ -f_1(\alpha)\Gamma^{\beta}_{\beta\beta}(\alpha, \beta)z_1^2, \\ \dot{z}_2 = -f_2(\alpha)\left[\Gamma^{\alpha}_{\alpha\alpha}(\alpha, \beta) + \tfrac{d\ln|f_2(\alpha)|}{d\alpha}\right]z_2^2 - f_1(\alpha)\cdot 2\Gamma^{\alpha}_{\alpha\beta}(\alpha, \beta)z_1z_2 - \\ -\tfrac{f_1^2(\alpha)}{f_2(\alpha)}\Gamma^{\alpha}_{\beta\beta}(\alpha, \beta)z_1^2, \end{aligned} \tag{17}$$

and under the conditions (16) will simplify:

$$\begin{aligned} \dot{z}_1 = -f_2(\alpha)\left[2\Gamma^{\beta}_{\alpha\beta}(\alpha, \beta) + \tfrac{d\ln|f_1(\alpha)|}{d\alpha}\right]z_1z_2, \\ \dot{z}_2 = -f_2(\alpha)\left[\Gamma^{\alpha}_{\alpha\alpha}(\alpha, \beta) + \tfrac{d\ln|f_2(\alpha)|}{d\alpha}\right]z_2^2 - \tfrac{f_1^2(\alpha)}{f_2(\alpha)}\Gamma^{\alpha}_{\beta\beta}(\alpha, \beta)z_1^2, \end{aligned} \tag{18}$$

and the Eq. (15) geodesics are almost everywhere equivalent to a composite system (14), (18) on the manifold $TM^2\{z_2, z_1; \alpha, \beta\}$ with new coordinates z_1, z_2 on the tangent space.

To fully integrate the system (14), (18) it is necessary to know, generally speaking, three independent tensor invariants: either the first three integrals, or three independent differential forms, or some combination of integrals and forms. At the same time, of course, the first integrals (in particular, for geodesic equations) can be searched for in a more general form than discussed below.

In [6, 8] examples of geodesic systems on a two-dimensional sphere with various metrics are considered, and in [12] examples of geodesic systems on two-dimensional surfaces of rotation and on the Lobachevsky plane are considered too.

Theorem 1 *If the following conditions are satisfied*

$$\begin{aligned} f_1^2(\alpha)\Gamma^{\alpha}_{\beta\beta}(\alpha, \beta) + f_2^2(\alpha)\left[2\Gamma^{\beta}_{\alpha\beta}(\alpha, \beta) + \tfrac{d\ln|f_1(\alpha)|}{d\alpha}\right] \equiv 0, \\ \Gamma^{\alpha}_{\alpha\alpha}(\alpha, \beta) + \tfrac{d\ln|f_2(\alpha)|}{d\alpha} \equiv 0, \end{aligned} \tag{19}$$

$$\Gamma^{\beta}_{\alpha\beta}(\alpha, \beta) = \Gamma^{\beta}_{\alpha\beta}(\alpha), \tag{20}$$

then the system (14), (18) has a complete set consisting of the first three integrals of the form

$$\Phi_1(z_2, z_1) = z_1^2 + z_2^2 = C_1^2 = const, \tag{21}$$

$$\Phi_2(z_1; \alpha) = z_1\Phi_0(\alpha) = C_2 = const,\ \Phi_0(\alpha) = f_1(\alpha)\exp\left\{2\int_{\alpha_0}^{\alpha}\Gamma^{\beta}_{\alpha\beta}(b)db\right\}, \tag{22}$$

$$\Phi_3(\alpha, \beta) = \beta \mp \int\limits_{\alpha_0}^{\alpha} \frac{C_2 f_1(b)}{f_2(b)\sqrt{C_1^2 \Phi_0^2(b) - C_2^2}} db = C_3 = const. \tag{23}$$

Moreover, after some reduction of that system, replacing the independent variable

$$\frac{d}{dt} = f_2(\alpha)\frac{d}{d\tau}, \tag{24}$$

and phase one

$$z_1^* = \ln|z_1|, \tag{25}$$

the phase flow of the system (14), (18) preserves the volume on the tangent bundle $TM^2\{z_2, z_1^*; \alpha, \beta\}$, i.e. the corresponding differential form is preserved:

$$dz_2 \wedge dz_1^* \wedge d\alpha \wedge d\beta. \tag{26}$$

The system (19) can be interpreted as the possibility of converting the quadratic form of the metric to a canonical form with the law of conservation of energy (21) (or see below (30)) depending on the problem under consideration. The history and current state of consideration of this more general problem are quite extensive (we note only the works of [12, 13]). Well, the search for both the integral (21) and (22) relies on the presence of additional symmetry groups in the system [5, 6].

5 Invariants of Potential Systems

We modify the system somewhat (14), (18), introducing into it a conservative smooth force field in projections on the axis $\dot{z}_1$, $\dot{z}_2$, respectively:

$$\tilde{F}(z_2, z_1; \alpha) = \begin{pmatrix} F_1(\beta) f_1(\alpha) \\ F_2(\alpha) f_2(\alpha) \end{pmatrix}. \tag{27}$$

The system under consideration on the tangent bundle $TM^2\{z_2, z_1; \alpha, \beta\}$ will take the form

$$\begin{cases} \dot{\alpha} = z_2 f_2(\alpha), \\ \dot{z}_2 = F_2(\alpha) f_2(\alpha) - f_2(\alpha)\left[\Gamma^{\alpha}_{\alpha\alpha}(\alpha, \beta) + \frac{d\ln|f_2(\alpha)|}{d\alpha}\right] z_2^2 - \\ \qquad - \frac{f_1^2(\alpha)}{f_2(\alpha)} \Gamma^{\alpha}_{\beta\beta}(\alpha, \beta) z_1^2, \\ \dot{z}_1 = F_1(\beta) f_1(\alpha) - f_2(\alpha)\left[2\Gamma^{\beta}_{\alpha\beta}(\alpha, \beta) + \frac{d\ln|f_1(\alpha)|}{d\alpha}\right] z_1 z_2, \\ \dot{\beta} = z_1 f_1(\alpha), \end{cases} \tag{28}$$

and it is almost everywhere equivalent to the following system:

$$\begin{aligned} \ddot{\alpha} - F_2(\alpha) f_2(\alpha) + \Gamma^{\alpha}_{\alpha\alpha}(\alpha, \beta)\dot{\alpha}^2 + \Gamma^{\alpha}_{\beta\beta}(\alpha, \beta)\dot{\beta}^2 = 0, \\ \ddot{\beta} - F_1(\beta) f_1(\alpha) + 2\Gamma^{\beta}_{\alpha\beta}(\alpha, \beta)\dot{\alpha}\dot{\beta} = 0, \end{aligned} \tag{29}$$

on the tangent bundle $TM^2\{\dot{\alpha}, \dot{\beta}; \alpha, \beta\}$.

Theorem 2 *If the conditions (19), (20) are satisfied, then the system (28) has a complete set consisting of the first three integrals of the form*

$$\Phi_1(z_2, z_1) = z_1^2 + z_2^2 + V(\alpha, \beta) = C_1 = const, \tag{30}$$

$$V(\alpha, \beta) = V_2(\alpha) + V_1(\beta) = -2\int\limits_{\alpha_0}^{\alpha} F_2(a)da - 2\int\limits_{\beta_0}^{\beta} F_1(b)db, \tag{31}$$

and also with $F_1(\beta) \equiv 0$—by the first integral (22) and

$$\Phi_3(\alpha, \beta) = \beta \mp \int\limits_{\alpha_0}^{\alpha} \frac{C_2 f_1(b)}{f_2(b)\sqrt{\Phi_0^2(b)[C_1 - V(b, \beta_0)] - C_2^2}} db = C_3 = const. \tag{32}$$

Moreover, after some reduction of that system, i.e. replacing the independent variable

$$\frac{d}{dt} = f_2(\alpha)\frac{d}{d\tau}, \tag{33}$$

and phase one

$$z_1^* = \ln|z_1|, \tag{34}$$

the phase flow of the system (28) preserves the volume on the tangent bundle $TM^2\{z_2, z_1^*; \alpha, \beta\}$, i.e. the corresponding differential form is preserved:

$$dz_2 \wedge dz_1^* \wedge d\alpha \wedge d\beta. \tag{35}$$

6 Invariants of Systems with Alternating Dissipation

Next, we modify the system somewhat (28) by introducing a smooth force field with dissipation into it. Its presence (generally speaking, alternating signs) characterizes not only the coefficient $b\delta(\alpha)$, $b > 0$, in the first equation of the system (37) (unlike the system (28)), but also the following dependence of the (external) force field in projections on the axis $\dot{z}_1$, $\dot{z}_2$, respectively:

$$\tilde{F}(z_2, z_1; \alpha, \beta) = \begin{pmatrix} F_1(\beta) f_1(\alpha) \\ F_2(\alpha) f_2(\alpha) \end{pmatrix} + \begin{pmatrix} z_1 F_1^1(\alpha) \\ z_2 F_2^1(\alpha) \end{pmatrix}. \tag{36}$$

The system under consideration on the tangent bundle $TM^2\{z_2, z_1; \alpha, \beta\}$ will take the form

$$\begin{cases} \dot{\alpha} = z_2 f_2(\alpha) + b\delta(\alpha), \\ \dot{z}_2 = F_2(\alpha) f_2(\alpha) - f_2(\alpha)\left[\Gamma^{\alpha}_{\alpha\alpha}(\alpha, \beta) + \frac{d\ln|f_2(\alpha)|}{d\alpha}\right] z_2^2 - \\ \quad -\frac{f_1^2(\alpha)}{f_2(\alpha)} \Gamma^{\alpha}_{\beta\beta}(\alpha, \beta) z_1^2 + z_2 F_2^1(\alpha), \\ \dot{z}_1 = F_1(\beta) f_1(\alpha) - f_2(\alpha)\left[2\Gamma^{\beta}_{\alpha\beta}(\alpha, \beta) + \frac{d\ln|f_1(\alpha)|}{d\alpha}\right] z_1 z_2 + z_1 F_1^1(\alpha), \\ \dot{\beta} = z_1 f_1(\alpha), \end{cases} \tag{37}$$

and it is almost everywhere equivalent to the following system:

$$\begin{cases} \ddot{\alpha} - \left\{b\tilde{\delta}(\alpha) + F_2^1(\alpha) + b\delta(\alpha)\left[2\Gamma^{\alpha}_{\alpha\alpha}(\alpha, \beta) + \frac{d\ln|f_2(\alpha)|}{d\alpha}\right]\right\}\dot{\alpha} - \\ -F_2(\alpha) f_2^2(\alpha) + b\delta(\alpha) F_2^1(\alpha) + b^2\delta^2(\alpha)\left[\Gamma^{\alpha}_{\alpha\alpha}(\alpha, \beta) + \frac{d\ln|f_2(\alpha)|}{d\alpha}\right] + \\ \quad +\Gamma^{\alpha}_{\alpha\alpha}(\alpha, \beta)\dot{\alpha}^2 + \Gamma^{\alpha}_{\beta\beta}(\alpha, \beta)\dot{\beta}^2 = 0, \\ \ddot{\beta} - \left\{F_1^1(\alpha) + b\delta(\alpha)\left[2\Gamma^{\beta}_{\alpha\beta}(\alpha, \beta) + \frac{d\ln|f_1(\alpha)|}{d\alpha}\right]\right\}\dot{\beta} - \\ \quad -F_1(\beta) f_1^2(\alpha) + 2\Gamma^{\beta}_{\alpha\beta}(\alpha, \beta)\dot{\alpha}\dot{\beta} = 0, \end{cases} \tag{38}$$

on the tangent bundle $TM^2\{\dot{\alpha}, \dot{\beta}; \alpha, \beta\}$. Here, as above,

$$\tilde{\delta}(\alpha) = \frac{d\delta(\alpha)}{d\alpha}. \tag{39}$$

We will integrate the fourth-order system (37) when performing the properties (19), (20), as well as when $F_1(\beta) \equiv 0$. At the same time, an independent subsystem of the third order is separated:

$$\begin{cases} \dot{\alpha} = z_2 f_2(\alpha) + b\delta(\alpha), \\ \dot{z}_2 = F_2(\alpha) f_2(\alpha) - \frac{f_1^2(\alpha)}{f_2(\alpha)} \Gamma^{\alpha}_{\beta\beta}(\alpha) z_1^2 + z_2 F_2^1(\alpha), \\ \dot{z}_1 = \frac{f_1^2(\alpha)}{f_2(\alpha)} \Gamma^{\alpha}_{\beta\beta}(\alpha) z_1 z_2 + z_1 F_1^1(\alpha), \end{cases} \tag{40}$$

if there is also a fourth equation

$$\dot{\beta} = z_1 f(\alpha). \tag{41}$$

We will also assume that for some $\kappa \in \mathbf{R}$ the equality is satisfied

$$\Gamma^{\alpha}_{\beta\beta}(\alpha)\frac{f_1^2(\alpha)}{f_2^2(\alpha)} = \kappa\frac{d}{d\alpha}\ln|\Delta(\alpha)| = \kappa\frac{\tilde{\Delta}(\alpha)}{\Delta(\alpha)},\ \tilde{\Delta}(\alpha) = \frac{d\Delta(\alpha)}{d\alpha},\ \Delta(\alpha) = \frac{\delta(\alpha)}{f_2(\alpha)}, \tag{42}$$

and for some $\lambda_2^0, \lambda_k^1 \in \mathbf{R}$, $k = 1, 2$, the equalities must be met

$$\begin{gathered} F_2(\alpha) = \lambda_2^0 \frac{d}{d\alpha}\frac{\Delta^2(\alpha)}{2} = \lambda_2^0\tilde{\Delta}(\alpha)\Delta(\alpha); \\ F_k^1(\alpha) = f_2(\alpha)\frac{d}{d\alpha}\Delta(\alpha) = \lambda_k^1\tilde{\Delta}(\alpha)f_2(\alpha),\ k = 1, 2. \end{gathered} \tag{43}$$

Condition (42) let's call it 'geometric', and the conditions from the group (43)—'energetic'.

Condition (42) it is called geometric, among other things, because it imposes a condition on the key coefficient of connectivity $\Gamma^{\alpha}_{\beta\beta}$, bringing the corresponding coefficients of the system to a homogeneous form with respect to the function $\Delta(\alpha)$. The conditions of the group (43) are called energetic, among other things, because the forces become, in a sense, 'potential' with respect to the functions of $\Delta^2(\alpha)/2$ and $\Delta(\alpha)$, bringing the corresponding coefficients of the system to a homogeneous form also with respect to the function $\Delta(\alpha)$ (see also [9]).

Theorem 3 *Let the conditions (42) and (43) be satisfied. Then the system (40), (41) has three independent, generally speaking, transcendental [4, 10] first integrals.*

In general, the first integrals are written out cumbersomely (since it is necessary to integrate the Abel equation [14]). In particular, if $\kappa = -1$, $\lambda_1^1 = \lambda_2^1$, the explicit form of the key first integral is:

$$\begin{gathered} \Theta_1(z_2, z_1; \alpha) = G_1\left(\frac{z_2}{\Delta(\alpha)}, \frac{z_1}{\Delta(\alpha)}\right) = \\ = \frac{f_2^2(\alpha)(z_2^2+z_1^2)+(b-\lambda_1^1)z_2\delta(\alpha)f_2(\alpha)-\lambda_2^0\delta^2(\alpha)}{z_1\delta(\alpha)f_2(\alpha)} = C_1 = \text{const}. \end{gathered} \tag{44}$$

In this case , the additional first integrals have the following structures:

$$\Theta_2(z_2, z_1; \alpha) = G_2\left(\Delta(\alpha), \frac{z_2}{\Delta(\alpha)}, \frac{z_1}{\Delta(\alpha)}\right) = C_2 = \text{const}, \tag{45}$$

$$\Theta_3(z_2, z_1; \alpha, \beta) = G_3\left(\Delta(\alpha), \beta, \frac{z_2}{\Delta(\alpha)}, \frac{z_1}{\Delta(\alpha)}\right) = C_3 = \text{const}. \tag{46}$$

The expression of functions (44)–(46) through a finite combination of elementary functions also depends on the explicit form of the function $\Delta(\alpha)$. So, for example, with $\kappa = -1$, $\lambda_1^1 = \lambda_2^1$ the additional first integral of the system (40) is found from the differential relation

$$
\begin{gathered}
d\ln|\Delta(\alpha)| = \frac{(b+u_2)du_2}{U_2(C_1,u_2)},\ u_2 = \frac{z_2}{\Delta(\alpha)},\ u_1 = \frac{z_1}{\Delta(\alpha)},\\
U_1(u_2) = u_2^2 + (b-\lambda_1^1)u_2 - \lambda_2^0,\\
U_2(C_1,u_2) = 2U_1(u_2) - \frac{C_1}{2}\left\{C_1 \pm \sqrt{C_1^2 - 4U_1(u_2)}\right\},\ C_1 \neq 0.
\end{gathered} \tag{47}
$$

The right part of this relation is expressed in terms of a finite combination of elementary functions, and the left—depending on the function $\Delta(\alpha)$.

Theorem 4 *If for systems of the form (40), (41) there are the first integrals of the form (44) to (46), then it also has the following three functionally independent invariant differential forms with transcendental coefficients:*

$$
\begin{gathered}
\rho_1(z_2,z_1;\alpha)dz_2\wedge dz_1\wedge d\alpha,\\
\rho_1(z_2,z_1;\alpha) = \exp\left\{(b+\lambda_1^1)\int\frac{du_2}{U_2(C_1,u_2)}\right\}\cdot\frac{u_2^2+u_1^2-(b-\lambda_1^1)u_2-\lambda_2^0}{u_1},\\
\rho_2(z_2,z_1;\alpha)dz_2\wedge dz_1\wedge d\alpha,\\
\rho_2(z_2,z_1;\alpha) = \Delta(\alpha)\exp\left\{(b+\lambda_1^1)\int\frac{du_2}{U_2(C_1,u_2)}\right\}\cdot\exp\left\{-\int\frac{(b+u_2)du_2}{U_2(C_1,u_2)}\right\},\\
\rho_3(z_2,z_1;\alpha,\beta)dz_2\wedge dz_1\wedge d\alpha\wedge d\beta,\\
\rho_3(z_2,z_1;\alpha,\beta) = \exp\left\{(b+\lambda_1^1)\int\frac{du_2}{U_2(C_1,u_2)}\right\}\cdot G_3\left(\Delta(\alpha),\beta,\frac{z_2}{\Delta(\alpha)},\frac{z_1}{\Delta(\alpha)}\right),
\end{gathered} \tag{48}
$$

but dependent with the first integrals (44)–(46).

For the complete integrability of the system (40), (41), you can use either the first three integrals, or three independent differential forms, or some combination (only independent elements) of integrals and forms (cf. with [2, 3, 15]).

On the structure of the first integrals for the systems under consideration with dissipation, see also [5, 6, 8]. Note only that for systems with dissipation, the transcendence of functions (in the sense of having essentially singular points) as the first integrals, it is inherited from the presence of attracting or repelling limit sets in the system.

In conclusion, we can refer to numerous applications concerning the integration of systems with dissipation, on the tangent bundle to a two-dimensional sphere, as well as more general systems on the bundle of two-dimensional surfaces of rotation and the Lobachevsky plane [15, 16].

7 Spatial Pendulum in the Flow of the Incoming Medium

Let us briefly describe the problem of a physical pendulum on a spherical hinge in the flow of an incoming medium, started in [8]. The position space of such a pendulum is a two-dimensional sphere $\mathbf{S}^2\{0 \le \xi \le \pi,\ \eta \bmod 2\pi\}$, phase space—tangent bundle $T\mathbf{S}^2\{\dot\xi,\dot\eta;\ 0 \le \xi \le \pi,\ \eta \bmod 2\pi\}$ to it.

Under the considered model assumptions, the equations of motion of such a pendulum are written out. Further, the statement is proved that the dynamical system describing the behavior of such a pendulum is trajectorically topologically equivalent to the following dynamical system on the tangent bundle of a two-dimensional sphere (the angle ξ is measured "along the flow"):

$$\begin{cases} \ddot{\xi} + b\dot{\xi}\cos\xi + \sin\xi\cos\xi - \dot{\eta}^2\frac{\sin\xi}{\cos\xi} = 0, \\ \ddot{\eta} + b\dot{\eta}\cos\xi + \dot{\xi}\dot{\eta}\frac{1+\cos^2\xi}{\cos\xi\sin\xi} = 0, \ b > 0. \end{cases} \tag{49}$$

The system (49) is almost everywhere equivalent to the system

$$\begin{cases} \dot{\xi} = -w_2 - b\sin\xi, \\ \dot{w}_2 = \sin\xi\cos\xi - w_1^2\frac{\cos\xi}{\sin\xi}, \\ \dot{w}_1 = w_1 w_2\frac{\cos\xi}{\sin\xi}, \end{cases} \tag{50}$$

$$\dot{\eta} = w_1\frac{\cos\xi}{\sin\xi}, \tag{51}$$

on the tangent bundle $T_*\mathbf{S}^2\{(w_2, w_1; \xi, \eta_1) \in \mathbf{R}^4 : \ 0 \le \xi \le \pi, \ \eta_1 \bmod 2\pi\}$ of two-dimensional sphere $\mathbf{S}^2\{(\xi, \eta_1) \in \mathbf{R}^2 : \ 0 \le \xi \le \pi, \ \eta_1 \bmod 2\pi\}$.

It can be seen that in the fourth-order system (50), (51), due to the cyclicity of the variable η, an independent third-order subsystem (50) is allocated, which can be independently considered on its three-dimensional manifold.

The key first integral of the system (50), (51) has the following form:

$$\Theta_1(w_2, w_1; \xi) = \frac{w_2^2 + w_1^2 + bw_2\sin\xi + \sin^2\xi}{w_1\sin\xi} = C_1 = \text{const}. \tag{52}$$

Remark 1 Consider a system (50) with variable dissipation with zero mean [5, 6, 8] becoming conservative at $b = 0$:

$$\begin{aligned} \dot{\xi} &= -w_2, \\ \dot{w}_2 &= \sin\xi\cos\xi - w_1^2\frac{\cos\alpha}{\sin\alpha}, \\ \dot{w}_1 &= w_1 w_2\frac{\cos\xi}{\sin\xi}. \end{aligned} \tag{53}$$

It has two analytic first integrals of the form

$$w_2^2 + w_1^2 + \sin^2\xi = C_1^* = \text{const}, \tag{54}$$

$$w_1\sin\xi = C_2^* = \text{const}. \tag{55}$$

Obviously, the ratio of two integrals (54), (55) it is also the first integral of the system (53). But with $b \neq 0$ each of the functions

$$w_2^2 + w_1^2 + bw_2 \sin\xi + \sin^2\xi \tag{56}$$

and (55) separately is not the first integral of the system (50). However, the ratio of functions (56), (55) is the first integral of the system (50) for any b.

The additional first integral of the system (50) is expressed in terms of a finite combination of elementary functions and has the following form (due to the bulkiness, we will write out the structural form):

$$\Theta_2(w_2, w_1; \xi) = G\left(\sin\xi, \frac{w_2}{\sin\xi}, \frac{w_1}{\sin\xi}\right) = C_2 = \text{const.} \tag{57}$$

Another (additional) first integral that 'binds' the Eq. (51) can be represented as

$$\Theta_3(w_2, w_1; \xi, \eta) = -\eta \pm \frac{1}{2}\text{arctg}\frac{w_1^2 - w_2^2 - bw_2\sin\xi - \sin^2\xi}{w_1(2w_2 + b\sin\xi)} = C_3 = \text{const.} \tag{58}$$

In the case under consideration, the system of dynamic equations (50), (51) has the first three integrals expressed by the relations (52), (57), (58), which are transcendental functions of phase variables (in the sense of complex analysis) and expressed in terms of a finite combination of elementary functions.

It is also possible to present invariant differential forms for the system of dynamic equations under consideration:

$$\begin{gathered}
\rho_1(w_2, w_1; \xi)dw_2 \wedge dw_1 \wedge d\xi,\\
\rho_1(w_2, w_1; \xi) = \exp\left\{b\int \frac{du_2}{U_2(C_1, u_2)}\right\} \cdot \frac{u_2^2 + u_1^2 + bu_2 + 1}{u_1},\\
\rho_2(w_2, w_1; \xi)dw_2 \wedge dw_1 \wedge d\xi,\\
\rho_2(w_2, w_1; \xi) = \sin\xi \exp\left\{b\int \frac{du_2}{U_2(C_1, u_2)}\right\} \cdot \exp\left\{-\int \frac{(b+u_2)du_2}{U_2(C_1, u_2)}\right\},\\
\rho_3(w_2, w_1; \xi, \eta)dw_2 \wedge dw_1 \wedge d\xi \wedge d\eta,\\
\rho_3(w_2, w_1; \xi, \eta) = \exp\left\{b\int \frac{du_2}{U_2(C_1, u_2)}\right\} \cdot \Theta_3(w_2, w_1; \xi, \eta),\\
u_2 = \frac{w_2}{\sin\xi},\ u_1 = \frac{w_1}{\sin\xi},\\
U_1(u_2) = u_2^2 + bu_2 + 1,\\
U_2(C_1, u_2) = 2U_1(u_2) - \frac{C_1}{2}\left\{C_1 \pm \sqrt{C_1^2 - 4U_1(u_2)}\right\},\ C_1 \neq 0.
\end{gathered} \tag{59}$$

References

1. Poincaré H (1912) Calcul des probabilités. Gauthier-Villars, Paris
2. Kolmogorov AN (1953) On dynamical systems with an integral invariant on the torus. Dokl Akad Nauk SSSR 93(5):763–766
3. Kozlov VV (2019) Tensor invariants and integration of differential equations. Russ Math Surv 74(1):111–140
4. Shamolin MV (1998) On integrability in transcendental functions. Russ Math Surv 53(3):637–638
5. Shamolin MV (2020) New cases of homogeneous integrable systems with dissipation on tangent bundles of twodimensional manifolds. Dokl Math 102(2):443–448
6. Shamolin MV (2020) New cases of integrable odd-order systems with dissipation. Dokl Math 101(2):158–164
7. Samsonov VA, Shamolin MV (1989) Body motion in a resisting medium. Moscow Univ Mech Bull 44(3):16–20
8. Shamolin MV (2009) Dynamical systems with variable dissipation: approaches, methods, and applications. J Math Sci 162(6):741–908
9. Kozlov VV (2015) Rational integrals of quasi-homogeneous dynamical systems. J Appl Math Mech 79(3):209–216
10. Shabat BV (1992) Introduction to complex analysis. American Mathematical Society, Providence, R.I
11. Trofimov VV (1984) Symplectic structures on groups of automorphisms of symmetric spaces. Vestn Mosk Gos Univ Ser 1: Mat Mekh (6):31–33
12. Klein F (2006) Vorlesungen uber nicht-euklidische Geometrie. VDM, Muller, Saarbr ucken
13. Weyl H (2016) Symmetry. Princeton University Press, Princeton, NJ
14. Kamke E (1959) Gewohnliche Differentialgleichungen, 5th edn. Akademie-Verlag, Leipzig
15. Trofimov VV, Shamolin MV (2012) Geometric and dynamical invariants of integrable Hamiltonian and dissipative systems. J Math Sci 180(4):365–530
16. Trofimov VV, Fomenko AT (1980) A method for constructing Hamiltonian flows on symmetric spaces and the integrability of some hydrodynamical systems. Sov Math Dokl 22:617–621

B-subharmonic Functions

Elina Shishkina

Abstract Considering different problems with Bessel operator we inevitably should obtain the main theorems of harmonic analysis for Laplace–Bessel operator. In this article we obtain condition of B-subharmonicity using the second Green's formula for the Laplace–Bessel operator.

Keywords B-subharmonic functions · Weighted spherical mean · B-harmonic functions · Laplace–Bessel operator

1 Introduction

Subharmonic functions have been introduced in the analysis Hartogs [1]. The systematic study of subharmonic functions began with the work of Riesz [2, 3], Privalov [4] and Radó [5]. It is widely known that subharmonic functions are used in the theory of surfaces of nonpositive Gaussian curvature [6], in solving boundary value problems [7], in the theory of random processes [8] and in studying analytic functions of a complex variable [4]. Now the theory of subharmonic functions is an actively developing area of modern mathematics.

In this article we introduce and proof B-subharmonicity condition. This is a part of B-harmonic analysis which provides a mathematical theory to deal with the singular Bessel differential operator of the form

$$B_{\gamma_j} = \frac{1}{x_j^{\gamma_j}} \frac{\partial}{\partial x_j} x_j^{\gamma_j} \frac{\partial}{\partial x_j} = \frac{\partial^2}{\partial x_j^2} + \frac{\gamma_j}{x_j} \frac{\partial}{\partial x_j}, \qquad j = 1, \ldots, n.$$

E. Shishkina (✉)
Voronezh State University, Universitetskaya pl., 1, Voronezh 394018, Russia
e-mail: ilina_dico@mail.ru

Belgorod State National Research University ("BelGU"), Pobedy Street, 85, Belgorod 308015, Russia

V. Vasilyev (ed.), *Differential Equations, Mathematical Modeling and Computational Algorithms*, Springer Proceedings in Mathematics & Statistics 423,
https://doi.org/10.1007/978-3-031-28505-9_13

We will use notation $\Delta_\gamma = (\Delta_\gamma)_x = \sum\limits_{k=1}^{n} (B_{\gamma_k})_{x_k}$. For Δ_γ the term *Laplace–Bessel operator* is used. A function $u = u(x) = u(x_1, \ldots, x_n)$ defined in a domain $\Omega \subset R^n$ is said to be *B–harmonic* if $u \in C^2(\Omega)$, $\frac{\partial u}{\partial x_j}|_{x_j=0} = 0$ for all $j = 1, \ldots, n$ and satisfies the Laplace–Bessel equation $\Delta_\gamma u = 0$ at every point of the domain Ω.

One can say that a function defined and continuous in some domain is B-subhartnonic if the value of this function at each point of the domain under consideration is less than or equal to its weighted spherical mean. It will be shown that B-subharmonicity of function in some domain follows from inequality $\Delta_\gamma u(x) \geq 0$ which is satisfied at all points of the considered domain.

In classical theory, the definition of subharmonic functions is often given in terms of the positivity of the Laplace operator, and then a generalized mean value theorem is derived with inequality instead of equality. For our case with the Laplace-Bessel operator, we rearrange this order and define subharmonic functions through the generalized mean value theorem with inequalities, and then derive for them a theorem about the non-negativity of the Laplace-Bessel operator.

2 Definitions

Suppose that R^n is the n-dimensional Euclidean space,

$$R^n_+ = \{x = (x_1, \ldots, x_n) \in R^n, \ \ x_1{>}0, \ldots, x_n{>}0\},$$

$$\overline{R}^n_+ = \{x = (x_1, \ldots, x_n) \in R^n, \ \ x_1{\geq}0, \ldots, x_n{\geq}0\},$$

$\gamma = (\gamma_1, \ldots, \gamma_n)$ is a multi-index consisting of positive fixed real numbers γ_i, $i = 1, \ldots, n$, and $|\gamma| = \gamma_1 + \ldots + \gamma_n$.

Let Ω be finite or infinite open set in R^n symmetric with respect to each hyperplane x_i=0, i=1, ..., n, $\Omega_+ = \Omega \cap R^n_+$ and $\overline{\Omega}_+ = \Omega \cap \overline{R}^n_+$.

We deal with the class $C^m(\Omega_+)$ consisting of m times differentiable on Ω_+ functions and denote by $C^m(\overline{\Omega}_+)$ the subset of functions from $C^m(\Omega_+)$ such that all derivatives of these functions with respect to x_i for any $i = 1, \ldots, n$ are continuous up to x_i=0. Class $C^m_{ev}(\overline{\Omega}_+)$ consists of all functions from $C^m(\overline{\Omega}_+)$ such that $\frac{\partial^{2k+1} f}{\partial x_i^{2k+1}}|_{x_i=0} = 0$ for all non-negative integer $k \leq \frac{m-1}{2}$ (see [9], p. 21).

In the following, we will denote $C^m_{ev}(\overline{R}^n_+)$ by C^m_{ev}. We set

$$C^\infty_{ev}(\overline{\Omega}_+) = \bigcap_{m=0}^{\infty} C^m_{ev}(\overline{\Omega}_+)$$

with intersection taken for all finite m and $C^\infty_{ev}(\overline{R}_+) = C^\infty_{ev}$.

The class $C_{ev}(\overline{\Omega}_+)$ is the restriction of the class of even continuous on Ω functions to $\overline{\Omega}_+$.

We will use notation $\overset{\circ}{C}{}^{\infty}_{ev}(\overline{\Omega}_+)$ for the space of all functions $f\in C^{\infty}_{ev}(\overline{\Omega}_+)$ with a compact support. We will use notations $\overset{\circ}{C}{}^{\infty}_{ev}(\overline{\Omega}_+)=\mathcal{D}_+(\overline{\Omega}_+)$ and $\overset{\circ}{C}{}^{\infty}_{ev}(\overline{R}_+)=\overset{\circ}{C}{}^{\infty}_{ev}$.

The multidimensional generalized translation is defined by the equality

$$({}^{\gamma}\mathbf{T}^{y}_{x}f)(x)={}^{\gamma}\mathbf{T}^{y}_{x}f(x)=({}^{\gamma_1}T^{y_1}_{x_1}\ldots{}^{\gamma_n}T^{y_n}_{x_n}f)(x), \tag{1}$$

where each of one-dimensional generalized translation ${}^{\gamma_i}T^{y_i}_{x_i}$ acts for i=1, ..., n according to (see [10])

$$({}^{\gamma_i}T^{y_i}_{x_i}f)(x)=\frac{\Gamma\left(\frac{\gamma_i+1}{2}\right)}{\sqrt{\pi}\,\Gamma\left(\frac{\gamma_i}{2}\right)}$$

$$\times\int\limits_0^{\pi}f(x_1,\ldots,x_{i-1},\sqrt{x_i^2+\tau_i^2-2x_iy_i\cos\varphi_i},x_{i+1},\ldots,x_n)\ \sin^{\gamma_i-1}\varphi_i\,d\varphi_i.$$

Next we will use notation

$$C(\gamma)=\pi^{-\frac{n}{2}}\prod_{i=1}^{n}\frac{\Gamma\left(\frac{\gamma_i+1}{2}\right)}{\Gamma\left(\frac{\gamma_i}{2}\right)}.$$

Part of the sphere of radius r with center at the origin belonging to R^n_+ we will denote $S^+_r(n)$:

$$S^+_r(n)=\{x\in\overline{R}{}^n_+:|x|=r\}\cup\{x\in\overline{R}{}^n_+:x_i=0,|x|\le r,i=1,\ldots,n\}.$$

For the weighed integral by the $S^+_1(n)$ we have formula [11], formula 107, p. 49

$$|S^+_1(n)|_{\gamma}=\int\limits_{S^+_1(n)}x^{\gamma}dS=\frac{\prod\limits_{i=1}^{n}\Gamma\left(\frac{\gamma_i+1}{2}\right)}{2^{n-1}\Gamma\left(\frac{n+|\gamma|}{2}\right)}. \tag{2}$$

3 *B*-harmonic Functions

In this section we will consider *B*-harmonic functions i.e. functions annihilated by the Laplace–Bessel operator in domain $\overline{\Omega}_+=\Omega\cap\overline{R}{}^n_+$.

A function $u=u(x)=u(x_1,\ldots,x_n)$ defined in a domain $\overline{\Omega}_+$ is said to be *B–harmonic* if $u\in C^2_{ev}(\overline{\Omega}_+)$ and satisfies the Laplace–Bessel equation $\Delta_{\gamma}u=0$ at every point of the domain $\overline{\Omega}_+$.

Theorem 1 *Let $x \in R^n_+$, $n > 1$ and*

$$E(x) = \begin{cases} \frac{1}{|S_1^+(n)|_\gamma} \ln |x|, & n+|\gamma| = 2; \\ \frac{|x|^{2-n-|\gamma|}}{(2-n-|\gamma|)|S_1^+(n)|_\gamma}, & n+|\gamma| > 2, \end{cases}$$

where $|S_1^+(n)|_\gamma$ is (2). Then for $|x| > \varepsilon\ \forall\varepsilon > 0$ we obtain that $E(x)$ is B-harmonic:

$$\Delta_\gamma E(x) = 0.$$

Proof Let consider first the case $n + |\gamma| > 2$. We can write

$$\Delta_\gamma E(x) = \sum_{j=1}^n B_{\gamma_j} E(x) = \sum_{j=1}^n \frac{1}{x_j^{\gamma_j}} \frac{\partial}{\partial x_j} x_j^{\gamma_j} \frac{\partial}{\partial x_j} E(x) =$$

$$= \frac{1}{(2-n-|\gamma|)|S_n^+|_\gamma} \sum_{j=1}^n \frac{1}{x_j^{\gamma_j}} \frac{\partial}{\partial x_j} x_j^{\gamma_j} \frac{\partial}{\partial x_j} |x|^{2-n-|\gamma|} =$$

$$= \frac{1}{(2-n-|\gamma|)|S_n^+|_\gamma} \sum_{j=1}^n \frac{1}{x_j^{\gamma_j}} \frac{\partial}{\partial x_j} x_j^{\gamma_j} \frac{(2-n-|\gamma|)}{2} |x|^{-n-|\gamma|} 2x_j =$$

$$= \frac{1}{|S_n^+|_\gamma} \sum_{j=1}^n \frac{1}{x_j^{\gamma_j}} \frac{\partial}{\partial x_j} |x|^{-n-|\gamma|} x_j^{1+\gamma_j} =$$

$$= \frac{1}{|S_n^+|_\gamma} \sum_{j=1}^n \frac{1}{x_j^{\gamma_j}} \left[\frac{(-n-|\gamma|)}{2} |x|^{-n-|\gamma|-2} 2x_j^{2+\gamma_j} + (1+\gamma_j)|x|^{-n-|\gamma|} x_j^{\gamma_j} \right] =$$

$$= \frac{1}{|S_n^+|_\gamma} \sum_{j=1}^n [(-n-|\gamma|)|x|^{-n-|\gamma|-2} x_j^2 + (1+\gamma_j)|x|^{-n-|\gamma|}] =$$

$$= \frac{1}{|S_n^+|_\gamma} [(-n-|\gamma|)|x|^{-n-|\gamma|} + (n+|\gamma|)|x|^{-n-|\gamma|}] = 0.$$

Now consider the case $n + |\gamma| = 2$:

$$\Delta_\gamma E(x) = \sum_{j=1}^n B_{\gamma_j} E(x) = \sum_{j=1}^n \frac{1}{x_j^{\gamma_j}} \frac{\partial}{\partial x_j} x_j^{\gamma_j} \frac{\partial}{\partial x_j} E(x) =$$

$$= \frac{1}{|S_n^+|_\gamma} \sum_{j=1}^{n} \frac{1}{x_j^{\gamma_j}} \frac{\partial}{\partial x_j} x_j^{\gamma_j} \frac{\partial}{\partial x_j} \ln|x| = \frac{1}{|S_n^+|_\gamma} \sum_{j=1}^{n} \frac{1}{x_j^{\gamma_j}} \frac{\partial}{\partial x_j} |x|^{-2} x_j^{1+\gamma_j} =$$

$$= \frac{1}{|S_n^+|_\gamma} \sum_{j=1}^{n} \frac{1}{x_j^{\gamma_j}} [-2|x|^{-4} x_j^{2+\gamma_j} + (1+\gamma_j)|x|^{-2} x_j^{\gamma_j}] =$$

$$= \frac{1}{|S_n^+|_\gamma} \sum_{j=1}^{n} [-2|x|^{-4} x_j^2 + (1+\gamma_j)|x|^{-2}] =$$

$$= \frac{1}{|S_n^+|_\gamma} [-2|x|^{-2} + (n+|\gamma|)|x|^{-2}] = 0,$$

because $n + |\gamma| = 2$.

4 Weighted Spherical Mean

In *B*-harmonic analysis when constructing a weighted spherical mean, instead of the usual shift, a multidimensional generalized translation (1) is used.

Weighted spherical mean (see [11–13]) of function $u(x)$, $x \in \overline{R}^n_+$ for $n \geq 2$ is

$$(M_r^\gamma u)(x) = (M_r^\gamma)_x [u(x)] = \frac{1}{|S_1^+(n)|_\gamma} \int\limits_{S_1^+(n)} {}^\gamma\mathbf{T}_x^{r\theta} u(x) \theta^\gamma dS, \tag{3}$$

where $\theta^\gamma = \prod\limits_{i=1}^{n} \theta_i^{\gamma_i}$.

Weighted spherical mean has properties

$$(M_r^\gamma u)(x)|_{r=0} = u(x), \qquad \frac{\partial}{\partial r} (M_r^\gamma u)(x) \bigg|_{r=0} = 0. \tag{4}$$

In the classical case, the transition from integration over a unit sphere centered at the origin to a sphere centered at a point x^0 of radius r is carried out by a simple linear change of coordinates. In our case, the presence of a generalized translation significantly complicates such a transition. Let's consider this point in more detail.

We will transform $(M_t^\gamma u)(x)$ so that the center of the part of the sphere over which the integration takes place moves. In this case, the dimension of the space will double. We have

$$(M_r^\gamma u)(x) = \frac{C(\gamma)}{|S_1^+(n)|_\gamma} \times$$

$$\times \int\limits_{S_1^+(n)} \int\limits_0^{\pi} \ldots \int\limits_0^{\pi} u(\sqrt{x_1^2 - 2rx_1\theta_1 \cos\beta_1 + r^2\theta_1^2}, \ldots, \sqrt{x_n^2 - 2rx_n\theta_n \cos\beta_1 + r^2\theta_n^2}) \times$$

$$\times \prod_{i=1}^{n} \sin^{\gamma_i - 1} \beta_i \, d\beta \theta^{\gamma} dS.$$

One can convert this integral into integral by the part of sphere in R^{2n} by using formulas

$$\begin{gathered} \widetilde{\theta}_1 = r\theta_1 \cos\beta_1, \qquad \widetilde{\theta}_2 = r\theta_1 \sin\beta_1, \\ \widetilde{\theta}_3 = r\theta_2 \cos\beta_2, \qquad \widetilde{\theta}_4 = r\theta_2 \sin\beta_2, \ldots, \\ \widetilde{\theta}_{2n-1} = r\theta_n \cos\beta_n, \qquad \widetilde{\theta}_{2n} = r\theta_n \sin\beta_n. \end{gathered} \tag{5}$$

We obtain

$$(M_r^{\gamma} u)(x) = \frac{C(\gamma)}{|S_1^+(n)|_{\gamma} r^{n+|\gamma|-1}} \times$$

$$\times \int\limits_{\widetilde{S}_r^+(2n)} u(\sqrt{(x_1 - \widetilde{\theta}_1)^2 + \widetilde{\theta}_2^2}, \ldots, \sqrt{(x_n - \widetilde{\theta}_{2n-1})^2 + \widetilde{\theta}_{2n}^2}) \prod_{i=1}^{n} \widetilde{\theta}_{2i}^{\gamma_i - 1} d\widetilde{S} =$$

$$= \frac{C(\gamma)}{|S_1^+(n)|_{\gamma} r^{n+|\gamma|-1}} \int\limits_{\widetilde{S}_{r,x}^+(2n)} u(\sqrt{z_1^2 + \widetilde{\theta}_2^2}, \ldots, \sqrt{z_{2n-1}^2 + \widetilde{\theta}_{2n}^2}) \prod_{i=1}^{n} \widetilde{\theta}_{2i}^{\gamma_i - 1} d\widetilde{S}',$$

where we put $\{\widetilde{\theta}_{2i-1} - x_i = z_{2i-1}, i = 1, \ldots, n\}$. Here $\widetilde{\theta}_{2i} > 0, i = 1, \ldots, n,$

$$\widetilde{S}_r^+(2n) = \{\widetilde{\theta} \in R^{2n} : |\widetilde{\theta}| = r\}$$

and

$$\widetilde{S}_{r,x}^+(2n) =$$

$$= \{(z_1, \widetilde{\theta}_2, \ldots, z_{2n-1}, \widetilde{\theta}_{2n}) \in R^{2n} : (z_1 - x_1)^2 + \widetilde{\theta}_2^2 + \cdots + (z_{2n-1} - x_n)^2 + \widetilde{\theta}_{2n}^2 = r^2\},$$

differentials $d\widetilde{S}$ and $d\widetilde{S}'$ mean that we are integrating over a surfaces $\widetilde{S}_r^+(2n)$ and $\widetilde{S}_{r,x}^+(2n)$ respectively.

Let now $z_{2i-1} = \theta_i \cos\beta_i$ $\widetilde{\theta}_{2i} = \theta_i \sin\beta_i$, $i = 1, \ldots, n$. We can write

$$(M_r^{\gamma} u)(x) = \frac{C(\gamma)}{|S_1^+(n)|_{\gamma} r^{n+|\gamma|-1}} \int\limits_0^{\pi} \ldots \int\limits_0^{\pi} \left(\int\limits_{\widetilde{\widetilde{S}}_{r,x}^+(n)} u(\theta) \theta^{\gamma} dS \right) \prod_{i=1}^{n} \sin^{\gamma_i - 1} \beta_i \, d\beta, \tag{6}$$

where $\widetilde{\widetilde{S^+_{r,x}}}(n)$ is a sphere (or a part of sphere) $(\theta_1\cos\beta_1-x_1)^2+\theta_1^2\sin^2\beta_1+\cdots+(\theta_n\cos\beta_n-x_n)^2+\theta_n^2\sin^2\beta_n=r^2$. To simplify the right part of (6) we introduce the next notation

$$\int\limits_{{}^\gamma\mathbf{T}^x_\theta S^+_r(n)} u(\theta)\theta^\gamma dS=C(\gamma)\int\limits_0^\pi\ldots\int\limits_0^\pi\left(\int\limits_{\widetilde{\widetilde{S^+_{r,x}}}(n)} u(\theta)\theta^\gamma dS\right)\prod_{i=1}^n\sin^{\gamma_i-1}\beta_i\,d\beta$$

so we can write

$$(M^\gamma_r u)(x)=\frac{1}{|S^+_1(n)|_\gamma r^{n+|\gamma|-1}}\int\limits_{{}^\gamma\mathbf{T}^x_\theta S^+_{r,x}(n)} u(\theta)\theta^\gamma dS. \tag{7}$$

5 *B*-subharmonic Functions

In this section we define the *B*-subharmonic function and prove that if Laplace-Bessel operator of a sufficiently smooth function is non-negative in domain then this function is *B*-subharmonic.

Let $u\in C_{ev}(\overline{\Omega}_+)$. We say that a function u is B-subharmonic if

$$u(x^0)\le(M^\gamma_r u)(x^0)=\frac{1}{|S^+_1(n)|_\gamma}\int\limits_{S^+_1(n)} {}^\gamma\mathbf{T}^{r\theta}_{x^0}u(x^0)\theta^\gamma dS$$

whenever the part of the sphere $\{x\in R^n_+:|x-x_0|\le r\}$ is contained in $\overline{\Omega}_+$.

Theorem 2 *Suppose $u\in C^2_{ev}(\overline{\Omega}_+)$ and $\Delta_\gamma u(x)\ge 0$ for all $x\in\overline{\Omega}_+$, then $u(x)$ B-subhartnonic at all points of $\overline{\Omega}_+$.*

Proof Let x^0 is any point of $\overline{\Omega}_+$,

$$v(x)=\begin{cases}-\ln|x-x^0|+\ln r, & \text{n+}|\gamma|=2s;\\ |x-x^0|^{2-n-|\gamma|}-r^{2-n-|\gamma|}, & \text{n+}|\gamma|>2,\end{cases}$$

is *B*-harmonic function by Theorem 1 in $\overline{\Omega}_+$: $\Delta_\gamma v=0$, $v(x)\ge 0$.

We consider $\theta\in R^n_+$,

$$I(x)=C(\gamma)\int\limits_0^\pi\ldots\int\limits_0^\pi\left(\int\limits_{\widetilde{G}^+}(u(\theta)\Delta_\gamma v(\theta)-v(\theta)\Delta_\gamma u(\theta))\,\theta^\gamma d\theta\right)\prod_{i=1}^n\sin^{\gamma_i-1}\beta_i\,d\beta,$$

where $\widetilde{G}^+$ the shell domain between

$$(\theta_1 \cos\beta_1 - x_1^0)^2 + \theta_1^2 \sin^2\beta_1 + \dots + (\theta_n \cos\beta_n - x_n^0)^2 + \theta_n^2 \sin^2\beta_n = \varepsilon^2$$

and

$$(\theta_1 \cos\beta_1 - x_1^0)^2 + \theta_1^2 \sin^2\beta_1 + \dots + (\theta_n \cos\beta_n - x_n^0)^2 + \theta_n^2 \sin^2\beta_n = r^2.$$

Numbers ε and r satisfy inequalities $0 < \varepsilon < r$ chosen so that set $\widetilde{G}^+$ lies entirely in $\overline{\Omega}_+$. The boundary of $\widetilde{G}^+$ can include parts of the coordinate plains.

Since $\Delta_\gamma v = 0$, $v(x) \geq 0$ and $\Delta_\gamma u(x) \geq 0$ for all $x \in \overline{\Omega}_+$ and $\widetilde{G}^+ \subseteq \overline{\Omega}_+$ we get

$$0 \geq I(x) = C(\gamma) \int_0^\pi \dots \int_0^\pi \left(\int_{\widetilde{G}^+} (u(\theta)\Delta_\gamma v(\theta) - v(\theta)\Delta_\gamma u(\theta))\, \theta^\gamma d\theta \right) \prod_{i=1}^n \sin^{\gamma_i - 1} \beta_i \, d\beta,$$

The second Green's formula for the Laplace–Bessel operator (see [14]) is

$$0 \geq I = C(\gamma) \int_0^\pi \dots \int_0^\pi \left(\int_{\partial\widetilde{G}^+} \left(u\frac{\partial v}{\partial \nu} - v\frac{\partial u}{\partial \nu} \right) \theta^\gamma \, dS \right) \prod_{i=1}^n \sin^{\gamma_i - 1} \beta_i \, d\beta,$$

where $\partial\widetilde{G}^+$ the boundary of $\widetilde{G}^+$, ν is a normal vector of the surface $\partial\widetilde{G}^+$.

In new coordinates

$$z_1 = \theta_1 \cos\beta_1, \qquad z_2 = \theta_1 \sin\beta_1,$$

$$z_3 = \theta_2 \cos\beta_2, \qquad z_4 = \theta_2 \sin\beta_2, \dots,$$

$$z_{2n-1} = \theta_n \cos\beta_n, \qquad z_{2n} = \theta_n \sin\beta_n,$$

such that $z_{2i} > 0$, $i = 1, ..., n$, we can write

$$0 \geq I = C(\gamma) \int_{\partial\widetilde{W}^+} \left(\widetilde{u}\frac{\partial \widetilde{v}}{\partial \widetilde{\nu}} - \widetilde{v}\frac{\partial \widetilde{u}}{\partial \widetilde{\nu}} \right) \prod_{i=1}^n z_{2i}^{\gamma_i - 1} \, d\widetilde{S},$$

where $\widetilde{u} = u\left(\sqrt{z_1^2 + z_2^2}, ..., \sqrt{z_{2n-1}^2 + z_{2n}^2}\right)$, $\widetilde{v} = v\left(\sqrt{z_1^2 + z_2^2}, ..., \sqrt{z_{2n-1}^2 + z_{2n}^2}\right)$, $\partial\widetilde{W}^+$ is a surface consisted of two spheres (or a parts of spheres in R^{2n}) with center at $\xi \in R^{2n}$, $\xi = (x_1, 0, x_2, 0, ..., x_{2n-1}, 0)$ of radii ε and r such that $0 < \varepsilon < r$:

$$\widetilde{S}^+_{\varepsilon,\xi}(2n) =$$

$$= \{z \in R^{2n} : (z_1 - x_1)^2 + z_2 + \dots + (z_{2n-1} - x_n)^2 + z_{2n}^2 = \varepsilon^2\},$$

$$\widetilde{S}^+_{r,\xi}(2n) =$$

$$= \{z \in R^{2n} : (z_1 - x_1)^2 + z_2 + \cdots + (z_{2n-1} - x_n)^2 + z_{2n}^2 = r^2\}$$

and possibly parts of coordinate plains, $\widetilde{\nu}$ is is a normal vector of the surface $\partial\widetilde{W}^+$, $d\widetilde{S}$ is the element of the surface $\partial\widetilde{W}^+$. Therefore,

$$0 \geq I = C(\gamma)\left[\left(\int\limits_{\widetilde{S}^+_{\varepsilon,\xi}(2n)} + \int\limits_{\widetilde{S}^+_{r,\xi}(2n)}\right)\widetilde{u}\frac{\partial\widetilde{v}}{\partial\widetilde{\nu}}\prod_{i=1}^{n} z_{2i}^{\gamma_i-1}\,d\widetilde{S} - \right.$$

$$\left. - \left(\int\limits_{\widetilde{S}^+_{\varepsilon,\xi}(2n)} + \int\limits_{\widetilde{S}^+_{r,\xi}(2n)}\right)\widetilde{v}\frac{\partial\widetilde{u}}{\partial\widetilde{\nu}}\prod_{i=1}^{n} z_{2i}^{\gamma_i-1}\,d\widetilde{S}\right].$$

On $\widetilde{S}^+_{r,\xi}(2n)$ we have $\widetilde{v} = 0$. Also, since $\Delta_\gamma u \geq 0$ and $\widetilde{\nu}$ is directed toward the center of the $\widetilde{S}^+_{\varepsilon,\xi}$ we get that $\int\limits_{\widetilde{S}^+_{\varepsilon,\xi}} \widetilde{v}\frac{\partial\widetilde{u}}{\partial\widetilde{\nu}}\prod\limits_{i=1}^{n} z_{2i}^{\gamma_i-1}\,d\widetilde{S} \leq 0$.

That means that

$$0 \geq C(\gamma)\left(\int\limits_{\widetilde{S}^+_{\varepsilon,\xi}(2n)} + \int\limits_{\widetilde{S}^+_{r,\xi}(2n)}\right)\widetilde{u}\frac{\partial\widetilde{v}}{\partial\widetilde{\nu}}\prod_{i=1}^{n} z_{2i}^{\gamma_i-1}\,d\widetilde{S}.$$

For $n + |\gamma| = 2$ we get

$$0 \geq C(\gamma)\left(\int\limits_{\widetilde{S}^+_{\varepsilon,\xi}(2n)} - \int\limits_{\widetilde{S}^+_{r,\xi}(2n)}\right)\frac{\widetilde{u}(z)}{|z-\xi|}\prod_{i=1}^{n} z_{2i}^{\gamma_i-1}\,d\widetilde{S}$$

and for $n + |\gamma| > 2$ we get

$$0 \geq C(\gamma)(n + |\gamma| - 2)\left(\int\limits_{\widetilde{S}^+_{\varepsilon,\xi}(2n)} - \int\limits_{\widetilde{S}^+_{r,\xi}(2n)}\right)\frac{\widetilde{u}(z)}{|z-\xi|^{n+|\gamma|-1}}\prod_{i=1}^{n} z_{2i}^{\gamma_i-1}\,d\widetilde{S},$$

where $\xi \in R^{2n}$, $\xi = (x_1, 0, x_2, 0, ..., x_{2n-1}, 0)$. In either case,

$$C(\gamma)\int\limits_{\widetilde{S}^+_{\varepsilon,\xi}(2n)}\frac{\widetilde{u}(z)}{|z-\xi|^{n+|\gamma|-1}}\prod_{i=1}^{n} z_{2i}^{\gamma_i-1}\,d\widetilde{S} \leq C(\gamma)\int\limits_{\widetilde{S}^+_{r,\xi}(2n)}\frac{\widetilde{u}(z)}{|z-\xi|^{n+|\gamma|-1}}\prod_{i=1}^{n} z_{2i}^{\gamma_i-1}\,d\widetilde{S}$$

or

$$\frac{C(\gamma)}{\varepsilon^{n+|\gamma|-1}}\int\limits_{\widetilde{S}^+_{\varepsilon,\xi}(2n)}\widetilde{u}(z)\prod_{i=1}^{n} z_{2i}^{\gamma_i-1}\,d\widetilde{S}\leq\frac{C(\gamma)}{r^{n+|\gamma|-1}}\int\limits_{\widetilde{S}^+_{r,\xi}(2n)}\widetilde{u}(z)\prod_{i=1}^{n} z_{2i}^{\gamma_i-1}\,d\widetilde{S}.$$

Returning to coordinates $\theta_1, ..., \theta_n$ by formulas $z_{2i-1} = \theta_i \cos\beta_i$ $\widetilde{\theta}_{2i} = \theta_i \sin\beta_i$, $i = 1, ..., n$ we obtain

$$\frac{1}{|S_1^+(n)|_\gamma \varepsilon^{n+|\gamma|-1}}\int\limits_{{}^\gamma\mathbf{T}^x_\theta S^+_{\varepsilon,x^0}(n)} u(\theta)\theta^\gamma dS\leq\frac{1}{|S_1^+(n)|_\gamma r^{n+|\gamma|-1}}\int\limits_{{}^\gamma\mathbf{T}^x_\theta S^+_{r,x^0}(n)} u(\theta)\theta^\gamma dS$$

or, using (7),

$$(M^\gamma_\varepsilon u)(x^0)=\frac{1}{|S_1^+(n)|_\gamma}\int\limits_{S_1^+(n)} {}^\gamma\mathbf{T}^{\varepsilon\theta}_{x^0} u(x^0)\theta^\gamma dS\leq$$

$$\leq\frac{1}{|S_1^+(n)|_\gamma}\int\limits_{S_1^+(n)} {}^\gamma\mathbf{T}^{r\theta}_{x^0} u(x^0)\theta^\gamma dS=(M^\gamma_r u)(x^0).$$

Letting ε tend to 0 the left side tends to $u(x^0)$ by (4) and we obtain inequality

$$u(x^0)\leq(M^\gamma_r u)(x^0).$$

Notes and Comments. There are a lot of properties of B-subharmonic functions need to prove. For example, it is interesting to consider the maximum principle, criterion of B-harmonicity in terms of B-subharmonic functions, the Perron method for solving the Dirichlet problem for Laplace-Bessel operator, the connection to the B-potential theory (for the B-potential theory see [15, 16]), Harnack inequality for singular equations and other.

References

1. Hartogs F (1906) Zur Theorie der analytischen Funktionen mehrerer unabhängiger Veränderlichen, insbesondere über die Darstellung derselben durch Reihen welche nach Potenzen einer Veränderlichen fortschreiten. Math Ann 62:1–88
2. Riesz F (1926) Sur les fonctions sousharmoniques et leur rapport à la théorie du potentiel I. Acta Math 48:329–343
3. Riesz F (1930) Sur les fonctions sousharmoniques et leur rapport à la théorie du potentiel II. Acta Math 54:321–360
4. Privalov II (1937) Subharmonic functions. ONTI NKTP USSR, Moscow-Leningrad
5. Radó T (1937) Subharmonic functions. Springer, Berlin, Heidelberg
6. Greene RE, Wu H (1974) Integrals of subharmonic functions on manifolds of nonnegative curvature. Invent Math 27(4):265–298. https://doi.org/10.1007/BF01425500

7. Gilbarg D, Trudinger NS (2001) Elliptic partial differential equations of second order, 2nd edn. Springer, Berlin, Heidelberg
8. Doob JL (1953) Stochastic Processes. Wiley, New York
9. Kipriyanov IA (1997) Singular elliptic boundary value problems. Nauka, Moscow
10. Levitan BM (1951) Expansion in Fourier series and integrals with Bessel functions. (Russian) Uspehi Matem Nauk (N.S.) 6(2)(42):102–143
11. Shishkina EL, Sitnik SM (2020) Transmutations, singular and fractional differential equations with applications to mathematical physics. Elsevier, Amsterdam
12. Lyakhov LN, Polovinkin IP, Shishkina EL (2014) On a Kipriyanov problem for a singular ultra-hyperbolic equation. Differ Equ 50(4):513–525. https://doi.org/10.1134/S0012266114040090
13. Lyakhov LN, Polovinkin IP, Shishkina EL (2014) Formulas for the solution of the Cauchy problem for a singular wave equation with Bessel time operator. Dokl Akad Nauk 459(5):533–538. https://doi.org/10.1134/S106456241407028X
14. Shishkina EL (2019) Generalized divergent theorem and second Green identity of B-elliptic and B-hyperbolic operators. Belgorod State Univ Sci Bull Math Phys 51(4):506–513 (2019). https://doi.org/10.18413/2075-4639-2019-51-4-506-513
15. Lyakhov LN (1992) Inversion of Riesz B-potentials. Dokl Math 44(3):717–720
16. Lyakhov LN (1991) A class of hypersingular integrals. Dokl Math 42(3):765–769

Some Multi–dimensional Modified G- and H-Integral Transforms on $\mathfrak{L}_{\overline{\nu},\overline{r}}$-Spaces

S. M. Sitnik, O. V. Skoromnik, and M. V. Papkopvich

Abstract This paper is devoted to the study of three classes of multidimensional integral transformations with Fox' H-function and the Meijer's G-function in kernels in weighted spaces integrable functions in the domain $R_+^n = R_+^1 \times R_+^1 \times \cdots \times R_+^1$. Mapping properties such as the boundedness, the rang, the representation and the inversion of the considered transforms are established.

Keywords Multidimensional integral transformations with Meijer's G-function and Fox' H-function in the kernels · Multidimensional Mellin transform · Weighted space of summable functions · Fractional integrals and derivatives

MSC Primary 44A30 · Secondary 33C60 · 35A22

1 Introduction

Multidimensional integral transformations are considered (see [1], formula (40); [2], formulas (1.1)–(1.2):

$$\left(\mathbf{H}^1_{\sigma,\kappa} f\right)(\mathbf{x}) = \mathbf{x}^{\sigma} \int\limits_{\mathbf{0}}^{\mathbf{x}} \mathbf{H}^{\mathbf{m},\mathbf{n}}_{\mathbf{p},\mathbf{q}}\left[\frac{\mathbf{x}}{\mathbf{t}} \middle| \begin{array}{c} (\mathbf{a}_i, \alpha_i)_{1,p} \\ (\mathbf{b}_j, \beta_j)_{1,q} \end{array}\right] \mathbf{t}^{\kappa} f(\mathbf{t}) \frac{\mathbf{dt}}{\mathbf{t}} (\mathbf{x} > 0); \tag{1}$$

$$\left(\mathbf{G}^1_{\sigma,\kappa} f\right)(\mathbf{x}) = \mathbf{x}^{\sigma} \int\limits_{\mathbf{0}}^{\mathbf{x}} \mathbf{G}^{\mathbf{m},\mathbf{n}}_{\mathbf{p},\mathbf{q}}\left[\frac{\mathbf{x}}{\mathbf{t}} \middle| \begin{array}{c} (\mathbf{a}_i)_{1,p} \\ (\mathbf{b}_j)_{1,q} \end{array}\right] \mathbf{t}^{\kappa} f(\mathbf{t}) \frac{\mathbf{dt}}{\mathbf{t}} (\mathbf{x} > 0); \tag{2}$$

S. M. Sitnik (✉)
Belgorod State National Research University (BSU), Belgorod, Russia
e-mail: sitnik@bsu.edu.ru

O. V. Skoromnik · M. V. Papkopvich
Polotsk State University, Navapolatsk, Belorussia
e-mail: mpapkovich@yandex.by

V. Vasilyev (ed.), *Differential Equations, Mathematical Modeling and Computational Algorithms*, Springer Proceedings in Mathematics & Statistics 423,
https://doi.org/10.1007/978-3-031-28505-9_14

$$\left(\mathbf{G}^{1}_{\sigma,\kappa;\,\delta}f\right)(\mathbf{x}) = \mathbf{x}^{\sigma}\int\limits_{\mathbf{0}}^{\mathbf{x}} \mathbf{G}^{\mathbf{m},\,\mathbf{n}}_{\mathbf{p},\,\mathbf{q}}\left[\frac{\boldsymbol{x}^{\delta}}{\boldsymbol{t}^{\delta}}\,\middle|\,\begin{matrix}(\mathbf{a}_i)_{1,p}\\(\mathbf{b}_j)_{1,q}\end{matrix}\right]\mathbf{t}^{\kappa}f(\mathbf{t})\frac{\mathbf{dt}}{\mathbf{t}}(\mathbf{x}>0); \tag{3}$$

here (see, for example, [1, 2]; [3, Sect. 28.4]; [4, 5]) $\mathbf{x} = (x_1, x_2, ..., x_n) \in \mathrm{R}^n$; $\mathbf{t} = (t_1, t_2, ..., t_n) \in \mathrm{R}^n$, R^n be the n-dimensional Euclidean space; $\mathbf{x}\cdot\mathbf{t} = \sum\limits_{n=1}^{n} x_n t_n$ denotes their scalar product; in particular, $\mathbf{x}\cdot\mathbf{1} = \sum\limits_{n=1}^{n} x_n$ for $\mathbf{1}=(1,1,...,1)$. The expression $\mathbf{x} > \mathbf{t}$ means that $x_1 > t_1, x_2 > t_2, ..., x_n > t_n$, similarly for signs $\geq$, $<$, $\leq$; $\int\limits_{\mathbf{0}}^{\mathbf{x}} = \int\limits_{0}^{x_1}\int\limits_{0}^{x_2}\cdots\int\limits_{0}^{x_n}$; $\int\limits_{\mathbf{0}}^{\infty} = \int\limits_{0}^{\infty}\int\limits_{0}^{\infty}\cdots\int\limits_{0}^{\infty}$; by $\mathrm{N} = \{1, 2, ...\}$ we denote the set of positive integers, $\mathrm{N}_0 = \mathrm{N}\bigcup\{0\}$, $\mathrm{N}_0^n = \mathrm{N}_0\times\mathrm{N}_0\times ...\times\mathrm{N}_0$; $\mathbf{k} = (k_1, k_2, ..., k_n) \in \mathrm{N}_0^n = \mathrm{N}_0 \times ... \times \mathrm{N}_0$ ($k_i \in \mathrm{N}_0$, $i = 1, 2, ..., n$) is a multi-index with $\mathbf{k}! = k_1!\cdots k_n!$ and $|\mathbf{k}| = k_1 + k_2 + ... + k_n$; $\mathrm{R}^n_+ = \{\mathbf{x}\in\mathrm{R}^n,\ \mathbf{x} > 0\}$; for $l = (l_1, l_2, ..., l_n) \in \mathrm{R}^n_+$ $\mathbf{D}^l = \frac{\partial^{|l|}}{(\partial x_1)^{l_1}\cdots(\partial x_n)^{l_n}}$; $\mathbf{dt} = dt_1\cdot dt_2\cdots dt_n$; $\mathbf{t}^l = t^{l_1}t^{l_2}\cdots t^{l_n}$; $f(\mathbf{t}) = f(t_1, t_2, ..., t_n)$. Let C^n ($n\in\mathrm{N}$) be the n-dimensional space of n complex numbers $z = (z_1, z_2, \cdots, z_n)$ ($z_j \in \mathrm{C}$, $j = 1, 2, \cdots, n$);

$\mathbf{m} = (m_1, m_2, ..., m_n) \in \mathrm{N}_0^n$ and $m_1 = m_2 = ... = m_n$; $\mathbf{n} = (\overline{n}_1, \overline{n}_2, ..., \overline{n}_n) \in \mathrm{N}_0^n$ and $\overline{n}_1 = \overline{n}_2 = ... = \overline{n}_n$; $\mathbf{p} = (p_1, p_2, ..., p_n) \in \mathrm{N}_0$ and $p_1 = p_2 = ... = p_n$; $\mathbf{q} = (q_1, q_2, ..., q_n) \in \mathrm{N}_0$ and $q_1 = q_2 = ... = q_n$) ($0 \leq \mathbf{m} \leq \mathbf{q}$, $0 \leq \mathbf{n} \leq \mathbf{p}$);

$\sigma = (\sigma_1, \sigma_2, ..., \sigma_n) \in \mathrm{C}^n$; $\kappa = (\kappa_1, \kappa_2, ..., \kappa_n) \in \mathrm{C}^n$; $\delta = (\delta_1, \delta_2, ..., \delta_n) \in \mathrm{R}^n_+$;

$\mathbf{a}_i = (a_{i_1}, a_{i_2}, ..., a_{i_n})$, $1 \leq i \leq p$, $a_{i_1}, a_{i_2}, ..., a_{i_n} \in \mathrm{C}$ ($1 \leq i_1 \leq p_1, ..., 1 \leq i_n \leq p_n$);

$\mathbf{b}_j = (b_{j_1}, b_{j_2}, ..., b_{j_n})$, $1 \leq j \leq q$, $b_{j_1}, b_{j_2}, ..., b_{j_n} \in \mathrm{C}$ ($1 \leq j_1 \leq q_1, ..., 1 \leq j_n \leq q_n$);

$\alpha_i = (\alpha_{i_1}, \alpha_{i_2}, ..., \alpha_{i_n})$, $1 \leq i \leq p$, $\alpha_{i_1}, \alpha_{i_2}, ..., \alpha_{i_n} \in \mathrm{R}^+_1$ ($1 \leq i_1 \leq p_1, ..., 1 \leq i_n \leq p_n$);

$\beta_j = (\beta_{j_1}, \beta_{j_2}, ..., \beta_{j_n})$, $1 \leq j \leq q$, $\beta_{j_1}, \beta_{j_2}, ..., \beta_{j_n} \in \mathrm{R}^+_1$ ($1 \leq j_1 \leq q_1, ..., 1 \leq j_n \leq q_n$).

We introduce the function

$$\mathrm{H}^{\mathbf{m},\,\mathbf{n}}_{\mathbf{p},\,\mathbf{q}}\left[\frac{\mathbf{x}}{\mathbf{t}}\,\middle|\,\begin{matrix}(\mathbf{a}_i,\alpha_i)_{1,p}\\(\mathbf{b}_j,\beta_j)_{1,q}\end{matrix}\right] = \prod_{k=1}^{n} H^{m_k,\,\overline{n}_k}_{p_k,\,q_k}\left[\frac{x_k}{t_k}\,\middle|\,\begin{matrix}(a_{i_k},\alpha_{i_k})_{1,p_k}\\(b_{j_k},\beta_{j_k})_{1,q_k}\end{matrix}\right], \tag{4}$$

which is the product of H-functions $H^{m,\,n}_{p,\,q}[z]$:

$$H^{m,n}_{p,q}[z] \equiv H^{m,n}_{p,q}\left[z\,\middle|\,\begin{matrix}(a_i,\alpha_i)_{1,p}\\(b_j,\beta_j)_{1,q}\end{matrix}\right] = \frac{1}{2\pi i}\int\limits_{L}\mathcal{H}^{m,n}_{p,q}(s)z^{-s}ds,\ \ z\neq 0, \tag{5}$$

where

$$\mathcal{H}^{m,n}_{p,q}(s) \equiv \mathcal{H}^{m,n}_{p,q}\left[\begin{matrix}(a_i,\alpha_i)_{1,p}\\(b_j,\beta_j)_{1,q}\end{matrix}\middle| s\right] = \frac{\prod\limits_{j=1}^{m}\Gamma(b_j+\beta_j s)\prod\limits_{i=1}^{n}\Gamma(1-a_i-\alpha_i s)}{\prod\limits_{i=n+1}^{p}\Gamma(a_i+\alpha_i s)\prod\limits_{j=m+1}^{q}\Gamma(1-b_j-\beta_j s)}; \tag{6}$$

and the function $\mathbf{G}^{\mathbf{m},\mathbf{n}}_{\mathbf{p},\mathbf{q}}\left[\mathbf{z}\middle|\begin{matrix}(\mathbf{a}_i)_{1,p}\\(\mathbf{b}_j)_{1,q}\end{matrix}\right] = \prod\limits_{k=1}^{n} G^{m_k,\bar{n}_k}_{p_k,q_k}\left[z_k\middle|\begin{matrix}(a_{i_k})_{1,p_k}\\(b_{j_k})_{1,q_k}\end{matrix}\right]$, which is a product of G- functions $G^{m,n}_{p,q}[z]$:

$$G^{m,n}_{p,q}[z] \equiv G^{m,n}_{p,q}\left[z\middle|\begin{matrix}(a_i)_{1,p}\\(b_j)_{1,q}\end{matrix}\right] = \frac{1}{2\pi i}\int\limits_L \mathcal{G}^{m,n}_{p,q}(s) z^{-s} ds, \; z \neq 0, \tag{7}$$

where

$$\mathcal{G}^{m,n}_{p,q}(s) \equiv \mathcal{G}^{m,n}_{p,q}\left[\begin{matrix}(a_i)_{1,p}\\(b_j)_{1,q}\end{matrix}\middle| s\right] = \frac{\prod\limits_{j=1}^{m}\Gamma(b_j+s)\prod\limits_{i=1}^{n}\Gamma(1-a_i-s)}{\prod\limits_{i=n+1}^{p}\Gamma(a_i+s)\prod\limits_{j=m+1}^{q}\Gamma(1-b_j-s)}. \tag{8}$$

In (5) and (7) L is a specially chosen infinite contour and empty product, if it occurs, being taken to be one. H-function is the most general of known special functions and includes as private cases elementary functions, Meijer special functions hypergeometric and Bessel types, as well as the G- function (7) obtained from H-function (5) with $\alpha_1 = \alpha_2 = ... = \alpha_p = \beta_1 = \beta_2 = ... = \beta_q = 1$ ([6, Sect. 2.9]). Modern theory of H- and G- functions (5), (7) is presented in Chaps. 1–2 of the monograph [6]. With elements of theory H-function and its special cases can also be found in books [7–10].

The theory of integral transformations has been intensively developing recently. This is due to the fact that integral transformations often arise both in problems of mathematics and in applied problems of physics, mechanics and other natural sciences. The use of integral transformations in the theory of differential and integral equations, operational calculus, and the theory of boundary value problems makes it possible to find their solutions in a closed form and study their structural properties.

The classical Fourier, Mellin and Hankel transformations are the most studied [11–14]. Such transformations are widely used in various problems of mathematical physics and applied mathematics [15–18]. They also find application in solving various model problems for partial differential equations.

Since the 70s of the twentieth century, the solution of various applied non-specific problems has led to the presentation of their solutions in the form of integral transformations with special functions in kernels. The interest in such transformations is also caused by the study of the corresponding integral equations of the first kind and the so-called pair and triple integral equations, which are often found in applications. The results in this direction are presented in books ([3, Chap. 7]; [19–22]).

Integral transformations with various special functions in kernels have been studied by many authors. They were mainly studied in the spaces L_1 and L_2 [11–13, 23, 24] and in some spaces of generalized functions [25, 26]. These monographs are devoted to one-dimensional integral transformations. Note also that separate results for multidimensional classical integral transformations and some transformations with special functions in kernels are presented in the monograph [27].

Most integral transformations with special functions in the kernels contain hypergeometric and Bessel type functions. In particular, the solution of the problems of axisymmetric potential theory is presented in the form of integral transformations with special Bessel functions of the second kind $\mathrm{Y}_\eta(z)$ and Struve $\mathrm{H}_\eta(z)$ in the kernels. For the first time, such constructions were considered by E. Titchmarsh [11] as a pair of mutually inverse transformations within the L_2- spaces. The questions of the action of these transformations in the spaces of r - summable functions $\mathfrak{L}_{\nu,r}$, $1 \leq r < \infty$, on the real semi-axis with power weight were studied by P. Rooney [28–30], Heywood and Rooney [31, 32] based on the theory of Mellin transformations. At the same time, they obtained analogues of the Parseval equality and gave a description of the spaces of functions represented by such integral operators. Similar questions for integral transformations with the Meijer G-function were studied by P. Rooney [33].

In 1993–1998 Kilbas and Saigo developed a theory integral H–transformations with special functions of general type in kernels, namely H–functions, in spaces $\mathfrak{L}_{\nu,r}$ of summable functions. Applying the technique of Mellin transformation and taking into account the asymptotic properties of the H- function, was constructed a theory of integral H- transformations with such functions in kernels in the spaces $\mathfrak{L}_{\nu,r}$ of summable functions with weight. Results presented in [6].

This paper presents properties of multidimensional transformations (1)–(3) in weighted spaces $\mathfrak{L}_{\overline{\nu},\overline{r}}$ of integrable functions $f(\mathbf{x}) = f(x_1, x_2, ..., x_n)$ on $\mathbf{R}^{\mathbf{n}}_+$, for which

$$\|f\|_{\overline{\nu},\overline{r}} = \left\{ \int_{R^1_+} x_n^{\nu_n \cdot r_n - 1} \left\{ \cdots \left\{ \int_{R^1_+} x_2^{\nu_2 \cdot r_2 - 1} \right.\right.\right.$$

$$\left.\left.\left. \times \left[\int_{R^1_+} x_1^{\nu_1 \cdot r_1 - 1} |f(x_1, ..., x_n)|^{r_1} dx_1 \right]^{r_2/r_1} dx_2 \right\}^{r_3/r_2} \cdots \right\}^{r_n/r_{n-1}} dx_n \right\}^{1/r_n} < \infty$$

($\overline{r} = (r_1, r_2, ..., r_n) \in \mathbf{R}^n, 1 < \overline{r} < \infty, r_1 = r_2 = ... = r_n; \overline{\nu} = (\nu_1, \nu_2, ..., \nu_n) \in \mathbf{R}^n$, $\nu_1 = \nu_2 = ... = \nu_n$).

In particular, for $\overline{\nu} = (\nu_1, \nu_2, ..., \nu_n) \in \mathbf{R}^n$, $\nu_1 = \nu_2 = ... = \nu_n$, and $\overline{2} = (2, 2, ..., 2)$ by $\mathfrak{L}_{\overline{\nu},\overline{2}}$ denote the weighted space of integrable functions $f(\mathbf{x}) = f(x_1, x_2, ..., x_n)$ on $\mathbf{R}^{\mathbf{n}}_+$ (see [1, 2]):

$$\|f\|_{\overline{\nu},\overline{2}} = \left\{ \int_{R^1_+} x_n^{\nu_n \cdot 2 - 1} \left\{ \cdots \right.\right.$$

$$\left\{\int_{R^1_+} x_2^{\nu_2 \cdot 2-1}\left[\int_{R^1_+} x_1^{\nu_1 \cdot 2-1}|f(x_1, ..., x_n)|^2 dx_1\right] dx_2\right\} \cdots \left.\right\} dx_n \Big\}^{1/2} < \infty.$$

For the transformations under consideration, various integral representations and inversion formulas are derived.

Research results for transformations (1), (2) generalize those obtained earlier for the corresponding one-dimensional cases ([6, Chaps. 5 and 6]).

2 Preliminaries

Denote by $[X, Y]$ a set of bounded linear operators acting from a Banach space X into a Banach space Y.

Multidimensional Mellin integral transform $(\mathfrak{M} f)(\mathbf{x})$ of function $f(\mathbf{x}) = f(x_1, x_2, ..., x_n)$, $\mathbf{x} = (x_1, x_2, ..., x_n) \in \mathrm{R}^n_+$, is determined by the formula

$$(\mathfrak{M} f)(\mathbf{s}) = \int_0^\infty f(\mathbf{t})\mathbf{t}^{\mathbf{s}-1}\mathbf{dt}, \;\; \mathrm{Re}(\mathbf{s}) = \overline{\nu}, \tag{9}$$

$\mathbf{s} = (s_1, s_2, ..., s_n) \in \mathrm{C}^n$; the inverse Mellin transform is given for $\mathbf{x} \in \mathrm{R}^n_+$ by the formula

$$(\mathfrak{M}^{-1} g)(\mathbf{x}) = \mathfrak{M}^{-1}[g(\mathbf{p})](\mathbf{x}) = \frac{1}{(2\pi i)^n}\int_{\gamma_1 - i\infty}^{\gamma_1 + i\infty} \cdots \int_{\gamma_n - i\infty}^{\gamma_n + i\infty} \mathbf{x}^{-\mathbf{s}} g(\mathbf{s})\mathbf{ds}, \tag{10}$$

$\gamma_j = \mathrm{Re}(s_j)$ $(j = 1, \cdots, n)$. The theory of multidimensional integral transformations (9), (10) can be recognized, for example, in books ([27]; [5, Chap. 1]). Let $\mathbf{N}_\varrho$ be elementary operator (see [5, Chap. 1]):

$$(\mathbf{N}_\varrho f)(\mathbf{x}) = f(\mathbf{x}^\varrho) \;\; (\mathbf{x} \in \mathrm{R}^n, \;\; \varrho = (\varrho_1, \varrho_2, ..., \varrho_n) \in \mathrm{R}^n, \;\; \varrho \neq 0). \tag{11}$$

This operator has the properties [[1]; [2], Lemma 2.1].

Lemma 2.1 *Let* $\overline{\nu} = (\nu_1, \nu_2, ..., \nu_n) \in \mathrm{R}^n$ $(\nu_1 = \nu_2 = ... = \nu_n)$ *and* $1 \le \overline{r} < \infty$. $\mathbf{N}_\varrho$ *is a bounded isomorphism* $\mathfrak{L}_{\overline{\nu},\overline{r}}$ *on* $\mathfrak{L}_{\varrho\overline{\nu},\overline{r}}$; *if* $f \in \mathfrak{L}_{\overline{\nu},\overline{r}}$ $(1 \le \overline{r} \le 2)$, *then*

$$(\mathfrak{M}\mathbf{N}_\varrho f)(\mathbf{s}) = \frac{1}{|\varrho|}(\mathfrak{M} f)(\frac{\mathbf{s}}{\varrho}) \;\; (\mathrm{Re}(\mathbf{s}) = \varrho\overline{\nu}).$$

Let $\mathrm{I}^\alpha_{0+;\,\sigma,\eta}$ *and* $\mathrm{I}^\alpha_{-;\,\sigma,\eta}$ *be the multidimensional Erdelyi-Kober operators of fractional integration, defined for* $\alpha = (\alpha_1, \alpha_2, ..., \alpha_n) \in \mathrm{C}^n$ $(\mathrm{Re}(\alpha) > 0)$, $\sigma > 0$, $\eta \in \mathrm{C}^n$ *by (see [1]; [2]):*

$$\left(\mathbf{I}^{\alpha}_{0+;\,\sigma,\,\eta}f\right)(\mathbf{x}) = \frac{\sigma\mathbf{x}^{-\sigma(\alpha+\eta)}}{\Gamma(\alpha)}\int\limits_{0}^{\mathbf{x}}\left(\mathbf{x}^{\sigma}-\mathbf{t}^{\sigma}\right)^{\alpha-1}\mathbf{t}^{\sigma\eta+\sigma-1}f(\mathbf{t})\mathbf{dt}\ (\mathbf{x}>0); \qquad (12)$$

$$\left(\mathbf{I}^{\alpha}_{-;\,\sigma,\,\eta}f\right)(\mathbf{x}) = \frac{\sigma\mathbf{x}^{\sigma\eta}}{\Gamma(\alpha)}\int\limits_{\mathbf{x}}^{\infty}\left(\mathbf{t}^{\sigma}-\mathbf{x}^{\sigma}\right)^{\alpha-1}\mathbf{t}^{\sigma(1-\alpha-\eta)-1}f(\mathbf{t})\mathbf{dt}\ (\mathbf{x}>0). \qquad (13)$$

3 $\mathfrak{L}_{\overline{\nu},\overline{r}}$-Theory And the Inversion Formulas for the Modified $\mathbf{H}^{1}_{\sigma,\kappa}$-Transform

To formulate the results presented $\mathfrak{L}_{\overline{\nu},\overline{2}}$-, $\mathfrak{L}_{\overline{\nu},\overline{r}}$-theories and the inversion formulas of the $\mathrm{H}^{1}_{\sigma,\kappa}$-transform (1) we need the following constants ([1], (57)–(60)), analogical for one-dimensional case defined via the parameters of the H - function (5) ([6], (3.4.1), (3.4.2), (1.1.7), (1.1.8), (1.1.10)):

$\overline{\alpha} = (\overline{\alpha}_1, \overline{\alpha}_2, ..., \overline{\alpha}_n)$ and $\overline{\beta} = (\overline{\beta}_1, \overline{\beta}_2, ..., \overline{\beta}_n)$, where

$$\overline{\alpha}_1 = \begin{cases} -\min\limits_{1\le j_1\le m_1}\left[\frac{\mathrm{Re}(b_{j_1})}{\beta_{j_1}}\right], & m_1>0,\\ -\infty, & m_1=0,\end{cases} \qquad \overline{\beta}_1 = \begin{cases} \min\limits_{1\le i_1\le \overline{n}_1}\left[\frac{1-\mathrm{Re}(a_{i_1})}{\alpha_{i_1}}\right], & \overline{n}_1>0,\\ \infty, & \overline{n}_1=0,\end{cases}$$

$$\overline{\alpha}_2 = \begin{cases} -\min\limits_{1\le j_2\le m_2}\left[\frac{\mathrm{Re}(b_{j_2})}{\beta_{j_2}}\right], & m_2>0,\\ -\infty, & m_2=0,\end{cases} \qquad \overline{\beta}_2 = \begin{cases} \min\limits_{1\le i_2\le \overline{n}_2}\left[\frac{1-\mathrm{Re}(a_{i_2})}{\alpha_{i_2}}\right], & \overline{n}_2>0,\\ \infty, & \overline{n}_2=0,\end{cases}$$

and so on

$$\overline{\alpha}_n = \begin{cases} -\min\limits_{1\le j_n\le m_n}\left[\frac{\mathrm{Re}(b_{j_n})}{\beta_{j_n}}\right], & m_n>0,\\ -\infty, & m_2=0,\end{cases} \qquad \overline{\beta}_n = \begin{cases} \min\limits_{1\le i_n\le \overline{n}_n}\left[\frac{1-\mathrm{Re}(a_{i_n})}{\alpha_{i_n}}\right], & \overline{n}_n>0,\\ \infty, & \overline{n}_n=0;\end{cases} \qquad (14)$$

$$a_1^* = \sum_{i=1}^{\overline{n}_1}\alpha_{i_1} - \sum_{i=\overline{n}_1+1}^{p_1}\alpha_{i_1} + \sum_{j=1}^{m_1}\beta_{j_1} - \sum_{j=m_1+1}^{q_1}\beta_{j_1},\quad \Delta_1 = \sum_{j=1}^{q_1}\beta_{j_1} - \sum_{i=1}^{p_1}\alpha_{i_1},$$

$$a_2^* = \sum_{i=1}^{\overline{n}_2}\alpha_{i_2} - \sum_{i=\overline{n}_2+1}^{p_2}\alpha_{i_2} + \sum_{j=1}^{m_2}\beta_{j_2} - \sum_{j=m_1+1}^{q_2}\beta_{j_2},\quad \Delta_2 = \sum_{j=1}^{q_2}\beta_{j_2} - \sum_{i=1}^{p_2}\alpha_{i_2},$$

and so on

$$a_n^* = \sum_{i=1}^{\overline{n}_n} \alpha_{i_n} - \sum_{i=\overline{n}_n+1}^{p_n} \alpha_{i_n} + \sum_{j=1}^{m_n} \beta_{j_n} - \sum_{j=m_n+1}^{q_n} \beta_{j_n}, \quad \Delta_n = \sum_{j=1}^{q_n} \beta_{j_n} - \sum_{i=1}^{p_n} \alpha_{i_n}; \quad (15)$$

$$\overline{\mu} = (\mu_1, \mu_2, ..., \mu_n),$$

where

$$\mu_1 = \sum_{j=1}^{q_1} b_{j_1} - \sum_{i=1}^{p_1} a_{i_1} + \frac{p_1 - q_1}{2}, \mu_2 = \sum_{j=1}^{q_2} b_{j_2} - \sum_{i=1}^{p_2} a_{i_2} + \frac{p_2 - q_2}{2}, ...,$$

$$\mu_n = \sum_{j=1}^{q_n} b_{j_n} - \sum_{i=1}^{p_n} a_{i_n} + \frac{p_n - q_n}{2}; \quad (16)$$

$$\alpha_0^1 = \begin{cases} 1 + \max\limits_{m_1+1 \le j_1 \le q_1} \left[\frac{\text{Re}(b_{j_1})-1}{\beta_{j_1}} \right], & q_1 > m_1, \\ -\infty, & q_1 = m_1, \end{cases}$$

$$\beta_0^1 = \begin{cases} 1 + \min\limits_{\overline{n}_1+1 \le i_1 \le p_1} \left[\frac{\text{Re}(a_{i_1})}{\alpha_{i_1}} \right], & p_1 > \overline{n}_1, \\ \infty, & p_1 = \overline{n}_1, \end{cases}$$

$$\alpha_0^2 = \begin{cases} 1 + \max\limits_{m_2+1 \le j_2 \le q_2} \left[\frac{\text{Re}(b_{j_2})-1}{\beta_{j_2}} \right], & q_2 > m_2, \\ -\infty, & q_2 = m_2, \end{cases}$$

$$\beta_0^2 = \begin{cases} 1 + \min\limits_{\overline{n}_2+1 \le i_2 \le p_2} \left[\frac{\text{Re}(a_{i_2})}{\alpha_{i_2}} \right], & p_2 > \overline{n}_2, \\ \infty, & p_2 = \overline{n}_2, \end{cases}$$

and so on

$$\alpha_0^n = \begin{cases} 1 + \max\limits_{m_n+1 \le j_n \le q_n} \left[\frac{\text{Re}(b_{j_n})-1}{\beta_{j_n}} \right], & q_n > m_n, \\ -\infty, & q_n = m_n, \end{cases}$$

$$\beta_0^n = \begin{cases} 1 + \min\limits_{\overline{n}_n+1 \le i_n \le p_2} \left[\frac{\text{Re}(a_{i_n})}{\alpha_{i_n}} \right], & p_n > \overline{n}_n, \\ \infty, & p_n = \overline{n}_n. \end{cases} \quad (17)$$

The exceptional set $\mathcal{E}_{\overline{\mathcal{H}}}$ of a function $\overline{\mathcal{H}}_{\mathbf{p},\mathbf{q}}^{\mathbf{m},\mathbf{n}}(\mathbf{s})$:

$$\overline{\mathcal{H}}_{\mathbf{p},\mathbf{q}}^{\mathbf{m},\mathbf{n}}(\mathbf{s}) \equiv \overline{\mathcal{H}}_{\mathbf{p},\mathbf{q}}^{\mathbf{m},\mathbf{n}} \left[\begin{matrix} (\mathbf{a}_i, \alpha_i)_{1,\mathbf{p}} \\ (\mathbf{b}_j, \beta_j)_{1,\mathbf{q}} \end{matrix} \middle| \mathbf{s} \right] = \prod_{k=1}^{n} \mathcal{H}_{p_k, q_k}^{m_k, \overline{n}_k} \left[\begin{matrix} (a_{i_k}, \alpha_{i_k})_{1,p_k} \\ (b_{j_k}, \beta_{j_k})_{1,q_k} \end{matrix} \middle| s \right], \quad (18)$$

is called a set of vectors $\overline{\nu} = (\nu_1, \nu_2, ..., \nu_n) \in \mathrm{R}^n$ $(\nu_1 = \nu_2 = ... = \nu_n)$, such that $\overline{\alpha}_1 < 1 - \nu_1 < \overline{\beta}_1$, $\overline{\alpha}_2 < 1 - \nu_2 < \overline{\beta}_2$, ..., $\overline{\alpha}_n < 1 - \nu_n < \overline{\beta}_n$, and functions $\mathcal{H}^{m_1,\overline{n}_1}_{p_1,q_1}(s_1)$, $\mathcal{H}^{m_2,\overline{n}_2}_{p_2,q_2}(s_2)$,...,$\mathcal{H}^{m_n,\overline{n}_n}_{p_n,q_n}(s_n)$ have zeros on lines $\mathrm{Re}(s_1) < 1 - \nu_1$, $\mathrm{Re}(s_2) < 1 - \nu_2$, ..., $\mathrm{Re}(s_n) < 1 - \nu_n$, respectively (see [1, (61)]).

Applying multidimensional Mellin transform (9) to (1), taking into account the results for the one-dimensional case ([6, Formulae (5.1.14)]), in the work [1] we obtained:

$$(\mathfrak{M}\mathrm{H}^1_{\sigma,\kappa} f)(\mathbf{s}) = \overline{\mathcal{H}}^{\mathbf{m},\mathbf{n}}_{\mathbf{p},\mathbf{q}}\left[\begin{array}{c}(\mathbf{a}_i, \alpha_i)_{1,\mathbf{p}}\\ (\mathbf{b}_j, \beta_j)_{1,\mathbf{q}}\end{array}\middle| \mathbf{s} + \sigma\right](\mathfrak{M}f)(\mathbf{s} + \sigma + \kappa). \tag{19}$$

The following assertion presents the $\mathfrak{L}_{\overline{\nu},\overline{2}}$-theory of the modified H-transform (1). One dimensional case see in ([6, Theorem 5.37]).

Theorem 3.1 ([1, Theorem 9]) *Let*
$\overline{\alpha}_1 < \nu_1 - \mathrm{Re}(\kappa_1) < \overline{\beta}_1, \overline{\alpha}_2 < \nu_2 - \mathrm{Re}(\kappa_2) < \overline{\beta}_2, \ldots, \overline{\alpha}_n < \nu_n - \mathrm{Re}(\kappa_1) < \overline{\beta}_n$,
$\nu_1 = \nu_2 = ... = \nu_n$;

$$a_1^* = 0, a_2^* = 0, ..., a_n^* = 0; \Delta_1[\nu_1 - \mathrm{Re}(\kappa_1)] + \mathrm{Re}(\mu_1) \le 0,$$

$$\Delta_2[\nu_2 - \mathrm{Re}(\kappa_2)] + \mathrm{Re}(\mu_2) \le 0, ..., \Delta_n[\nu_n - \mathrm{Re}(\kappa_n)] + \mathrm{Re}(\mu_n) \le 0. \tag{20}$$

There hold the following assertions:

(a) There exists a one-to-one map $\mathrm{H}^1_{\sigma,\kappa} \in [\mathfrak{L}_{\overline{\nu},\overline{2}}, \mathfrak{L}_{\overline{\nu}-\mathrm{Re}(\kappa+\sigma),\overline{2}}]$ *such the relation (19) holds for* $f \in \mathfrak{L}_{\overline{\nu},\overline{2}}$ *and* $\mathrm{Re}(\mathbf{s}) = \overline{\nu} - \mathrm{Re}(\kappa + \sigma)$.
If $a_1^* = 0$, $a_2^* = 0$, ..., $a_n^* = 0$; $\Delta_1[\nu_1 - \mathrm{Re}(\kappa_1)] + \mathrm{Re}(\mu_1) = 0$, $\Delta_2[\nu_2 - \mathrm{Re}(\kappa_2)] + \mathrm{Re}(\mu_2) = 0$, ..., $\Delta_n[\nu_n - \mathrm{Re}(\kappa_n)] + \mathrm{Re}(\mu_n) = 0$ *and* $1 - \overline{\nu} + \mathrm{Re}(\kappa) \notin \mathcal{E}_{\overline{\mathcal{H}}}$, *then* $\mathrm{H}^1_{\sigma,\kappa}$ *maps* $\mathfrak{L}_{\overline{\nu},\overline{2}}$ *onto* $\mathfrak{L}_{\overline{\nu}-\mathrm{Re}(\kappa+\sigma),\overline{2}}$.

(b) The transform $\mathrm{H}^1_{\sigma,\kappa}$ *does not depend on* $\overline{\nu}$ *in the sense if* $\overline{\nu}$ *and* $\widetilde{\overline{\nu}}$ *satisfy Eq. (20) and if the transforms* $\mathrm{H}^1_{\sigma,\kappa}$ *and* $\widetilde{\mathrm{H}}^1_{\sigma,\kappa}$ *are defined in respective spaces* $\mathfrak{L}_{\overline{\nu},\overline{2}}$ *and* $\mathfrak{L}_{\widetilde{\overline{\nu}},\overline{2}}$ *by Eq. (19), then* $\mathrm{H}^1_{\sigma,\kappa} = \widetilde{\mathrm{H}}^1_{\sigma,\kappa}$ *for* $f \in \mathfrak{L}_{\widetilde{\overline{\nu}},\overline{2}} \bigcap \mathfrak{L}_{\overline{\nu},\overline{2}}$.

(c) If $a_1^* = 0$, $a_2^* = 0$, ..., $a_n^* = 0$; $\Delta_1[\nu_1 - \mathrm{Re}(\kappa_1)] + \mathrm{Re}(\mu_1) < 0$, $\Delta_2[\nu_2 - \mathrm{Re}(\kappa_2)] + \mathrm{Re}(\mu_2) < 0$, ..., $\Delta_n[\nu_n - \mathrm{Re}(\kappa_n)] + \mathrm{Re}(\mu_n) < 0$; *then for* $f \in \mathfrak{L}_{\overline{\nu},\overline{2}}$ *the transform* $\mathrm{H}^1_{\sigma,\kappa} f$ *is given by Eq. (1).*

(d) Let $\overline{\lambda} = (\lambda_1, ..., \lambda_n) \in \mathrm{C}^n$, $\overline{h} = (h_1, ..., h_n) > 0$, *and* $f \in \mathfrak{L}_{\overline{\nu},\overline{2}}$. *If* $\mathrm{Re}(\overline{\lambda}) > (\overline{\nu} - \mathrm{Re}(\kappa))\overline{h} - 1$, *then* $\mathrm{H}^1_{\sigma,\kappa} f$ *is represented in the form*

$$\left(\mathrm{H}^1_{\sigma,\kappa} f\right)(\mathbf{x}) = \overline{h}\mathbf{x}^{\sigma+1-(\overline{\lambda}+1)/\overline{h}} \frac{\mathbf{d}}{\mathbf{dx}} \mathbf{x}^{(\overline{\lambda}+1)/\overline{h}} \times$$

$$\times \int\limits_0^\infty \mathrm{H}^{\mathbf{m},\mathbf{n}+1}_{\mathbf{p}+1,\mathbf{q}+1}\left[\frac{\mathbf{x}}{\mathbf{t}}\middle| \begin{array}{c}(-\overline{\lambda}, \overline{h}), (\mathbf{a}_i, \alpha_i)_{1,\mathbf{p}}\\ (\mathbf{b}_j, \beta_j)_{1,\mathbf{q}}, (-\overline{\lambda} - 1, \overline{h})\end{array}\right]\mathbf{t}^{\kappa-1} f(\mathbf{t})\mathbf{dt}, \tag{21}$$

while for $\mathrm{Re}(\overline{\lambda}) < (\overline{\nu} - \mathrm{Re}(k))\overline{h} - 1$ *is given by*

$$(\mathrm{H}^1_{\sigma,\kappa} f)(\mathbf{x}) = -\overline{h}\mathbf{x}^{\sigma+1-(\overline{\lambda}+1)/\overline{h}} \frac{\mathbf{d}}{\mathbf{dx}} \mathbf{x}^{(\overline{\lambda}+1)/\overline{h}} \times$$

$$\times \int_0^\infty \mathrm{H}^{\mathbf{m+1,n}}_{\mathbf{p+1,q+1}} \left[\frac{\mathbf{x}}{\mathbf{t}} \middle| \begin{matrix} (\mathbf{a}_i, \alpha_i)_{1,\mathbf{p}}, (-\overline{\lambda}, \overline{h}) \\ (-\overline{\lambda}-1, \overline{h}), (\mathbf{b}_j, \beta_j)_{1,\mathbf{q}} \end{matrix} \right] \mathbf{t}^{\kappa-1} f(\mathbf{t})\mathbf{dt}. \tag{22}$$

(e) If $f \in \mathfrak{L}_{\overline{\nu},\overline{2}}$ *and* $g \in \mathfrak{L}_{1-\overline{\nu}+\mathrm{Re}(\kappa+\sigma),\overline{2}}$, *then there holds the relation :*

$$\int_0^\infty f(\mathbf{x})(\mathrm{H}^1_{\sigma,\kappa} g)(\mathbf{x})\mathbf{dx} = \int_0^\infty (\mathrm{H}^2_{\sigma,\kappa} f)(\mathbf{x}) g(\mathbf{x}) d\mathbf{x}, \tag{23}$$

where

$$(\mathrm{H}^2_{\sigma,\kappa} f)(\mathbf{x}) = \mathbf{x}^\sigma \int_0^\infty \mathrm{H}^{\mathbf{m,n}}_{\mathbf{p,q}} \left[\frac{\mathbf{t}}{\mathbf{x}} \middle| \begin{matrix} (\mathbf{a}_i, \alpha_i)_{1,\mathbf{p}} \\ (\mathbf{b}_j, \beta_j)_{1,\mathbf{q}} \end{matrix} \right] \mathbf{t}^\kappa f(\mathbf{t}) \frac{\mathbf{dt}}{\mathbf{x}}. \tag{24}$$

Taking into account the results for the multidimensional case in [4], Theorems 4.1–4.2 and Theorems 5.38–5.39 in [6], Lemma 2.1, Theorem 3.1, we present $\mathfrak{L}_{\overline{\nu},\overline{r}}$-*theory of the modified* $\mathrm{H}^1_{\sigma,\kappa}$-*transform for two cases, when* $a_1^* = a_2^* = \ldots = a_n^* = 0$, $\Delta_1 = \Delta_2 = \ldots = \Delta_n = 0$, $\mathrm{Re}(\mu_1) = \mathrm{Re}(\mu_2) = \ldots = \mathrm{Re}(\mu_n) = 0$ *and* $a_1^* = a_2^* = \ldots = a_n^* = 0$, $\Delta_1 = \Delta_2 = \ldots = \Delta_n = 0$, $\mathrm{Re}(\mu_1) < 0, \mathrm{Re}(\mu_2) < 0, \ldots, \mathrm{Re}(\mu_n) < 0$.

Theorem 3.2 *Let*

$$a_1^* = 0, a_2^* = 0, \ldots, a_n^* = 0; \Delta_1 = \Delta_2 = \ldots = \Delta_n = 0;$$

$$\mathrm{Re}(\mu_1) = \mathrm{Re}(\mu_2) = \ldots = \mathrm{Re}(\mu_n) = 0;$$

$$\overline{\alpha}_1 < \nu_1 - \mathrm{Re}(\kappa_1) < \overline{\beta}_1, \overline{\alpha}_2 < \nu_1 - \mathrm{Re}(\kappa_2) < \overline{\beta}_2, \ldots, \overline{\alpha}_n < \nu_n - \mathrm{Re}(\kappa_n) < \overline{\beta}_n,$$

$$\nu_1 = \nu_2 = \ldots = \nu_n; 1 < \overline{r} < \infty, r_1 = r_2 = \ldots = r_n.$$

There the following assertions are true:

(a) The transform $\mathrm{H}^1_{\sigma,\kappa}$ *defined on* $\mathfrak{L}_{\overline{\nu},\overline{2}}$ *can be extended to* $\mathfrak{L}_{\overline{\nu},\overline{r}}$ *as an element of* $\mathrm{H}^1_{\sigma,\kappa} \in [\mathfrak{L}_{\overline{\nu},\overline{r}}, \mathfrak{L}_{\overline{\nu}-\mathrm{Re}(\kappa+\sigma),\overline{r}}]$. *If* $1 < \overline{r} \leq \overline{2}$, *then the transform* $\mathrm{H}^1_{\sigma,\kappa}$ *is one-to-one and there holds the equality (19) for* $f \in \mathfrak{L}_{\overline{\nu},\overline{r}}$ *and* $\mathrm{Re}(s) = \overline{\nu} - \mathrm{Re}(\kappa + \sigma)$.

(b) If $f \in \mathfrak{L}_{\overline{\nu},\overline{r}}$, $\overline{\lambda} = (\lambda_1, \ldots, \lambda_n) \in \mathrm{C}^n$ *and* $\overline{h} = (h_1, \ldots, h_n) > 0$, *then* $\mathrm{H}^1_{\sigma,\kappa} f$ *is represented in the form (21) for* $\mathrm{Re}(\overline{\lambda}) > (\overline{\nu} - \mathrm{Re}(\kappa))\overline{h} - 1$ *and in the form (22) for* $\mathrm{Re}(\overline{\lambda}) < (\overline{\nu} - \mathrm{Re}(k))\overline{h} - 1$.

(c) If $f \in \mathfrak{L}_{\overline{\nu},\overline{r}}$ *and* $g \in \mathfrak{L}_{1-\overline{\nu}+\mathrm{Re}(\kappa+\sigma),\overline{r}'}$ *with* $\overline{r}' = \overline{r}/(\overline{r} - 1)$, *then the relation (23) holds.*

(d) If $1-\overline{\nu}+\mathrm{Re}(\kappa)\notin\mathcal{E}_{\overline{\mathcal{H}}}$, *then the transform* $\mathrm{H}^{1}_{\sigma,\kappa}$ *is one-to-one on* $\mathfrak{L}_{\overline{\nu},\overline{r}}$ *and its image is given by*

$$\mathrm{H}^{1}_{\sigma,\kappa}(\mathfrak{L}_{\overline{\nu},\overline{r}})=\mathfrak{L}_{\overline{\nu}-\mathrm{Re}(\kappa+\sigma),\overline{r}}. \tag{25}$$

Theorem 3.3 *Let*

$$a_1^*=0,a_2^*=0,...,\ a_n^*=0;\ \Delta_1=\Delta_2=...=\Delta_n=0;$$

$$\mathrm{Re}(\mu_1)<0,\mathrm{Re}(\mu_2)<0,...,\mathrm{Re}(\mu_n)<0;$$

$$\overline{\alpha}_1<\nu_1-\mathrm{Re}(\kappa_1)<\overline{\beta}_1,\overline{\alpha}_2<\nu_1-\mathrm{Re}(\kappa_2)<\overline{\beta}_2,\dots,\overline{\alpha}_n<\nu_n-\mathrm{Re}(\kappa_n)<\overline{\beta}_n,$$

$\nu_1=\nu_2=...=\nu_n$; $1<\overline{r}<\infty, r_1=r_2=...=r_n$; *and let* $\mathbf{m}>\mathbf{0}$ *or* $\mathbf{n}>\mathbf{0}$. *There the following assertions are true:*

(a) The transform $\mathrm{H}^{1}_{\sigma,\kappa}$ *defined on* $\mathfrak{L}_{\overline{\nu},\overline{2}}$ *can be extended to* $\mathfrak{L}_{\overline{\nu},\overline{r}}$ *as an element of* $\mathrm{H}^{1}_{\sigma,\kappa}\in[\mathfrak{L}_{\overline{\nu},\overline{r}},\mathfrak{L}_{\overline{\nu}-\mathrm{Re}(\kappa+\sigma),\mathbf{s}}]$ *for any* $\mathbf{s}=(s_1,s_2,...,s_n)$, $\mathbf{s}\geq\overline{r}$, *such that* $1/s_j=1/r_j+\mathrm{Re}(\mu_j)$, $j=1,2,...,n$. *If* $1<\overline{r}\leq\overline{2}$, *then the transform* $\mathrm{H}^{1}_{\sigma,\kappa}$ *is one-to-one and there holds the relation (19) for* $f\in\mathfrak{L}_{\overline{\nu},\overline{r}}$ *and* $\mathrm{Re}(s)=\overline{\nu}-\mathrm{Re}(\kappa+\sigma)$.

(b) If $f\in\mathfrak{L}_{\overline{\nu},\overline{r}}$ *and* $g\in\mathfrak{L}_{1-\overline{\nu}+\mathrm{Re}(\kappa+\sigma),\mathbf{s}}$ *with* $1<\mathbf{s}<\infty$ *and* $1\leq1/\overline{r}+1/\mathbf{s}<1-\mathrm{Re}(\overline{\mu})$, *then the relation (23) holds.*

(c) Let $\overline{k}=(k_1,k_2,...,k_n)>0$. *If* $1-\overline{\nu}+\mathrm{Re}(\kappa)\notin\mathcal{E}_{\overline{\mathcal{H}}}$, *then the transform* $\mathrm{H}^{1}_{\sigma,\kappa}$ *is one-to-one on* $\mathfrak{L}_{\overline{\nu},\overline{r}}$ *and there hold*

$$\mathrm{H}^{1}_{\sigma,\kappa}(\mathfrak{L}_{\overline{\nu},\overline{r}})=\mathrm{I}^{-\overline{\mu}}_{-;\overline{k},(\sigma-\overline{\alpha})/\overline{k}}(\mathfrak{L}_{\overline{\nu}-\mathrm{Re}(\kappa+\sigma),\overline{r}}) \tag{26}$$

for $\mathbf{m}>\mathbf{0}$, *and*

$$\mathrm{H}^{1}_{\sigma,\kappa}(\mathfrak{L}_{\overline{\nu},\overline{r}})=\mathrm{I}^{-\overline{\mu}}_{0+;\overline{k},(\overline{\beta}-\sigma)/\overline{k}-1}(\mathfrak{L}_{\overline{\nu}-\mathrm{Re}(\kappa+\sigma),\overline{r}}) \tag{27}$$

for $\mathbf{n}>\mathbf{0}$. *When* $1-\overline{\nu}+\mathrm{Re}(\kappa)\in\mathcal{E}_{\overline{\mathcal{H}}}$, $\mathrm{H}^{1}_{\sigma,\kappa}(\mathfrak{L}_{\overline{\nu},\overline{r}})$ *is a subset of right hand sides of (26) and (27) in respective cases.*

(d) If $f\in\mathfrak{L}_{\overline{\nu},\overline{r}}$, $\overline{\lambda}=(\lambda_1,...,\lambda_n)\in\mathrm{C}^n$ *and* $\overline{h}=(h_1,...,h_n)>0$, *then* $\mathrm{H}^{1}_{\sigma,\kappa}f$ *is given in (21) for* $\mathrm{Re}(\overline{\lambda})>(\overline{\nu}-\mathrm{Re}(\kappa))\overline{h}-1$, *while in (22) for* $\mathrm{Re}(\overline{\lambda})<(\overline{\nu}-\mathrm{Re}(k))\overline{h}-1$. *Furthermore* $\mathrm{H}^{1}_{\sigma,\kappa}f$ *is given in (1).*

In ([1], formulas (68), (69)) were obtained inversion formulas for transformation $\mathrm{H}^{1}_{\sigma,\kappa}f$ *(1), which generalize the corresponding one-dimensional case (see [6], (5.5.23) and (5.5.24)):*

$$f(\mathbf{x})=-\overline{h}\mathbf{x}^{(\overline{\lambda}+1)/\overline{h}-\kappa}\frac{\mathbf{d}}{\mathbf{dx}}\mathbf{x}^{-(\overline{\lambda}+1)/\overline{h}}\times$$

$$\int_0^\infty \mathbf{H}^{\mathbf{q-m,p-n+1}}_{\mathbf{p+1,q+1}}\left[\frac{\mathbf{t}}{\mathbf{x}}\middle| \begin{matrix} (-\overline{\lambda},\overline{h}),(1-\mathbf{a}_i-\alpha_i,\alpha_i)_{\mathbf{n+1,p}},(1-\mathbf{a}_i-\alpha_i,\alpha_i)_{\mathbf{1,n}} \\ (1-\mathbf{b}_j-\beta_j,\beta_j)_{\mathbf{m+1,q}},(1-\mathbf{b}_j-\beta_j,\beta_j)_{\mathbf{1,m}}\,(-\overline{\lambda}-1,\overline{h}) \end{matrix}\right]\times$$

$$\mathbf{t}^{-\sigma}(\mathbf{H}^1_{\sigma,\kappa}f)(\mathbf{t})\mathbf{dt} \tag{28}$$

or

$$f(\mathbf{x})=\overline{h}\mathbf{x}^{(\overline{\lambda}+1)/\overline{h}-1}\frac{\mathbf{d}}{\mathbf{dx}}\mathbf{x}^{-(\overline{\lambda}+1)/\overline{h}}\times$$

$$\int_0^\infty \mathbf{H}^{\mathbf{q-m+1,p-n}}_{\mathbf{p+1,q+1}}\left[\frac{\mathbf{t}}{\mathbf{x}}\middle| \begin{matrix} (1-\mathbf{a}_i-\alpha_i,\alpha_i)_{\mathbf{n+1,p}},(1-\mathbf{a}_i-\alpha_i,\alpha_i)_{\mathbf{1,n}},\ (-\overline{\lambda},\overline{h}) \\ (-\overline{\lambda}-1,\overline{h}),(1-\mathbf{b}_j-\beta_j,\beta_j)_{\mathbf{m+1,q}},(1-\mathbf{b}_j-\beta_j,\beta_j)_{\mathbf{1,m}} \end{matrix}\right]\times$$

$$\mathbf{t}^{-\sigma}(\mathbf{H}^1_{\sigma,\kappa}f)(\mathbf{t})\mathbf{dt}. \tag{29}$$

Condition for the validity of these formulas are given by the following assertion (one-dimensional case see in ([6], Theorem 5.47)).

Theorem 3.4 *Let* $a_1^*=0, a_2^*=0, ..., a_n^*=0;\quad \overline{\alpha}_1<\nu_1-\mathrm{Re}(\kappa_1)<\overline{\beta}_1,\ \overline{\alpha}_2<\nu_2-\mathrm{Re}(\kappa_2)<\overline{\beta}_2, ...,\quad \overline{\alpha}_n<\nu_n-\mathrm{Re}(\kappa_n)<\overline{\beta}_n;\quad \alpha_0^1<1-\nu_1+\mathrm{Re}(\kappa_1)<\beta_0^1,\ \alpha_0^2<1-\nu_2+\mathrm{Re}(\kappa_2)<\beta_0^2, ...,\ \alpha_0^n<1-\nu_n+\mathrm{Re}(\kappa_n)<\beta_0^n;$ *and let* $\overline{\lambda}\in C^n$, $\overline{h}>0$.

(a) If $\Delta_1[\nu_1-\mathrm{Re}(\kappa_1)]+\mathrm{Re}(\mu_1)=0,\quad \Delta_2[\nu_2-\mathrm{Re}(\kappa_2)]+\mathrm{Re}(\mu_2)=0, ..., \Delta_n[\nu_n-\mathrm{Re}(\kappa_n)]+\mathrm{Re}(\mu_n)=0$ *and* $f\in\mathfrak{L}_{\overline{\nu},\overline{2}}$ $(\nu_1=...=\nu_n)$, *then the inversion formulas (28) and (29) are valid for* $\mathrm{Re}(\overline{\lambda})>(1-\overline{\nu}+\mathrm{Re}(\kappa))\overline{h}-1$ *and for* $\mathrm{Re}(\overline{\lambda})<(1-\overline{\nu}+\mathrm{Re}(\kappa))\overline{h}-1$, *respectively.*

(b) If $\Delta_1=\Delta_2=...=\Delta_n=0;$ $\mathrm{Re}(\mu_1)=\mathrm{Re}(\mu_2)=...=\mathrm{Re}(\mu_n)=0$ *and* $f\in\mathfrak{L}_{\overline{\nu},\overline{r}}$ $(\nu_1=\nu_2=...=\nu_n)$, $1<\overline{r}<\infty, r_1=r_2=...=r_n$, *then the inversion formulas (28) and (29) are valid for* $\mathrm{Re}(\overline{\lambda})>(1-\overline{\nu}+\mathrm{Re}(\kappa))\overline{h}-1$ *and for* $\mathrm{Re}(\overline{\lambda})<(1-\overline{\nu}+\mathrm{Re}(\kappa))\overline{h}-1$, *respectively.*

4 $\mathfrak{L}_{\overline{\nu},\overline{r}}$-Theory and the Inversion Formulas of the Modified $\mathbf{G}^1_{\sigma,\kappa}$- transform

Modified $G^1_{\sigma,\kappa}$-transform (2) is a special case of the modified $H^1_{\sigma,\kappa}$-transformation (1) when parameters in (5), (6) are equal: $\alpha_1=\alpha_2=...=\alpha_p=\beta_1=\beta_2=...=\beta_q=1$. Therefore $\mathfrak{L}_{\overline{\nu},\overline{2}}$-theory of the $G^1_{\sigma,\kappa}$-transformation (2) follows from the corresponding results for the modified $H^1_{\sigma,\kappa}$-transformation (1) and it was presented in the work [2].

To formulate statements representing $\mathfrak{L}_{\overline{\nu},\overline{2}}$- and $\mathfrak{L}_{\overline{\nu},\overline{r}}$-theories and inversion formulas for the $G^1_{\sigma,\kappa}$-transform (2), we need the following multidimensional constants ([2], (3.1)–(3.5)):

$\overline{\alpha} = (\overline{\alpha}_1, \overline{\alpha}_2, ..., \overline{\alpha}_n)$ and $\overline{\beta} = (\overline{\beta}_1, \overline{\beta}_2, ..., \overline{\beta}_n)$, where

$$\overline{\alpha}_1 = \begin{cases} -\min\limits_{1\le j_1\le m_1}\Big[\mathrm{Re}(b_{j_1})\Big], & m_1 > 0, \\ -\infty, & m_1 = 0, \end{cases} \quad \overline{\beta}_1 = \begin{cases} \min\limits_{1\le i_1\le \overline{n}_1}\Big[1-\mathrm{Re}(a_{i_1})\Big], & \overline{n}_1 > 0, \\ \infty, & \overline{n}_1 = 0, \end{cases}$$

$$\overline{\alpha}_2 = \begin{cases} -\min\limits_{1\le j_2\le m_2}\Big[\mathrm{Re}(b_{j_2})\Big], & m_2 > 0, \\ -\infty, & m_2 = 0, \end{cases} \quad \overline{\beta}_2 = \begin{cases} \min\limits_{1\le i_2\le \overline{n}_2}\Big[1-\mathrm{Re}(a_{i_2})\Big], & \overline{n}_2 > 0, \\ \infty, & \overline{n}_2 = 0, \end{cases}$$

and so on

$$\overline{\alpha}_n = \begin{cases} -\min\limits_{1\le j_n\le m_n}\Big[\mathrm{Re}(b_{j_n})\Big], & m_n > 0, \\ -\infty, & m_2 = 0, \end{cases} \quad \overline{\beta}_n = \begin{cases} \min\limits_{1\le i_n\le \overline{n}_n}\Big[1-\mathrm{Re}(a_{i_n})\Big], & \overline{n}_n > 0, \\ \infty, & \overline{n}_n = 0; \end{cases} \tag{30}$$

$$a_1^* = 2(m_1+n_1) - p_1 - q_1,\ a_2^* = 2(m_2+n_2) - p_2 - q_2, ..., a_n^* = 2(m_n+n_n) - p_n - q_n; \tag{31}$$

$$\Delta_1 = q_1 - p_1,\ \Delta_2 = q_2 - p_2, ..., \Delta_n = q_n - p_n; \tag{32}$$

$$\mu_1 = \sum_{j=1}^{q_1} b_{j_1} - \sum_{i=1}^{p_1} a_{i_1} + \frac{p_1 - q_1}{2},\ \mu_2 = \sum_{j=1}^{q_2} b_{j_2} - \sum_{i=1}^{p_2} a_{i_2} + \frac{p_2 - q_2}{2}, ...,$$

$$\mu_n = \sum_{j=1}^{q_n} b_{j_n} - \sum_{i=1}^{p_n} a_{i_n} + \frac{p_n - q_n}{2}; \tag{33}$$

$$\alpha_0^1 = \begin{cases} 1 + \max\limits_{m_1+1\le j_1\le q_1}\Big[\mathrm{Re}(b_{j_1}) - 1\Big], & q_1 > m_1, \\ -\infty, & q_1 = m_1, \end{cases}$$

$$\beta_0^1 = \begin{cases} 1 + \min\limits_{\overline{n}_1+1\le i_1\le p_1}\Big[\mathrm{Re}(a_{i_1})\Big], & p_1 > \overline{n}_1, \\ \infty, & p_1 = \overline{n}_1, \end{cases}$$

$$\alpha_0^2 = \begin{cases} 1 + \max\limits_{m_2+1\le j_2\le q_2}\Big[\mathrm{Re}(b_{j_2}) - 1\Big], & q_2 > m_2, \\ -\infty, & q_2 = m_2, \end{cases}$$

$$\beta_0^2 = \begin{cases} 1 + \min\limits_{\overline{n}_2+1\le i_2\le p_2}\Big[\mathrm{Re}(a_{i_2})\Big], & p_2 > \overline{n}_2, \\ \infty, & p_2 = \overline{n}_2, \end{cases}$$

and so on

$$\alpha_0^n = \begin{cases} 1 + \max\limits_{m_n+1 \le j_n \le q_n} \Big[\mathrm{Re}(b_{j_n}) - 1 \Big], & q_n > m_n, \\ -\infty, & q_n = m_n, \end{cases}$$

$$\beta_0^n = \begin{cases} 1 + \min\limits_{\overline{n}_n+1 \le i_n \le p_2} \Big[\mathrm{Re}(a_{i_n}) \Big], & p_n > \overline{n}_n, \\ \infty, & p_n = \overline{n}_n. \end{cases} \tag{34}$$

The exceptional set $\mathcal{E}_{\overline{\mathcal{G}}}$ of a function $\overline{\mathcal{G}}^{\mathbf{m},\mathbf{n}}_{\mathbf{p},\mathbf{q}}(\mathbf{s})$:

$$\overline{\mathcal{G}}^{\mathbf{m},\mathbf{n}}_{\mathbf{p},\mathbf{q}}(\mathbf{s}) \equiv \overline{\mathcal{G}}^{\mathbf{m},\mathbf{n}}_{\mathbf{p},\mathbf{q}}\left[\begin{matrix} (\mathbf{a}_i)_{1,\mathbf{p}} \\ (\mathbf{b}_j)_{1,\mathbf{q}} \end{matrix} \middle| \mathbf{s} \right] = \prod_{k=1}^{n} \mathcal{G}^{m_k,\overline{n}_k}_{p_k,q_k}\left[\begin{matrix} (a_{i_k})_{1,p_k} \\ (b_{j_k})_{1,q_k} \end{matrix} \middle| s_k \right], \tag{35}$$

is called a set of vectors $\overline{\nu} = (\nu_1, \nu_2, ..., \nu_n) \in \mathrm{R}^n$ $(\nu_1 = \nu_2 = ... = \nu_n)$, such that $\overline{\alpha}_1 < 1 - \nu_1 < \overline{\beta}_1$, $\overline{\alpha}_2 < 1 - \nu_2 < \overline{\beta}_2$, ..., $\overline{\alpha}_n < 1 - \nu_n < \overline{\beta}_n$, and functions $\mathcal{G}^{m_1,\overline{n}_1}_{p_1,q_1}(s_1)$, $\mathcal{G}^{m_2,\overline{n}_2}_{p_2,q_2}(s_2)$,...,$\mathcal{G}^{m_n,\overline{n}_n}_{p_n,q_n}(s_n)$ have zeros on lines $\mathrm{Re}(s_1) < 1 - \nu_1$, $\mathrm{Re}(s_2) < 1 - \nu_2$, ..., $\mathrm{Re}(s_n) < 1 - \nu_n$, respectively (see [2], (3.6)).

Apply the multidimensional Mellin transform (9) to transformation $\mathrm{G}^1_{\sigma,\kappa} f$ (2) and, taking into account (19), we obtain:

$$(\mathfrak{M}\mathrm{G}^1_{\sigma,\kappa} f)(\mathbf{s}) = \overline{\mathcal{G}}^{\mathbf{m},\mathbf{n}}_{\mathbf{p},\mathbf{q}}\left[\begin{matrix} (\mathbf{a}_i)_{1,\mathbf{p}} \\ (\mathbf{b}_j)_{1,\mathbf{q}} \end{matrix} \middle| \mathbf{s} + \sigma \right] (\mathfrak{M} f)(\mathbf{s} + \sigma + \kappa), \tag{36}$$

where $\overline{\mathcal{G}}^{\mathbf{m},\mathbf{n}}_{\mathbf{p},\mathbf{q}}(\mathbf{s})$ is given by (35).

Theorem 4.1 ([2], Theorem 3.1) *Let*

$$\overline{\alpha}_1 < (\nu_1 - \mathrm{Re}(\kappa_1)) < \overline{\beta}_1, \overline{\alpha}_2 < (\nu_2 - \mathrm{Re}(\kappa_2)) < \overline{\beta}_2,$$

$$\ldots, \overline{\alpha}_n < (\nu_n - \mathrm{Re}(\kappa_1)) < \overline{\beta}_n, \nu_1 = \nu_2 = \ldots = \nu_n; \tag{37}$$

$$a_1^* = 0, a_2^* = 0, ..., a_n^* = 0; \Delta_1[\nu_1 - \mathrm{Re}(\kappa_1)] + \mathrm{Re}(\mu_1) \le 0,$$

$$\Delta_2[\nu_2 - \mathrm{Re}(\kappa_2)] + \mathrm{Re}(\mu_2) \le 0, ..., \Delta_n[\nu_n - \mathrm{Re}(\kappa_n)] + \mathrm{Re}(\mu_n) \le 0. \tag{38}$$

There hold the following assertions:

(a) There exists a one-to-one map $\mathrm{G}^1_{\sigma,\kappa} \in [\mathfrak{L}_{\overline{\nu},\overline{2}}, \mathfrak{L}_{\overline{\nu}-\mathrm{Re}(\kappa+\sigma),\overline{2}}]$ *such the relation (36) holds for* $f \in \mathfrak{L}_{\overline{\nu},\overline{2}}$ *and* $\mathrm{Re}(\mathbf{s}) = \overline{\nu} - \mathrm{Re}(\kappa + \sigma)$.
If $a_1^* = 0$, $a_2^* = 0$, ..., $a_n^* = 0$; $\Delta_1[\nu_1 - \mathrm{Re}(\kappa_1)] + \mathrm{Re}(\mu_1) = 0$, $\Delta_2[\nu_2 - \mathrm{Re}(\kappa_2)] + \mathrm{Re}(\mu_2) = 0$, ..., $\Delta_n[\nu_n - \mathrm{Re}(\kappa_n)] + \mathrm{Re}(\mu_n) = 0$ *and* $1 - \overline{\nu} + \mathrm{Re}(\kappa)) \notin \mathcal{E}_{\overline{\mathcal{G}}}$, *then the transform* $\mathrm{G}^1_{\sigma,\kappa}$ *maps* $\mathfrak{L}_{\overline{\nu},\overline{2}}$ *onto* $\mathfrak{L}_{\overline{\nu}-\mathrm{Re}(\kappa+\sigma),\overline{2}}$.

(b) The transform $\mathrm{G}^1_{\sigma,\kappa}$ *does not depend on* $\overline{\nu}$ *in the sense if* $\overline{\nu}$ *and* $\widetilde{\overline{\nu}}$ *satisfy Eqs. (37), (38) and if the transforms* $\mathrm{G}^1_{\sigma,\kappa}$ *and* $\widetilde{\mathrm{G}}^1_{\sigma,\kappa}$ *are defined in respective spaces* $\mathfrak{L}_{\overline{\nu},\overline{2}}$ *and* $\mathfrak{L}_{\widetilde{\overline{\nu}},\overline{2}}$ *by Eq. (36), then* $\mathrm{G}^1_{\sigma,\kappa} = \widetilde{\mathrm{G}}^1_{\sigma,\kappa}$ *for* $f \in \mathfrak{L}_{\widetilde{\overline{\nu}},\overline{2}} \bigcap \mathfrak{L}_{\overline{\nu},\overline{2}}$.

(c) If $a_1^* = 0,\ a_2^* = 0, \ldots, a_n^* = 0;\ \Delta_1[\nu_1 - \mathrm{Re}(\kappa_1)] + \mathrm{Re}(\mu_1) < 0,\ \Delta_2[\nu_2 - \mathrm{Re}(\kappa_2)] + \mathrm{Re}(\mu_2) < 0, \ldots, \Delta_n[\nu_n - \mathrm{Re}(\kappa_n)] + \mathrm{Re}(\mu_n) < 0$; *then for* $f \in \mathfrak{L}_{\overline{\nu},\overline{2}}$ *the transform* $\mathrm{G}^1_{\sigma,\kappa} f$ *is given by Eq. (2).*

(d) Let $\overline{\lambda} = (\lambda_1, \ldots, \lambda_n) \in \mathrm{C}^n$ *and* $f \in \mathfrak{L}_{\overline{\nu},\overline{2}}$. *If* $\mathrm{Re}(\overline{\lambda}) > \overline{\nu} - \mathrm{Re}(\kappa) - 1$, *then the transform* $\mathrm{G}^1_{\sigma,\kappa} f$ *is represented in the form*

$$\left(\mathrm{G}^1_{\sigma,\kappa} f\right)(\mathbf{x}) = \mathbf{x}^{\sigma-\overline{\lambda}} \frac{\mathbf{d}}{\mathbf{dx}} \mathbf{x}^{(\overline{\lambda}+1)} \int_0^{\mathbf{x}} G^{\mathbf{m},\mathbf{n}+1}_{\mathbf{p}+1,\mathbf{q}+1}\left[\frac{\mathbf{x}}{\mathbf{t}} \,\middle|\, \begin{matrix} -\overline{\lambda}, (\mathbf{a}_i)_{1,\mathbf{p}} \\ (\mathbf{b}_j)_{1,\mathbf{q}}, -\overline{\lambda}-1 \end{matrix}\right] \mathbf{t}^{\kappa-1} f(\mathbf{t})\mathbf{dt}, \tag{39}$$

while for $\mathrm{Re}(\overline{\lambda}) < \overline{\nu} - \mathrm{Re}(\kappa) - 1$ *the transform* $\mathrm{G}^1_{\sigma,\kappa} f$ *is given by*

$$\left(\mathrm{G}^1_{\sigma,\kappa} f\right)(\mathbf{x}) = -\mathbf{x}^{\sigma-\overline{\lambda}} \frac{\mathbf{d}}{\mathbf{dx}} \mathbf{x}^{(\overline{\lambda}+1)} \int_0^{\mathbf{x}} G^{\mathbf{m}+1,\mathbf{n}}_{\mathbf{p}+1,\mathbf{q}+1}\left[\frac{\mathbf{x}}{\mathbf{t}} \,\middle|\, \begin{matrix} (\mathbf{a}_i)_{1,\mathbf{p}}, -\overline{\lambda} \\ -\overline{\lambda}-1, (\mathbf{b}_j)_{1,\mathbf{q}} \end{matrix}\right] \mathbf{t}^{\kappa-1} f(\mathbf{t})\mathbf{dt}. \tag{40}$$

(e) If $f \in \mathfrak{L}_{\overline{\nu},\overline{2}}$ *and* $g \in \mathfrak{L}_{1-\overline{\nu}+\mathrm{Re}(\kappa+\sigma),\overline{2}}$, *then there holds the relation :*

$$\int_0^{\mathbf{x}} f(\mathbf{x})\left(\mathrm{G}^1_{\sigma,\kappa} g\right)(\mathbf{x})\mathbf{dx} = \int_0^{\mathbf{x}} \left(\mathrm{G}^2_{\sigma,\kappa} f\right)(\mathbf{x}) g(\mathbf{x}) d\mathbf{x}, \tag{41}$$

where

$$\left(\mathrm{G}^2_{\sigma,\kappa} f\right)(\mathbf{x}) = \mathbf{x}^{\sigma} \int_0^{\mathbf{x}} G^{\mathbf{m},\mathbf{n}}_{\mathbf{p},\mathbf{q}}\left[\frac{\mathbf{t}}{\mathbf{x}} \,\middle|\, \begin{matrix} (\mathbf{a}_i)_{1,\mathbf{p}} \\ (\mathbf{b}_j)_{1,\mathbf{q}} \end{matrix}\right] \mathbf{t}^{\kappa} f(\mathbf{t}) \frac{\mathbf{dt}}{\mathbf{x}}.$$

Now present $\mathfrak{L}_{\overline{\nu},\overline{r}}$*-theory of the* $\mathrm{G}^1_{\sigma,\kappa}$*-transform, which follows from Theorems 3.2 and 3.3 for* $\mathrm{H}^1_{\sigma,\kappa}$*- transform when* $\alpha_1 = \alpha_2 = \ldots = \alpha_p = \beta_1 = \beta_2 = \ldots = \beta_q = 1$.

Theorem 4.2 *Let*

$$a_1^* = 0, a_2^* = 0, \ldots, a_n^* = 0;\ \Delta_1 = \Delta_2 = \ldots = \Delta_n = 0;$$

$$\mathrm{Re}(\mu_1) = \mathrm{Re}(\mu_2) = \ldots = \mathrm{Re}(\mu_n) = 0;$$

$$\overline{\alpha}_1 < \nu_1 - \mathrm{Re}(\kappa_1) < \overline{\beta}_1, \overline{\alpha}_2 < \nu_1 - \mathrm{Re}(\kappa_2) < \overline{\beta}_2, \ldots, \overline{\alpha}_n < \nu_n - \mathrm{Re}(\kappa_n) < \overline{\beta}_n,$$

$$\nu_1 = \nu_2 = \ldots = \nu_n;\ 1 < \overline{r} < \infty, r_1 = r_2 = \ldots = r_n.$$

There the following assertions are true:

(a) The transform $\mathrm{G}^1_{\sigma,\kappa}$ *defined on* $\mathfrak{L}_{\overline{\nu},\overline{2}}$ *can be extended to* $\mathfrak{L}_{\overline{\nu},\overline{r}}$ *as an element of* $\mathrm{G}^1_{\sigma,\kappa} \in [\mathfrak{L}_{\overline{\nu},\overline{r}}, \mathfrak{L}_{\overline{\nu}-\mathrm{Re}(\kappa+\sigma),\overline{r}}]$. *If* $1 < \overline{r} \le \overline{2}$, *then the transform* $\mathrm{G}^1_{\sigma,\kappa}$ *is one-to-one on* $\mathfrak{L}_{\overline{\nu},\overline{r}}$ *and there holds the equality (36) for* $f \in \mathfrak{L}_{\overline{\nu},\overline{r}}$ *and* $\mathrm{Re}(s) = \overline{\nu} - \mathrm{Re}(\kappa+\sigma)$.

(b) If $f \in \mathfrak{L}_{\overline{\nu},\overline{r}}$ *and* $\overline{\lambda} = (\lambda_1, ..., \lambda_n) \in \mathbb{C}^n$*, then* $\mathrm{G}^1_{\sigma,\kappa} f$ *is given in (39) for* $\mathrm{Re}(\overline{\lambda}) > \overline{\nu} - \mathrm{Re}(\kappa) - 1$*, while in (40) for* $\mathrm{Re}(\overline{\lambda}) < \overline{\nu} - \mathrm{Re}(k) - 1$.

(c) If $f \in \mathfrak{L}_{\overline{\nu},\overline{r}}$ *and* $g \in \mathfrak{L}_{1-\overline{\nu}+\mathrm{Re}(\kappa+\sigma),\overline{r}'}$ *with* $\overline{r}' = \overline{r}/(\overline{r} - 1)$*, then the relation (41) holds.*

(d) If $1 - \overline{\nu} + \mathrm{Re}(\kappa) \notin \mathcal{E}_{\overline{\mathcal{G}}}$*, then the transform* $\mathrm{G}^1_{\sigma,\kappa}$ *is one-to-one on* $\mathfrak{L}_{\overline{\nu},\overline{r}}$ *and its image is given by*

$$\mathrm{G}^1_{\sigma,\kappa}(\mathfrak{L}_{\overline{\nu},\overline{r}}) = \mathfrak{L}_{\overline{\nu}-\mathrm{Re}(\kappa+\sigma),\overline{r}}. \tag{42}$$

Theorem 4.3 *Let*

$$a_1^* = 0, a_2^* = 0, ..., a_n^* = 0; \Delta_1 = \Delta_2 = ... = \Delta_n = 0;$$

$$\mathrm{Re}(\mu_1) < 0, \mathrm{Re}(\mu_2) < 0, ..., \mathrm{Re}(\mu_n) < 0;$$

$$\overline{\alpha}_1 < \nu_1 - \mathrm{Re}(\kappa_1) < \overline{\beta}_1, \overline{\alpha}_2 < \nu_1 - \mathrm{Re}(\kappa_2) < \overline{\beta}_2, \dots, \overline{\alpha}_n < \nu_n - \mathrm{Re}(\kappa_n) < \overline{\beta}_n,$$

$\nu_1 = \nu_2 = ... = \nu_n$*;* $1 < \overline{r} < \infty, r_1 = r_2 = ... = r_n$*; and let* $\mathbf{m} > \mathbf{0}$ *or* $\mathbf{n} > \mathbf{0}$*. There the following assertions are true:*

(a) The transform $\mathrm{G}^1_{\sigma,\kappa}$ *defined on* $\mathfrak{L}_{\overline{\nu},\overline{2}}$ *can be extended to* $\mathfrak{L}_{\overline{\nu},\overline{r}}$ *as an element of* $[\mathfrak{L}_{\overline{\nu},\overline{r}}, \mathfrak{L}_{\overline{\nu}-\mathrm{Re}(\kappa+\sigma),\mathbf{s}}]$ *for any* $\mathbf{s} = (s_1, s_2, ..., s_n)$*,* $\mathbf{s} \geq \overline{r}$*, such that* $1/s_j = 1/r_j + \mathrm{Re}(\mu_j)$*,* $j = 1, 2, ..., n$*. If* $1 < \overline{r} \leq \overline{2}$*, then the transform* $\mathrm{G}^1_{\sigma,\kappa}$ *is one-to-one on* $\mathfrak{L}_{\overline{\nu},\overline{r}}$ *and there holds the equality (36) for* $f \in \mathfrak{L}_{\overline{\nu},\overline{r}}$ *and* $\mathrm{Re}(s) = \overline{\nu} - \mathrm{Re}(\kappa + \sigma)$.

(b) If $f \in \mathfrak{L}_{\overline{\nu},\overline{r}}$ *and* $g \in \mathfrak{L}_{1-\overline{\nu}+\mathrm{Re}(\kappa+\sigma),\mathbf{s}}$ *with* $1 < \mathbf{s} < \infty$ *and* $1 \leq 1/\overline{r} + 1/\mathbf{s} < 1 - \mathrm{Re}(\overline{\mu})$*, then the relation (41) holds.*

(c) Let $\overline{k} = (k_1, k_2, ..., k_n) > 0$*. If* $1 - \overline{\nu} + \mathrm{Re}(\kappa) \notin \mathcal{E}_{\overline{\mathcal{G}}}$*, then the transform* $\mathrm{G}^1_{\sigma,\kappa}$ *is one-to-one on* $\mathfrak{L}_{\overline{\nu},\overline{r}}$ *and there hold*

$$\mathrm{G}^1_{\sigma,\kappa}(\mathfrak{L}_{\overline{\nu},\overline{r}}) = \mathrm{I}^{-\overline{\mu}}_{-;\overline{k},(\sigma-\overline{\alpha})/\overline{k}}(\mathfrak{L}_{\overline{\nu}-\mathrm{Re}(\kappa+\sigma),\overline{r}}) \tag{43}$$

for $\mathbf{m} > \mathbf{0}$*, and*

$$\mathrm{G}^1_{\sigma,\kappa}(\mathfrak{L}_{\overline{\nu},\overline{r}}) = \mathrm{I}^{-\overline{\mu}}_{0+;\overline{k},(\overline{\beta}-\sigma)/\overline{k}-1}(\mathfrak{L}_{\overline{\nu}-\mathrm{Re}(\kappa+\sigma),\overline{r}}) \tag{44}$$

for $\mathbf{n} > \mathbf{0}$*. When* $1 - \overline{\nu} + \mathrm{Re}(\kappa) \in \mathcal{E}_{\overline{\mathcal{G}}}$*,* $\mathrm{G}^1_{\sigma,\kappa}(\mathfrak{L}_{\overline{\nu},\overline{r}})$ *is a subset of right hand sides of (43) and (44) in respective cases.*

(d) If $f \in \mathfrak{L}_{\overline{\nu},\overline{r}}$*,* $\overline{\lambda} = (\lambda_1, ..., \lambda_n) \in \mathbb{C}^n$*, then* $\mathrm{G}^1_{\sigma,\kappa} f$ *is given in (39) for* $\mathrm{Re}(\overline{\lambda}) > \overline{\nu} - \mathrm{Re}(\kappa) - 1$*, while in (40) for* $\mathrm{Re}(\overline{\lambda}) < \overline{\nu} - \mathrm{Re}(k) - 1$*. Furthermore* $\mathrm{G}^1_{\sigma,\kappa} f$ *is given in (2).*

In ([2], formulas (3.14), (3.15)) were obtained inversion formulas for transformation $\mathrm{G}^1_{\sigma,\kappa} f$ *(2), which generalize the corresponding one-dimensional case (see [6], (6.6.23) and (6.6.24)):*

$f(\mathbf{x}) = -\mathbf{x}^{\overline{\lambda}+1-\kappa} \frac{\mathbf{d}}{\mathbf{dx}} \mathbf{x}^{-\overline{\lambda}-1} \times$

$$\int_0^\infty G^{q-m,p-n+1}_{p+1,q+1}\left[\frac{t}{x}\left|\begin{matrix}-\overline{\lambda}, -a_{n+1}, ..., -a_p, -a_1, ..., -a_n\\ -b_{m+1}, ..., -b_q, -b_1, ..., -b_m, -\overline{\lambda}-1\end{matrix}\right.\right]t^{-\sigma}(G^1_{\sigma,\kappa}f)(t)dt \tag{45}$$

or

$$f(x) = x^{\overline{\lambda}+1-\kappa}\frac{d}{dx}x^{-\overline{\lambda}-1}\times$$

$$\int_0^\infty G^{q-m+1,p-n}_{p+1,q+1}\left[\frac{t}{x}\left|\begin{matrix}-a_{n+1}, ..., -a_p, -a_1, ..., -a_n, -\overline{\lambda},\\ -\overline{\lambda}-1, -b_{m+1}, ..., -b_q, -b_1, ..., -b_m,\end{matrix}\right.\right]t^{-\sigma}(G^1_{\sigma,\kappa}f)(t)dt. \tag{46}$$

The conditions for the validity of these formulas are given by the following statement, which follows from ([1], Theorem 10) and Theorem 3.4.

Theorem 4.4 *Let* $a_1^* = 0, a_2^* = 0, ..., a_n^* = 0;$ $\overline{\alpha}_1 < \nu_1 - \mathrm{Re}(\kappa_1) < \overline{\beta}_1,$ $\overline{\alpha}_2 < \nu_2 - \mathrm{Re}(\kappa_2) < \overline{\beta}_2, ...,$ $\overline{\alpha}_n < \nu_n - \mathrm{Re}(\kappa_n) < \overline{\beta}_n;$ $\alpha_0^1 < 1 - \nu_1 + \mathrm{Re}(\kappa_1) < \beta_0^1,$ $\alpha_0^2 < 1 - \nu_2 + \mathrm{Re}(\kappa_2) < \beta_0^2, ..., \alpha_0^n < 1 - \nu_n + \mathrm{Re}(\kappa_n) < \beta_0^n;$ *and let* $\overline{\lambda} \in C^n$.

(a) If $\Delta_1[\nu_1 - \mathrm{Re}(\kappa_1)] + \mathrm{Re}(\mu_1) = 0,$ $\Delta_2[\nu_2 - \mathrm{Re}(\kappa_2)] + \mathrm{Re}(\mu_2) = 0, ...,$ $\Delta_n[\nu_n - \mathrm{Re}(\kappa_n)] + \mathrm{Re}(\mu_n) = 0$ $(\nu_1 = \nu_2 = ... = \nu_n)$ *and* $f \in \mathfrak{L}_{\overline{\nu},\overline{2}}$ *, then the inversion formulas (45) and (46) are valid for* $\mathrm{Re}(\overline{\lambda}) > -\overline{\nu} + \mathrm{Re}(\kappa)$ *and* $\mathrm{Re}(\overline{\lambda}) < -\overline{\nu} + \mathrm{Re}(\kappa)$, *respectively.*

(b) If $\Delta_1 = \Delta_2 = ... = \Delta_n = 0;$ $\mathrm{Re}(\mu_1) = \mathrm{Re}(\mu_2) = ... = \mathrm{Re}(\mu_n) = 0$ *and* $f \in \mathfrak{L}_{\overline{\nu},\overline{r}}$ $(\nu_1 = \nu_2 = ... = \nu_n)$, $1 < \overline{r} < \infty$, $r_1 = r_2 = ... = r_n$, *then the inversion formulas (45) and (46) are valid for* $\mathrm{Re}(\overline{\lambda}) > -\overline{\nu} + \mathrm{Re}(\kappa)$ *and for* $\mathrm{Re}(\overline{\lambda}) < -\overline{\nu} + \mathrm{Re}(\kappa)$, *respectively.*

5 $\mathfrak{L}_{\overline{\nu},\overline{r}}$-Theory and the Inversion Formulas of the Modified $G^1_{\sigma,\kappa;\,\delta}$-Transform

$G^1_{\sigma,\kappa;\delta}$-transformation (3) represent as a composition $G^1_{\sigma/\delta,\kappa/\delta}$-transformation (2) and elementary operators of the form (11) N_ϱ . Indeed, replacing in (2) x^δ into $x^{1/\delta}$ and changing variables $t^\delta = \o$, we have:

$$(G^1_{\sigma,\kappa;\delta}f)(x^{1/\delta}) = x^{\sigma/\delta}\int_0^x G^{m,n}_{p,q}\left[\frac{x}{t^\delta}\left|\begin{matrix}(a_i)_{1,p}\\ (b_j)_{1,q}\end{matrix}\right.\right]t^\kappa f(t)\frac{dt}{t} =$$

$$= \frac{1}{\delta}x^{\sigma/\delta}\int_0^x G^{m,n}_{p,q}\left[\frac{x}{\o}\left|\begin{matrix}(a_i)_{1,p}\\ (b_j)_{1,q}\end{matrix}\right.\right]\o^{\kappa/\delta}f(\o^{1/\delta})\frac{d\o}{\o}dt = \frac{1}{\delta}(G^1_{\sigma,\kappa}N_{1/\delta}f)(x^{1/\delta}). \tag{47}$$

Applying the operator N_δ to the last equality, in [2] we obtained next view of transform $G^1_{\sigma,\kappa;\,\delta}f$ (3):

$$(\mathbf{G}^1_{\sigma,\kappa;\,\delta} f)(\mathbf{x}) = \frac{1}{\delta}(\mathbf{N}_\delta \mathbf{G}^1_{\sigma/\delta,\,\kappa/\delta}\mathbf{N}_{1/\delta} f)(\mathbf{x}). \tag{48}$$

Then applying the Mellin transform (9) to (48), taking into account (47) and assertion of Lemma 2.1, we get :

$$(\mathfrak{M}\mathbf{G}^1_{\sigma,\kappa;\delta} f)(\mathbf{s}) = \left(\mathfrak{M}\left(\frac{1}{\delta}\mathbf{N}_\delta \mathbf{G}^1_{\sigma/\delta,\,\kappa/\delta}\mathbf{N}_{1/\delta} f\right)\right)(\mathbf{s}) =$$

$$\frac{1}{\delta^2}\left(\mathfrak{M}\left(\mathbf{G}^1_{\sigma/\delta,\,\kappa/\delta}\mathbf{N}_{1/\delta} f\right)\right)\left(\frac{\mathbf{s}}{\delta}\right) =$$

$$\frac{1}{\delta^2}\overline{\mathcal{G}}^{\mathbf{m},\mathbf{n}}_{\mathbf{p},\mathbf{q}}\left[\begin{matrix}(\mathbf{a}_i)_{1,\mathbf{p}}\\(\mathbf{b}_j)_{1,\mathbf{q}}\end{matrix}\middle|\frac{\mathbf{s}+\sigma}{\delta}\right](\mathfrak{M}N_{1/\delta} f)(\frac{\mathbf{s}+\sigma+\kappa}{\delta}) =$$

$$\frac{1}{\delta}\overline{\mathcal{G}}^{\mathbf{m},\mathbf{n}}_{\mathbf{p},\mathbf{q}}\left[\begin{matrix}(\mathbf{a}_i)_{1,\mathbf{p}}\\(\mathbf{b}_j)_{1,\mathbf{q}}\end{matrix}\middle|\frac{\mathbf{s}+\sigma}{\delta}\right](\mathfrak{M} f)(\mathbf{s}+\sigma+\kappa).$$

Thus, (see [2], formula 4.3)

$$(\mathfrak{M}\mathbf{G}^1_{\sigma,\kappa;\delta} f)(\mathbf{s}) = \frac{1}{\delta}\overline{\mathcal{G}}^{\mathbf{m},\mathbf{n}}_{\mathbf{p},\mathbf{q}}\left[\begin{matrix}(\mathbf{a}_i)_{1,\mathbf{p}}\\(\mathbf{b}_j)_{1,\mathbf{q}}\end{matrix}\middle|\frac{\mathbf{s}+\sigma}{\delta}\right](\mathfrak{M} f)(\mathbf{s}+\sigma+\kappa). \tag{49}$$

The next theorem gives $\mathfrak{L}_{\overline{\nu},\overline{2}}$-theory of transformation (3), which follows from the corresponding assertions of Theorem 4.1, Lemma 2.1 and representations (47)–(48).

Theorem 5.1 ([2], Theorem 4.1) *Let*

$$\overline{\alpha}_1 < (\nu_1 - \mathrm{Re}(\kappa_1))/\delta_1 < \overline{\beta}_1, \overline{\alpha}_2 < (\nu_2 - \mathrm{Re}(\kappa_2))/\delta_2 < \overline{\beta}_2,$$

$$\ldots, \overline{\alpha}_n < (\nu_n - \mathrm{Re}(\kappa_1))/\delta_n < \overline{\beta}_n, \tag{50}$$

$$\nu_1 = \ldots = \nu_n;\ a_1^* = 0, a_2^* = 0, \ldots,\ a_n^* = 0;\ \Delta_1[\nu_1 - \mathrm{Re}(\kappa_1)]/\delta_1 + \mathrm{Re}(\mu_1) \le 0,$$

$$\Delta_2[\nu_2 - \mathrm{Re}(\kappa_2)]/\delta_2 + \mathrm{Re}(\mu_2) \le 0, \ldots, \Delta_n[\nu_n - \mathrm{Re}(\kappa_n)]/\delta_2 + \mathrm{Re}(\mu_n) \le 0. \tag{51}$$

There hold the following assertions:

(a) There exists a one-to-one map $\mathrm{G}^1_{\sigma,\kappa;\delta} \in [\mathfrak{L}_{\overline{\nu},\overline{2}}, \mathfrak{L}_{\overline{\nu}-\mathrm{Re}(\kappa+\sigma),\overline{2}}]$ *such the relation (49) holds for* $f \in \mathfrak{L}_{\overline{\nu},\overline{2}}$ *and* $\mathrm{Re}(\mathbf{s}) = \overline{\nu} - \mathrm{Re}(\kappa+\sigma)$.
If $a_1^* = 0,\ a_2^* = 0, \ldots, a_n^* = 0;$ $\Delta_1[\nu_1 - \mathrm{Re}(\kappa_1)]/\delta_1 + \mathrm{Re}(\mu_1) = 0,\ \Delta_2[\nu_2 - \mathrm{Re}(\kappa_2)]/\delta_2 + \mathrm{Re}(\mu_2) = 0, \ldots, \Delta_n[\nu_n - \mathrm{Re}(\kappa_n)]/\delta_n + \mathrm{Re}(\mu_n) = 0$ *and* $1 - (\overline{\nu} - \mathrm{Re}(\kappa))/\delta \notin \mathcal{E}_{\overline{\mathcal{G}}}$, *then the transform* $\mathrm{G}^1_{\sigma,\kappa;\delta}$ *maps* $\mathfrak{L}_{\overline{\nu},\overline{2}}$ *onto* $\mathfrak{L}_{\overline{\nu}-\mathrm{Re}(\kappa+\sigma),\overline{2}}$.

(b) The transform $\mathrm{G}^1_{\sigma,\kappa;\delta} f$ *does not depend on* $\overline{\nu}$ *in the sense if* $\overline{\nu}$ *and* $\widetilde{\overline{\nu}}$ *satisfy Eq. (50) – (51) and if the transforms* $\mathrm{G}^1_{\sigma,\kappa;\delta} f$ *and* $\widetilde{\mathrm{G}}^1_{\sigma,\kappa;\delta} f$ *are defined in respective spaces* $\mathfrak{L}_{\overline{\nu},\overline{2}}$ *and* $\mathfrak{L}_{\widetilde{\overline{\nu}},\overline{2}}$ *by Eq. (49), then* $\mathrm{G}^1_{\sigma,\kappa;\delta} f = \widetilde{\mathrm{G}}^1_{\sigma,\kappa;\delta} f$ *for* $f \in \mathfrak{L}_{\widetilde{\overline{\nu}},\overline{2}} \bigcap \mathfrak{L}_{\overline{\nu},\overline{2}}$.

(c) If $a_1^* = 0,\ a_2^* = 0, ..., a_n^* = 0;\ \Delta_1[\nu_1 - \mathrm{Re}(\kappa_1)]/\delta_1 + \mathrm{Re}(\mu_1) < 0,\ \Delta_2[\nu_2 - \mathrm{Re}(\kappa_2)]/\delta_2 + \mathrm{Re}(\mu_2) < 0, ..., \Delta_n[\nu_n - \mathrm{Re}(\kappa_n)]/\delta_n + \mathrm{Re}(\mu_n) < 0;$ *then for* $f \in \mathfrak{L}_{\overline{\nu},\overline{2}}$ *the transform* $\mathrm{G}^1_{\sigma,\kappa;\delta} f$ *is given by Eq. (3).*

(d) Let $\overline{\lambda} = (\lambda_1, ..., \lambda_n) \in \mathrm{C}^n$ *and* $f \in \mathfrak{L}_{\overline{\nu},\overline{2}}$. *If* $\mathrm{Re}(\overline{\lambda}) > (\overline{\nu} - \mathrm{Re}(\kappa))/\delta - 1$, *then the transform* $\mathrm{G}^1_{\sigma,\kappa;\delta} f$ *is represented in the form*
$(\mathrm{G}^1_{\sigma,\kappa;\delta} f)(\mathbf{x}) =$

$$\frac{1}{\delta}\mathbf{x}^{\sigma+1-\delta(\overline{\lambda}+1)}\frac{\mathbf{d}}{\mathbf{dx}}\mathbf{x}^{\delta(\overline{\lambda}+1)}\int\limits_0^\infty \mathrm{G}^{\mathbf{m},\mathbf{n}+1}_{\mathbf{p}+1,\mathbf{q}+1}\left[\frac{\mathbf{x}^\delta}{\mathbf{t}^\delta}\,\middle|\,\begin{matrix}-\overline{\lambda}, (\mathbf{a}_i)_{1,\mathbf{p}}\\ (\mathbf{b}_j)_{1,\mathbf{q}}, -\overline{\lambda}-1\end{matrix}\right]\mathbf{t}^{\kappa-1}f(\mathbf{t})\mathbf{dt}, \tag{52}$$

while for $\mathrm{Re}(\overline{\lambda}) < (\overline{\nu} - \mathrm{Re}(\kappa))/\delta - 1$ *the transform* $\mathrm{G}^1_{\sigma,\kappa;\delta} f$ *is given by*
$(\mathrm{G}^1_{\sigma,\kappa;\delta} f)(\mathbf{x}) =$

$$-\frac{1}{\delta}\mathbf{x}^{\sigma+1-\delta(\overline{\lambda}+1)}\frac{\mathbf{d}}{\mathbf{dx}}\mathbf{x}^{\delta(\overline{\lambda}+1)}\int\limits_0^\infty \mathrm{G}^{\mathbf{m}+1,\mathbf{n}}_{\mathbf{p}+1,\mathbf{q}+1}\left[\frac{\mathbf{x}^\delta}{\mathbf{t}^\delta}\,\middle|\,\begin{matrix}(\mathbf{a}_i)_{1,\mathbf{p}}, -\overline{\lambda}\\ -\overline{\lambda}-1, (\mathbf{b}_j)_{1,\mathbf{q}}\end{matrix}\right]\mathbf{t}^{\kappa-1}f(\mathbf{t})\mathbf{dt}. \tag{53}$$

(e) If $f \in \mathfrak{L}_{\overline{\nu},\overline{2}}$ *and* $g \in \mathfrak{L}_{1-\overline{\nu}+\mathrm{Re}(\kappa+\sigma),\overline{2}}$, *then there holds the relation:*

$$\int\limits_0^\infty f(\mathbf{x})\left(\mathrm{G}^1_{\sigma,\kappa;\delta} g\right)(\mathbf{x})\mathbf{dx} = \int\limits_0^\infty \left(\mathrm{G}^2_{\kappa,\sigma;\delta} f\right)(\mathbf{x})g(\mathbf{x})d\mathbf{x}, \tag{54}$$

where

$$\left(\mathrm{G}^2_{\kappa,\sigma;\delta} f\right)(\mathbf{x}) = \mathbf{x}^\kappa \int\limits_0^\infty \mathrm{G}^{\mathbf{m},\mathbf{n}}_{\mathbf{p},\mathbf{q}}\left[\frac{\mathbf{t}^\delta}{\mathbf{x}^\delta}\,\middle|\,\begin{matrix}(\mathbf{a}_i)_{1,\mathbf{p}}\\ (\mathbf{b}_j)_{1,\mathbf{q}}\end{matrix}\right]\mathbf{t}^\sigma f(\mathbf{t})\frac{\mathbf{dt}}{\mathbf{x}}.$$

Now present $\mathfrak{L}_{\overline{\nu},\overline{r}}$*-theory of the* $\mathrm{G}^1_{\sigma,\kappa;\delta}$*- transform, which follows from Theorems 4.2 and 4.3 for* $\mathrm{G}^1_{\sigma,\kappa}$*- transform and Lemma 2.1.*

Theorem 5.2 *Let*

$$a_1^* = 0, a_2^* = 0, ..., a_n^* = 0;\ \Delta_1 = \Delta_2 = ... = \Delta_n = 0;$$

$$\mathrm{Re}(\mu_1) = \mathrm{Re}(\mu_2) = ... = \mathrm{Re}(\mu_n) = 0;$$

$$\overline{\alpha}_1 < (\nu_1 - \mathrm{Re}(\kappa_1))/\delta_1 < \overline{\beta}_1, \overline{\alpha}_2 < (\nu_2 - \mathrm{Re}(\kappa_2))/\delta_2 < \overline{\beta}_2, ...,$$

$$\overline{\alpha}_n < (\nu_n - \mathrm{Re}(\kappa_1))/\delta_n < \overline{\beta}_n, \nu_1 = \nu_2 = ... = \nu_n;\ 1 < \overline{r} < \infty, r_1 = r_2 = ... = r_n.$$

There the following assertions are true:

(a) The transform $\mathrm{G}^1_{\sigma,\kappa;\delta}$ *defined on* $\mathfrak{L}_{\overline{\nu},\overline{2}}$ *can be extended to* $\mathfrak{L}_{\overline{\nu},\overline{r}}$ *as an element of* $[\mathfrak{L}_{\overline{\nu},\overline{r}}, \mathfrak{L}_{\overline{\nu}-\mathrm{Re}(\kappa+\sigma),\overline{r}}]$. *If* $1 < \overline{r} \leq \overline{2}$, *then the transform* $\mathrm{G}^1_{\sigma,\kappa;\delta}$ *is one-to-one on* $\mathfrak{L}_{\overline{\nu},\overline{r}}$ *and there holds the equality (49) for* $f \in \mathfrak{L}_{\overline{\nu},\overline{r}}$ *and* $\mathrm{Re}(s) = \overline{\nu} - \mathrm{Re}(\kappa + \sigma)$.

(b) If $f \in \mathfrak{L}_{\overline{\nu},\overline{r}}$ *and* $\overline{\lambda} = (\lambda_1, ..., \lambda_n) \in \mathrm{C}^n$, *then* $\mathrm{G}^1_{\sigma,\kappa;\delta} f$ *is represented in the form (52) for* $\mathrm{Re}(\overline{\lambda}) > (\overline{\nu} - \mathrm{Re}(\kappa))/\delta - 1$ *and in the form (53) for* $\mathrm{Re}(\overline{\lambda}) < (\overline{\nu} - \mathrm{Re}(k))/\delta - 1$.

(c) If $f \in \mathfrak{L}_{\overline{\nu},\overline{r}}$ *and* $g \in \mathfrak{L}_{1-\overline{\nu}+\mathrm{Re}(\kappa+\sigma),\overline{r}'}$ *with* $\overline{r}' = \overline{r}/(\overline{r} - 1)$, *then the relation (54) holds.*

(d) If $1 - \overline{\nu} + \mathrm{Re}(\kappa) \notin \mathcal{E}_{\overline{\mathcal{G}}}$, *then the transform* $\mathrm{G}^1_{\sigma,\kappa;\delta}$ *is one-to-one on* $\mathfrak{L}_{\overline{\nu},\overline{r}}$ *and its image is given by*

$$\mathrm{G}^1_{\sigma,\kappa;\delta}(\mathfrak{L}_{\overline{\nu},\overline{r}}) = \mathfrak{L}_{\overline{\nu}-\mathrm{Re}(\kappa+\sigma),\overline{r}}. \tag{55}$$

Theorem 5.3 *Let*

$$a_1^* = 0, a_2^* = 0, ..., a_n^* = 0; \Delta_1 = \Delta_2 = ... = \Delta_n = 0;$$

$$\mathrm{Re}(\mu_1) < 0, \mathrm{Re}(\mu_2) < 0, ..., \mathrm{Re}(\mu_n) < 0;$$

$$\overline{\alpha}_1 < (\nu_1 - \mathrm{Re}(\kappa_1))/\delta_1 < \overline{\beta}_1, \overline{\alpha}_2 < (\nu_2 - \mathrm{Re}(\kappa_2))/\delta_2 < \overline{\beta}_2, ...,$$

$\overline{\alpha}_n < (\nu_n - \mathrm{Re}(\kappa_1))/\delta_n < \overline{\beta}_n, \nu_1 = \nu_2 = ... = \nu_n;\ 1 < \overline{r} < \infty, r_1 = r_2 = ... = r_n;$ *and let* $\mathbf{m} > \mathbf{0}$ *or* $\mathbf{n} > \mathbf{0}$.

The following assertions are true:

(a) The transform $\mathrm{G}^1_{\sigma,\kappa;\delta}$ *defined on* $\mathfrak{L}_{\overline{\nu},\overline{2}}$ *can be extended to* $\mathfrak{L}_{\overline{\nu},\overline{r}}$ *as an element of* $[\mathfrak{L}_{\overline{\nu},\overline{r}}, \mathfrak{L}_{\overline{\nu}-\mathrm{Re}(\kappa+\sigma),\mathbf{s}}]$ *for any* $\mathbf{s} = (s_1, s_2, ..., s_n)$, $\mathbf{s} \geq \overline{r}$, *such that* $1/s_j = 1/r_j + \mathrm{Re}(\mu_j)$, $j = 1, 2, ..., n$. *If* $1 < \overline{r} \leq \overline{2}$, *then the transform* $\mathrm{G}^1_{\sigma,\kappa;\delta}$ *is one-to-one on* $\mathfrak{L}_{\overline{\nu},\overline{r}}$ *and there holds the equality (49) for* $f \in \mathfrak{L}_{\overline{\nu},\overline{r}}$ *and* $\mathrm{Re}(s) = \overline{\nu} - \mathrm{Re}(\kappa + \sigma)$.

(b) If $f \in \mathfrak{L}_{\overline{\nu},\overline{r}}$ *and* $g \in \mathfrak{L}_{1-\overline{\nu}+\mathrm{Re}(\kappa+\sigma),\mathbf{s}}$ *with* $1 < \mathbf{s} < \infty$ *and* $1 \leq 1/\overline{r} + 1/\mathbf{s} < 1 - \mathrm{Re}(\overline{\mu})$, *then the relation (54) holds.*

(c) Let $\overline{k} = (k_1, k_2, ..., k_n) > 0$. *If* $1 - (\overline{\nu} - \mathrm{Re}(\kappa))/\delta \notin \mathcal{E}_{\overline{\mathcal{G}}}$, *then the transform* $\mathrm{G}^1_{\sigma,\kappa;\delta}$ *is one-to-one on* $\mathfrak{L}_{\overline{\nu},\overline{r}}$ *and there hold*

$$\mathrm{G}^1_{\sigma,\kappa;\delta}(\mathfrak{L}_{\overline{\nu},\overline{r}}) = \mathrm{I}^{-\overline{\mu}}_{-;\delta\overline{k},(\sigma/\delta-\overline{\alpha})/\overline{k}}(\mathfrak{L}_{\overline{\nu}-\mathrm{Re}(\kappa+\sigma),\overline{r}}) \tag{56}$$

for $\mathbf{m} > \mathbf{0}$, *and*

$$\mathrm{G}^1_{\sigma,\kappa;\delta}(\mathfrak{L}_{\overline{\nu},\overline{r}}) = \mathrm{I}^{-\overline{\mu}}_{0+;\delta\overline{k},(\overline{\beta}-\sigma/\delta)/\overline{k}-1}(\mathfrak{L}_{\overline{\nu}-\mathrm{Re}(\kappa+\sigma),\overline{r}}) \tag{57}$$

for $\mathbf{n} > \mathbf{0}$. *When* $1 - (\overline{\nu} - \mathrm{Re}(\kappa))/\delta \in \mathcal{E}_{\overline{\mathcal{G}}}$, $\mathrm{G}^1_{\sigma,\kappa;\delta}(\mathfrak{L}_{\overline{\nu},\overline{r}})$ *is a subset of right hand sides of (56) and (57) in respective cases.*

(d) If $f \in \mathfrak{L}_{\overline{\nu},\overline{r}}$, $\overline{\lambda} = (\lambda_1, ..., \lambda_n) \in \mathrm{C}^n$, *then* $\mathrm{G}^1_{\sigma,\kappa;\delta} f$ *is given in (52) for* $\mathrm{Re}(\overline{\lambda}) > (\overline{\nu} - \mathrm{Re}(\kappa))/\delta - 1$, *while in (53) for* $\mathrm{Re}(\overline{\lambda}) < (\overline{\nu} - \mathrm{Re}(k))/\delta - 1$. *Furthermore* $\mathrm{G}^1_{\sigma,\kappa;\delta} f$ *is given in (3).*

In ([2], formulas (4.10), (4.11)) inversion formulas were obtained for transformation $\mathrm{G}^1_{\sigma,\kappa;\delta}f$*:*

$$f(\mathbf{x}) = -\delta \mathbf{x}^{\delta-(\kappa-1)+\delta(\overline{\lambda}+1)} \frac{\mathbf{d}}{\mathbf{dx}} \mathbf{x}^{-\delta(\overline{\lambda}+1)} \times$$

$$\int_0^\infty \mathrm{G}^{\mathbf{q-m,p-n+1}}_{\mathbf{p+1,q+1}} \left[\frac{\mathbf{t}^\delta}{\mathbf{x}^\delta} \middle| \begin{matrix} -\overline{\lambda}, -\mathbf{a_{n+1}}, \ldots, -\mathbf{a_p}, -\mathbf{a}_1, \ldots, -\mathbf{a_n} \\ -\mathbf{b_{m+1}}, \ldots, -\mathbf{b_q}, -\mathbf{b}_1, \ldots, -\mathbf{b_m}, -\overline{\lambda} - 1 \end{matrix} \right] \times$$

$$\mathbf{t}^{\delta-\sigma-1} (\mathrm{G}^1_{\sigma,\kappa;\delta} f)(\mathbf{t}) \mathbf{dt} \tag{58}$$

or

$$f(\mathbf{x}) = \delta \mathbf{x}^{\delta-(\kappa-1)+\delta(\overline{\lambda}+1)} \frac{\mathbf{d}}{\mathbf{dx}} \mathbf{x}^{-\delta(\overline{\lambda}+1)} \times$$

$$\int_0^\infty \mathrm{G}^{\mathbf{q-m+1,p-n}}_{\mathbf{p+1,q+1}} \left[\frac{\mathbf{t}^\delta}{\mathbf{x}^\delta} \middle| \begin{matrix} -\mathbf{a_{n+1}}, \ldots, -\mathbf{a_p}, -\mathbf{a}_1, \ldots, -\mathbf{a_n}, -\overline{\lambda}, \\ -\overline{\lambda} - 1, -\mathbf{b_{m+1}}, \ldots, -\mathbf{b_q}, -\mathbf{b}_1, \ldots, -\mathbf{b_m}, \end{matrix} \right] \times$$

$$\mathbf{t}^{\delta-\sigma-1} (\mathrm{G}^1_{\sigma,\kappa;\delta} f)(\mathbf{t}) \mathbf{dt}. \tag{59}$$

The conditions for the validity of these formulas are given by the statement that follows from Theorem 4.4, Lemma 2.1, and ([1], Theorem 10).

Theorem 5.4 *Let* $a_1^* = 0, a_2^* = 0, \ldots, a_n^* = 0;$ $\overline{\alpha}_1 < (\nu_1 - \mathrm{Re}(\kappa_1))/\delta_1 < \overline{\beta}_1,$ $\overline{\alpha}_2 < (\nu_2 - \mathrm{Re}(\kappa_2))/\delta_2 < \overline{\beta}_2, \ldots,$ $\overline{\alpha}_n < (\nu_n - \mathrm{Re}(\kappa_n))/\delta_n < \overline{\beta}_n;$ $\alpha_0^1 < 1 - (\nu_1 - \mathrm{Re}(\kappa_1))/\delta_1 < \beta_0^1,$ $\alpha_0^2 < 1 - (\nu_2 - \mathrm{Re}(\kappa_2))/\delta_2 < \beta_0^2, \ldots,$ $\alpha_0^n < 1 - (\nu_n - \mathrm{Re}(\kappa_n))/\delta_n < \beta_0^n$*; and let* $\overline{\lambda} \in \mathbf{C}^n$.

(a) If $\Delta_1[\nu_1 - \mathrm{Re}(\kappa_1)]/\delta_1 + \mathrm{Re}(\mu_1) = 0,$ $\Delta_2[\nu_2 - \mathrm{Re}(\kappa_2)]/\delta_2 + \mathrm{Re}(\mu_2) = 0, \ldots,$ $\Delta_n[\nu_n - \mathrm{Re}(\kappa_n)]/\delta_n + \mathrm{Re}(\mu_n) = 0$ $(\nu_1 = \nu_2 = \ldots = \nu_n)$ *and* $f \in \mathfrak{L}_{\overline{\nu},\overline{2}}$ *, then the inversion formulas (58) and (59) are valid for* $\mathrm{Re}(\overline{\lambda}) > (-\overline{\nu} + \mathrm{Re}(\kappa))/\delta$ *and* $\mathrm{Re}(\overline{\lambda}) < (-\overline{\nu} + \mathrm{Re}(\kappa))/\delta$*, respectively.*

(b) If $\Delta_1 = \Delta_2 = \ldots = \Delta_n = 0;$ $\mathrm{Re}(\mu_1) = \mathrm{Re}(\mu_2) = \ldots = \mathrm{Re}(\mu_n) = 0$ *and* $f \in \mathfrak{L}_{\overline{\nu},\overline{r}}$ $(\nu_1 = \nu_2 = \ldots = \nu_n),$ $1 < \overline{r} < \infty, r_1 = r_2 = \ldots = r_n$*, then the inversion formulas (58) and (59) are valid for* $\mathrm{Re}(\overline{\lambda}) > (-\overline{\nu} + \mathrm{Re}(\kappa))/\delta$ *and* $\mathrm{Re}(\overline{\lambda}) < (-\overline{\nu} + \mathrm{Re}(\kappa))/\delta$*, respectively.*

References

1. Sitnik SM, Skoromnik OV (2020) One-dimensional and multi-dimensional integral transforms of Buschman–Erdelyi type with Legendre Functions in kernels, Transmutat Oper Appl Trends in Math 293–319
2. M.V. Papkovich , O.V. Skoromnik , *Multi-dimensional modified G- transformations and integral transformations with hypergeometric Gauss functions in kernels in weight spaces of summed functions.* Belorussia, Bulletin of the Vitebsk State university No 1 (114)(2022, in Russian), pp. 5–20
3. Samko SG, Kilbas AA, Marichev OI (1993) Fractional integrals and derivatives: theory and applications. Gordon and Breach Science Publishers, London
4. Sitnik SM, Skoromnik OV, Shlapakov SA (2019) Multidimensional general integral transformation with special functions in the kernel, Belorussia, Bulletin of the Vitebsk State university No 3 (104) (in Russian), pp 18–27
5. Kilbas AA, Srivastava HM, Trujillo JJ (2006) Theory and applications of fractional differential equations. Elsevier, Amsterdam
6. Kilbas AA, Saigo M (2004) H-Transforms. Theory and applications. Chapman and Hall, Boca Raton, 400 pp
7. Kiryakova V (1994) Generalized fractional calculus and applications. Wiley, New York, p 205
8. Prudnikov AP, Brychkov YA, Marichev OI (1990) Integrals and series. More special functions, vol 3. Gordon and Breach., New York
9. Mathai AM, Saxena RK (1978) The H-function with applications in statistics and other disciplines. Halsted Press, Wiley, New York, p 192
10. Srivastava HM, Gupta KC, Goyal SL (1982) The H-function of one and two variables with applications. South Asian Publishers., New Delhi, 415 pp
11. Titchmarsh E (1948) Introduction to the theory of Fourier integrals (Gostekhizdat M-L, in Russian), 334 pp
12. Ditkin VA, Prudnikov VP (1974) Integral transformations and operational calculus (M Nauka, in Russian), 542 pp
13. Sneddon IN (1951) Fourier transform (M. ILL., in Russian), 667 pp
14. Zayed AI (1996) Handbook of function and generalized function transformation. CRC Press, Boca Raton, New York, London, Tokyo, 540 pp
15. Antipov MY, Kolyshkin AA, Remi V (1993) Applied integral transforms, CRM monograph series, vol 2. American Mathematical Society, Providence, Rhode Island USA, 265 pp
16. Sneddon IN (1975) Application of integral transforms in the theory of elasticity. International Centre for Mathematical Sciences, Courses and Lectures. Springer, Vienna-New York, 220 pp
17. Sneddon IN (1966) Mixed boundary problems in potential theory . North - Holland Publ., Amsterdam, 283 pp
18. Sneddon IN (1974) The use of integral transformations, THM Edition. McGraw Hill, New Delhi, 187 pp
19. Virchenko NA (1969) Paired (triple) integral equations. Vishcha shkola, Kiev (in Russian), 160 pp
20. Uflyand YG (1977) The method of paired equations in problems of mathematical physics. Nauka, Leningrad department (in Russian), 220 pp
21. Sneddon IN (1962) Fractional integrals and derivatives and dual integral equations. North Carolina State College, The Applied Mathematics Research Group Ralligh, PSR–6, 40 pp
22. Srivastava HM, Buschman GG (1977) Convolution integral equations with special function kernels. Wiley Eastern, New Delhi, Banglore, p 164
23. Zayed AI (1968) Generalized integral transformations, Pure applied mathematics, vol 18. Interscience Publ., Wiley, New York, London, Sydney, 254 pp
24. Debnath L (2000) Integral transforms and their applications. CRC Press, Boca Raton, Florida, p 402
25. Brychkov YA, Prudnikov AP (1977) Integral transformations of generalized functions M. Nauka (in Russian), 287 pp

26. Pathak RS (1997) Intergral transforms of generalized functions and their applications. Gordon and Breach Science Publ., Amsterdam, 467 pp
27. Brychkov YA, Glaeske H-Y, Prudnikov AP, Tuan VK (1992) Multidimensional Integral Transformations. Gordon And Breach, Philadelphia, 404 p
28. Rooney PG (1989) On the and transformations. Can J Math 32:1021–1044
29. Rooney PG (1994) On the range of the integral transformation. Can Math Bull 37(4):545–548
30. Rooney PG (1995) On the representation of functions by the Hankel and some related transformations. Proc Roy Soc Edinb Sect A 125(3):449–463
31. Heywood PQ, Rooney PG (1988) On the Hankel and some related transformations. Can J Math 40:989–1009
32. Heywood PQ, Rooney PG (1994) On the Struve transformations. SIAM J Math Anal 25:450–461
33. Rooney PG (1982/83) On integral transformations with G- function kernels. Proc Roy Soc Edinb Sect A 93:265–297
34. Erdelyi A, Magnus W, Oberhettinger F, Tricomi FG (1953) Higher transcendental functions, vol 1. McGraw-Hill, New York

On Sufficient Conditions of the Faddeev–Marchenko Theorem

B. D. Koshanov and A. P. Soldatov

Abstract We study sufficient conditions, ensuring validity of the Faddeev–Marchenko fundamental theorem on restoration of the potential of the Sturm–Liouville equation on the entire axis along the given linear ratios between the Jost functions. These conditions are formulated in terms of so-called reflection coefficient within the framework of the corresponding weighted Holder spaces on the real line with power behavior at infinity.

Keywords Jost functions · Reflection coefficient · Markushevich's problem · Weighted Gelder spaces · Fredholm property · Index formula of the problem

We consider the following spectral Sturm–Liouville problem

$$-y'' + q(x)y = k^2 y, \quad x \in \mathbb{R}, \tag{1}$$

on the entire axis in the classical setting, when the real coefficient q satisfies the condition

$$\int_{\mathbb{R}} (1 + |x|)|q(x)|dx < \infty.$$

B. D. Koshanov
Department of Mathematics, Al-Farabi Kazakh National University, Al-Farabi Ave. 71, Almaty 050040, Kazakhstan
e-mail: koshanov@list.ru

International University of Information Technologies, st. Manas, 34A, Almaty 050040, Kazakhstan

A. P. Soldatov (✉)
Federal Research Center "Informatics and Control" RAS, Moscow, Russia
e-mail: soldatov48@gmail.com

NRU BelSU, Belgorod, Russia

V. Vasilyev (ed.), *Differential Equations, Mathematical Modeling and Computational Algorithms*, Springer Proceedings in Mathematics & Statistics 423,
https://doi.org/10.1007/978-3-031-28505-9_15

It is well-known [1], that in the class of functions, vanishing at infinity, this problem has a discrete spectrum at a finite number of points $\lambda = -æ_j^2$, $1 \le j \le n$, of the negative part of the real axis $\lambda = k^2$, and continuous spectrum fills the real half. In this case, all eigenvalues $-æ_j^2$ are simple.

We introduce the so-called Ghost functions $f_j(x, k)$, $j = 1, 2$, as solutions of the Eq. (1) with given asymptotics at infinity:

$$f_1(x,k) - e^{ikx} \to 0 \text{ as } x \to +\infty; \quad f_2(x,k) - e^{-ikx} \to 0 \text{ as } x \to -\infty.$$

They are defined unambiguously and represented by the following formulas

$$f_1(x,k) = e^{ikx} + \int_x^\infty A_1(x,t)e^{ikt}dt, \quad f_2(x,k) = e^{-ikx} + \int_x^\infty A_2(x,t)e^{-ikt}dt,$$

with certain real kernels $A_j(x, t)$, moreover, the integrals converge absolutely. Therefore,

$$\begin{aligned} f_1(x,k) &= e^{ikx}[1 + g_1(x,k)], \quad g_1(x,k) = \textstyle\int_0^\infty A_1(x, s+x)e^{iks}ds, \\ f_2(x,k) &= e^{-ikx}[1 + g_2(x,k)], \quad g_2(x,k) = \textstyle\int_0^\infty A_2(x, x-s)e^{iks}ds, \end{aligned} \tag{2}$$

where at every fixed x functions $B_1(s) = A_1(x, s + x)$ and $B_2(s) = A_2(x, x - s)$ are summable on the half axis $(0, \infty)$. In particular, $f_j(x, k)$ are continuously extended to the functions $f_j(x, \zeta)$, analytic in upper half plane $D_+ = \{\operatorname{Im} \zeta > 0\}$ and vanishing at infinity. With respect to the variable x, the corresponding functions $f_j(x, \zeta)$ satisfy analogous to (1) equation $-f_j'' + f_j = \zeta^2 f_j$.

At a fixed real $k \neq 0$ pairs of functions $\{f_j(x, k), \overline{f_j(x, k)}\}$ form two fundamental systems of solutions, and consequently, connected by the relation:

$$f_1(x,k) = b(k) f_2(x,k) + a(k)\overline{f_2(x,k)}$$

with some coefficients $a(k)$, $b(k)$. Properties of these coefficients a, b are clarified in detail (see, e.g., [1])

$$\begin{aligned} a(-k) = \overline{a(k)}, \quad b(-k) = \overline{b(k)}, \quad |a(k)|^2 = 1 + |b(k)|^2, \\ a(k) = 1 + O(k^{-1}), \; b(k) = O(k^{-1}) \;\text{ as }\; k \to \infty, \\ k[a(k) + b(k)] \to 0 \;\text{ as }\; k \to 0. \end{aligned} \tag{3}$$

Moreover, the function $a(k)$ continues analytically in the upper half-plane D_+, has simple zeros at points $\zeta = iæ_j$, $1 \le j \le n$, and excluding them, everywhere is different from zero, moreover,

$$\widetilde{a}(\zeta) = \zeta a(\zeta) \in C(B_1), \quad B_1 = \{|\zeta| \le 1, \; \operatorname{Im} \zeta \ge 0\}, \tag{4}$$

and $a(\zeta) \to 1$ as $\zeta \to \infty$.

In explicit form $2ika(k) = (f_1'f_2 - f_2'f_1)(x, k)$, where due to (1), the Wronskian in the right-hand side of this equality does not depend on x. Hence, a similar expression is also valid for analytic extension

$$2i\zeta a(\zeta) = f_1'(x, \zeta)f_2(x, \zeta) - f_2'(x, \zeta)f_1(x, \zeta)$$

of this function.

Obviously, the function $a(\zeta)$ can be represented in the form

$$a(\zeta) = a_1(\zeta)\left(\frac{\zeta - i\text{æ}_1}{\zeta + i\text{æ}_1}\right)\cdots\left(\frac{\zeta - i\text{æ}_n}{\zeta + i\text{æ}_n}\right), \tag{5}$$

where the function $a_1(\zeta)$ is everywhere different from zero. Consequently, in the upper half plane, the analytic function $\ln a_1(\zeta)$ is defined, which vanishes at infinity and continuous in its closure (except the point $\zeta = 0$).

The inverse problem of scattering theory is in restoring the coefficient q of the Sturm–Liouville equation for the given set $\text{æ}_1, \ldots, \text{æ}_n$ of positive numbers and pairs of the functions a, b with the properties (3)–(5). Let's describe the central result of this theory developed by Faddeev [2] and Marchenko [3].

Main theorem of Faddeev–Marchenko *Suppose that set of positive numbers $\text{æ}_1, \ldots, \text{æ}_n$, continuous at $k \neq 0$ pairs of functions a, b with the properties (3)–(5) and sets of numbers $m_{1,l}, \ldots, m_{n,l}$, $l = 1, 2$, with properties $m_{j,1}m_{j,2} = -[a'(i\text{æ}_j)]^2$, $1 \le j \le n$ are given. Let the functions*

$$\begin{aligned} F_1(x) &= \sum_{j=1}^{n} \frac{1}{m_{j,1}} e^{-\text{æ}_j x} - \frac{1}{2\pi}\int_{\mathbb{R}} \frac{b(-k)}{a(k)} e^{ikx}\,dk, \\ F_2(x) &= \sum_{j=1}^{n} \frac{1}{m_{j,2}} e^{\text{æ}_j x} + \frac{1}{2\pi}\int_{\mathbb{R}} \frac{b(k)}{a(k)} e^{-ikx}\,dk, \end{aligned} \tag{6}$$

be continuous, differentiable and for any $x_0 \in \mathbb{R}$, satisfy the condition

$$F_1(x),\ F_1'(x), xF_1'(x) \in L^1(x_0, +\infty);\ F_2(x),\ F_2'(x), xF_2'(x) \in L^1(-\infty, x_0). \tag{7}$$

Then in the integral representation (2), kernels A_j are uniquely defined as a solution to the Gelfand–Levitan integral equations

$$\begin{aligned} A_1(x, y) + F_1(x + y) + \int_x^{\infty} A_1(x, t)F_1(t + y)dt &= 0, \quad y \ge x, \\ A_2(x, y) + F_2(x + y) + \int_{-\infty}^{x} A_2(x, t)F_2(t + y)dt &= 0, \quad x \ge y, \end{aligned}$$

and the functions $f_j(x, k)$ satisfy the Sturm–Liouville equation with the coefficient

$$q(x) = -2[A_1(x, x)]' = 2[A_2(x, x)]'.$$

Often, instead of the coefficient b, the ratio $r = b/a$ is used, which is called the reflection coefficient. From (3) it can be seen that it should satisfy the conditions

$$r(-k) = \overline{r(k)}, \quad |r(k)| < 1, \; k \neq 0; \quad r(k) = O(k^{-1}) \text{ as } k \to \infty, \tag{8}$$

$$|a(k)|^{-2} = 1 - |r(k)|^2. \tag{9}$$

Moreover, in designation (4) we can write $k[a(k) + b(k)] = \widetilde{a}(k)[1 + r(k)]$, so when $\widetilde{a}(0) = 0$ condition (3c) holds automatically. On the other hand, this condition leads to the limit

$$\lim_{k \to 0} r(k) = -1 \quad \text{for} \quad \widetilde{a}(0) \neq 0.$$

We note that at $\widetilde{a}(0) = 0$ behavior of $r(k)$ as $k \to 0$ is not fully clarified.

Obviously, in the equality (9) we can replace a to the function a_1, figured in (5). In particular, for the boundary value $\ln a_1^+(t)$, $t \in \mathrm{R}$, of the analytic function $\ln a_1(\zeta)$ we have the relation $-2 \ln a_1^+ = \ln(1 - |r|^2)$. Therefore, this function with an accuracy up to an imaginary constant can be restored by using the Schwartz formula [4]:

$$-2 \ln a_1(\zeta) = \frac{1}{\pi i} \int_{\mathrm{I}} \mathrm{R} \frac{\ln[1 - |r(t)|^2] dt}{t - \zeta}, \quad z \in D_+. \tag{10}$$

The following question arises: for which reflection coefficients $r(k)$ and coefficients a, b built from them, the functions (6) satisfy conditions (7) of the main theorem. This issue was discussed in detail by Levitan [5]. His results have been completed in [6], based on the classical properties of the Cauchy-type integral:

$$(I\varphi)(\zeta) = \frac{1}{2\pi i} \int_{\mathbb{R}} \frac{\varphi(t) dt}{t - \zeta}, \; \zeta \in D_\pm, \tag{11}$$

defining an analytic function in the half plane $D_\pm = \{\pm \mathrm{Im}\, \zeta > 0\}$, and the singular Cauchy integral

$$(S\varphi)(t_0) = \frac{1}{\pi i} \int_{\mathrm{I}} \mathrm{R} \frac{\varphi(t) dt}{t - t_0}, \; t_0 \in \mathbb{R}. \tag{12}$$

Relationship between the last integral and boundary values $(I\varphi)^\pm$ of the function $I\varphi$ is caried out by the Sokhotski–Plemelj formulas

$$2(I\varphi)^\pm = \pm\varphi + S\varphi. \tag{13}$$

In this paper, within the weighted Holder spaces, we describe the class of reflection coefficients $r(k)$, which ensures validity of the main theorem.

Assume that $C^{\mu}(G)$, $0 < \mu < 1$, means the Holder class on the closed set $G \subseteq \mathbb{C}$. We recall that it consists of all functions $\varphi(z)$, $z \in G$, for which the norm

$$|\varphi| = |\varphi|_0 + [\varphi]_{\mu}, \tag{14}$$

is finite, where

$$|\varphi|_0 = \sup_{z \in G} |\varphi(z)|, \quad [\varphi]_{\mu} = \sup_{z_1, z_2 \in G} \frac{|\varphi(z_1) - \varphi(z_2)|}{|z_1 - z_2|^{\mu}}.$$

For $0 < \mu < \nu$, embedding of Banach spaces $C^{\nu}(G) \subseteq C^{\mu}(G)$ holds. If the set G is bounded, then, according to the Ascoli's theorem, this embedding is compact.

For unbounded sets G, they also introduce [7] the space of functions $\varphi(z)$, that satisfy Holder conditions with some exponent μ in relation to the metric of the Riemann sphere $\mathbb{C} \cup \{\infty\}$. This condition can be expressed in the form

$$|\varphi(z_1) - \varphi(z_2)| \leq \frac{C|z_1 - z_2|^{\mu}}{(1 + |z_1|)^{\mu}(1 + |z_2|)^{\mu}}, \quad z_1, z_2 \in G, \tag{15}$$

with some constant $C > 0$, it is equivalent to the fact that $\varphi(z)$ as well as the function $\underset{\sim}{\varphi}(1/z)$ satisfies the Gelder conditions on any compact subsets, consequently, $\overline{G}$ and $\tilde{G} = \{z, \ 1/z \in \overline{G}\}$. This class we denote by $H(G)$, and its elements φ, obviously, allow the limit $\varphi(\infty) = \lim \varphi(z)$ as $z \to \infty$. Condition $\varphi(\infty) = 0$ distinguishes in it a class, which we denote as $\overset{\circ}{H} (\overline{D}_{+})$.

Let the weight function $\rho_{\lambda}(z) = (1 + |z|^2)^{\lambda/2}$ be with real exponent λ. We denote by $C^{\mu}_{\lambda}(G, \infty)$ a weighted space of functions $\varphi(z)$, for which the following norm is finite:

$$|\varphi| = |\rho_{-\lambda}\varphi|_0 + [\rho_{\mu-\lambda}\varphi]_{\mu}. \tag{16}$$

In the case when G is a closed domain, the space $C^{n,\mu}(G)$ of differentiable functions, all derivatives of which up to the nth order, belong to $C^{\mu}(G)$, can be inductively defined by the conditions φ, $\varphi' \in C^{n-1,\mu}(G)$. By analogy with it the space $C^{n,\mu}_{\lambda}(G, \infty)$ is inductively introduced by the conditions

$$\varphi \in C^{n-1,\mu}_{\lambda}, \quad \varphi' \in C^{n-1,\mu}_{\lambda-1}, \tag{17}$$

where φ' is understood as a pair of private derivatives.

All these spaces are particular case of analogous spaces with more general weight functions, introduced and studied in [8]. In particular, multiplication as a bilinear mapping is bounded $C^{n,\mu}_{\lambda'} \times C^{n,\mu}_{\lambda''} \to C^{n,\mu}_{\lambda'+\lambda''}$, the space $C^{n,\mu}_{0}$ is a Banach algebra by multiplication, and the operator $\varphi \to \rho_{\delta}\varphi$ performs isomorphism of Banach spaces $C^{n,\mu}_{\lambda} \to C^{n,\mu}_{\lambda+\delta}$. Moreover, family of Banach spaces $(C^{n,\mu}_{\lambda})$ decreases monotonically (in the sense of embedding between Banach spaces) with respect to the parameter

μ and monotonically increases with respect to λ. At the same time, for a function φ, given in the set G and vanishing at infinity, condition (15) is equivalent to the fact that φ belongs to the space $C^{\mu}_{-\mu}(G,\infty)$. In particular, we get the embedding $C^{\mu}_{\lambda} \subseteq C^{\varepsilon}_{-\varepsilon}$ for $0 < \varepsilon \le \min(\mu, -\lambda)$ and the equality

$$\overset{\circ}{H}(G) = \cup_{0<\varepsilon<1} C^{\varepsilon}_{-\varepsilon}(G,\infty).$$

We also note the embedding of Banach spaces

$$C^{\mu}(G) \subseteq C^{\mu}_{\mu}(G,\infty),$$

which directly follows from definition (16), since

$$C^{\mu}(G) \subseteq C^{\mu}_{\mu}(G,\infty) = \{\varphi, \mid [\varphi]_{\mu} < \infty\}.$$

Furthermore, as a set G we will consider mainly the line $\mathbb{R}$ and the half-plane $D_{\pm}$. In the first case, φ' in (17) is understood as ordinary derivative, and in the second case, $C^{n,\mu}_{\lambda}$ is understood as a space of analytical in $D_{\pm}$ functions. We special study the space $C^{n,\mu}_{0}(\mathbb{R},\infty)$.

Lemma 1 *Let $\varphi \in C^{n,\mu}_{0}(\mathbb{R},\infty)$ and a function f be analytic in a neighbourhood of some compact K, containing a set of values φ.*

Then the superposition $f \circ \varphi \in C^{n,\mu}_{0}(\mathbb{R},\infty)$.

Proof First, consider the case $n = 0$. On the compact K function f satisfies the Lipschitz conditions

$$|f(z_1) - f(z_2)| \le L|z_1 - z_2|$$

with some constant $L > 0$. Therefore, as applied to the above-introduced semi-norm $\{\varphi\}_{\mu}$ we have obvious estimate $\{f \circ \varphi\}_{\mu} \le L\{\varphi\}_{\mu}$, hence, $f \circ \varphi \in C^{\mu}_{0}$.

In general case use induction by n. Then, due to (17), functions $f \circ \varphi$ and $\rho_1 f' \circ \varphi$ belong to $C^{n-1,\mu}_{0}$. Therefore, in the right side of the equality

$$\rho_1(f \circ \varphi)' = (f' \circ \varphi)(\rho_1 \varphi')$$

both cofactors belong to $C^{n-1,\mu}_{0}$, thus, by definition $f \circ \varphi \in C^{n,\mu}_{0}$.

Based on the negative non-integer weight order λ, we introduce the space $\widetilde{C}^{\mu}_{\lambda}(\overline{D}_{\pm},\infty)$ as a finite-dimensional expansion $C^{\mu}_{\lambda}(\overline{D}_{\pm},\infty)$ by polynomials of the variable $u = (\zeta \pm i)^{-1}$. Obviously, for $-1 < \lambda < 0$ it coincides with $C^{\mu}_{\lambda}(\overline{D}_{\pm},\infty)$, for $-2 < \lambda < -1$ its elements uniquely represented in the form

$$\phi(\zeta) = \alpha(\zeta \pm i)^{-1} + \phi_0(\zeta), \quad \phi_0 \in C^{n,\mu}_{\lambda}(\overline{D}_{\pm},\infty),$$

with some $\alpha \in \mathbb{C}$. Similarly, when $-3 < \lambda < -2$ elements of this space are uniquely represented as

$$\phi(\zeta) = \alpha(\zeta \pm i)^{-1} + \beta(\zeta \pm i)^{-2} + \phi_0(\zeta), \quad \phi_0 \in C_\lambda^\mu(\overline{D}_\pm, \infty),$$

with some α, $\beta \in \mathbb{C}$ and etc.

In general case, the space $\widetilde{C}_\lambda^\mu(\overline{D}_\pm, \infty)$ for $-s-1 < \lambda < -s$ is an extension of $C_\lambda^\mu(\overline{D}_\pm, \infty)$ on s measurements.

Similar extension $\widetilde{C}_\lambda^\mu(\mathbb{R}, \infty)$ we can introduce in relation to polynomials of the variable $v = t/(t^2+1)$, $t \in \mathbb{R}$, or, equivalently, in relation to polynomials of the variable $u = (t \pm i)^{-1}$.

At last, corresponding spaces $\widetilde{C}_\lambda^{n,\mu}$ of differentiable functions are determined inductively by conditions analogous to (17)

$$\varphi \in \widetilde{C}_\lambda^{n-1,\mu}, \quad \varphi' \in \widetilde{C}_{\lambda-1}^{n-1,\mu}.$$

From the definitions, it is clear that the operation $\phi \to \phi^\pm$ acts $\widetilde{C}_\lambda^{n,\mu}(D_\pm, \infty) \to \widetilde{C}_\lambda^{n,\mu}(\mathbb{R}, \infty)$.

Let's turn to the Cauchy-type integral (11) and the associated singular integral (12).

Theorem 1 *For any non-integer negative λ operator I is bounded and invertible $\widetilde{C}_\lambda^{n,\mu}(\mathbb{R}, \infty) \to \widetilde{C}_\lambda^{n,\mu}(\overline{D}_\pm, \infty)$, moreover, for a piecewise analytic function ϕ from $\widetilde{C}_\lambda^{n,\mu}(\overline{D}_\pm, \infty)$ in the half-plane $D_\pm$, equality $\phi^+ - \phi^- = \varphi$ is equivalent to $\phi = I\varphi$.*

Singular operator S is bounded in $\widetilde{C}_\lambda^{n,\mu}(\mathbb{R}, \infty)$ and coincides with its inverse, i.e. $S^2 = 1$, where 1 means the unit operator.

Proof Let us first make sure that the operator I is bounded $C_\lambda^{n,\mu}(\mathbb{R}, \infty) \to \widetilde{C}_\lambda^{n,\mu}(\overline{D}_\pm, \infty)$. For definiteness, we consider only the case of upper half plane, and first consider the case $-1 < \lambda < 0$, when the wave in space notation can be omitted. For $n = 0$ this statement is established in [8], moreover, it is easy to see that for $\phi \in C_\lambda^\mu(\overline{D}_+, \infty)$ the Cauchy formula is valid:

$$\phi(\zeta) = (I\phi^+)(\zeta), \quad \zeta \in D_+. \tag{18}$$

Let $\varphi \in C_\lambda^{n,\mu}$, $n \geq 1$, then by using the method of integration by parts, we obtain the equality:

$$(I\varphi)'(\zeta) = (I\varphi')(\zeta), \quad \zeta \in D_+. \tag{19}$$

Since

$$\frac{1}{t-\zeta} + \frac{1}{\zeta+i} = \frac{t+i}{\zeta+i}\,\frac{1}{t-\zeta}, \quad \zeta \in D_+, \tag{20}$$

and integral from φ' over $\mathbb{R}$ is equal to zero, it follows that

$$(\zeta+i)(I\varphi)'(\zeta)=\frac{1}{2\pi i}\int_{\mathrm{I}}\mathrm{R}\frac{(t+i)\varphi'(t)dt}{t-\zeta}.$$

On the basis of the Sokhotski–Plemelj formula, taking into account the relation $[(I\varphi)^+]'=[(I\varphi)']^+$, following from (19), we obtain the corresponding equality for the singular integral:

$$(t_0+i)(S\varphi)'(t_0)=\frac{1}{\pi i}\int_{\mathrm{I}}\mathrm{R}\frac{(t+i)\varphi'(t)dt}{t-t_0}.$$

Therefore, in relation to the operation $(\mathcal{D}\phi)(\zeta)=(\zeta+i)\phi'(\zeta)$ and a similar operation on the line we get equality $\mathcal{D}I=I\mathcal{D},\ \mathcal{D}S=S\mathcal{D}$. Since (17) can be rewritten in the form $\phi,\mathcal{D}\phi\in C^{n-1,\mu}_{\lambda}(\overline{D}_+,\infty)$ and similarly for spaces on the line, From here, by induction, we establish rightness of the theorem and for the space $C^{n,\mu}_{\lambda},\ -1<\lambda<0$.

For $-2<\lambda<-1$ equality (20) shows that

$$(\zeta+i)(I\varphi)(\zeta)=\frac{1}{2\pi i}\int_{\mathrm{I}}\mathrm{R}\varphi(t)dt+(I\varphi_0)(\zeta)$$

with the function $\varphi_0(t)=(t+i)\varphi(t)\in C^{n,\mu}_{\lambda+1}(\mathbb{R},\infty)$. It remains to note that, as proved above, the operator I is bounded $C^{n,\mu}_{\lambda+1}(\mathbb{R},\infty)\to C^{n,\mu}_{\lambda+1}(\overline{D}_+,\infty)$.

Further, let $-3<\lambda<-2$. Then, due to (20), we can write

$$\frac{t+i}{\zeta+i}\,\frac{1}{t-\zeta}+\frac{t+i}{(\zeta+i)^2}=\left(\frac{t+i}{\zeta+i}\right)^2\frac{1}{t-\zeta},$$

thus

$$\frac{1}{t-\zeta}+\frac{1}{\zeta+i}+\frac{t+i}{(\zeta+i)^2}=\left(\frac{t+i}{\zeta+i}\right)^2\frac{1}{t-\zeta}.$$

Similarly to the previous case, from here we get boundedness of the operator I for the considered case of weight orders λ.

Continuing this process, we establish validity of the considered statement for all non-integer negative λ. On the other hand, the Cauchy formula (18) can be applied to polynomial of variable $u=(\zeta+i)^{-1}$, which, according to the definition, leads to boundedness of the operator $I:\ \widetilde{C}^{n,\mu}_{\lambda}(\mathbb{R},\infty)\to\widetilde{C}^{n,\mu}_{\lambda}(\overline{D}_+,\infty)$.

Let, further, $\phi\in\widetilde{C}^{n,\mu}_{\lambda}(\widehat{D},\infty)$. Applying the Sokhotski–Plemelj formulas (13) to the function $\phi_0=\phi-I(\phi^+-\phi^-)$, we come to equality $\phi_0^+=\phi_0^-$. Therefore, the function ϕ_0 is analytic over the entire plane and vanishes at infinity, what is possible only for $\phi_0=0$.

Thus, the first part of Theorem is established. Statement about boundedness of the operator S in the space $\widetilde{C}^{n,\mu}_{\lambda}(\mathbb{R},\infty)$ follows from the first part of Theorem and formulas (13). Equality $S^2=1$ is proved by usual way [7], by writing the Cauchy formula (18) for $\phi=I\varphi$ and again using (13).

We return to description of the reflection coefficients $r(k)$, ensuring validity of the main theorem.

Theorem 2 *Let a function* $r(t)$ *with properties (8) belong to the class* $C^{2,\nu}_{\delta-\nu}(\mathbb{R},\infty)$, $0<\nu<1$, $-3<\delta<-2$, *and be subordinated to one of the following two assumptions:*

(*i*) $|r(0)|<1$;

(*ii*) $r(0)=-1$, $(|r|^2)''(0)\neq 0$, *moreover, the function* $r_0(t)=t^{-2}[1-|r(t)|^2]$ *belongs to the class* $C^{2,\nu}$ *in the neighbourhood* $t=0$.

Then functions a, *defined by formulas (5), (10), and* $b=ra$ *satisfy all conditions (3)–(4), moreover,*

$$a^{-1}-1\in\widetilde{C}^{\nu}_{\delta}(\overline{D}_{+},\infty), \tag{21}$$

and functions F_1, F_2, *defined by formulas (6), satisfy conditions (7) of the main theorem.*

Proof Each of two suggestions (i), (ii) we will consider separately.

(i) In this case, the function $\ln(1-|r|^2)$ can be represented as $|r|^2h(|r|^2)$, where $h(s)=s^{-1}\ln(1-s)$, $|s|<1$. Since $|r(t)|^2=r(t)\overline{r(t)}\in C^{2,\nu}_{2\delta-2\nu}\subseteq C^{2,\nu}_{0}$, according to Lemma 1, we conclude that $h(|r|^2)\in C^{2,\nu}_{0}(\mathbb{R},\infty)$ and, consequently,

$$\ln(1-|r|^2)\in C^{2,\nu}_{2\delta-2\nu}(\mathbb{R},\infty)\subseteq C^{2,\nu}_{\delta}(\mathbb{R},\infty). \tag{22}$$

Therefore, due to (10) and Theorem 1, the function $\ln a_1\in\widetilde{C}^{2,\nu}_{\delta}(\overline{D}_{+},\infty)$. According to Lemma 1, as above, we conclude that

$$e^{\pm\ln a_1}-1\in\widetilde{C}^{2,\nu}_{\delta}(\overline{D}_{+},\infty)$$

and, consequently,

$$a-1\in\widetilde{C}^{2,\nu}_{\delta}(\mathbb{R},\infty),\quad a^{-1}-1\in\widetilde{C}^{2,\nu}_{\delta}(\overline{D}_{+},\infty).$$

Let us turn to verification of condition (7) of the main theorem. For the first terms in the right side of (8), this condition is obviously done. Therefore, it is enough to check it for the Fourier transform of the function

$$r(t)=b(t)/a(t),\quad \widetilde{r}(t)=b(-t)/a(t),$$

figured in (6).

We denote by $M(\mathbb{R})$ image of $L^1(\mathbb{R})$ in the Fourier transform. Then taking into account well-known properties, it is enough to prove that all functions $r(t)$, $r'(t)$, $tr(t)$ and $\widetilde{r}(t)$, $\widetilde{r}'(t)$, $t\widetilde{r}(t)$ belong to $M(\mathbb{R})$. We use the fact that class of differentiable functions, which, together with their derivatives, belong to $L^2(\mathbb{R})$, contain in $M(\mathbb{R})$. Hence, the case is reduced to proof that

$$\varphi(t), \varphi'(t), \varphi''(t), t\varphi(t), t\varphi'(t) \in M(\mathbb{R}). \tag{23}$$

for each of functions $\varphi = r, \widetilde{r}$.

For $\varphi = r \in C^{2,\nu}_{\delta-\nu}(\mathbb{R}, \infty)$, taking into account the inequality $\delta < -3/2$, these conditions obviously hold. For function $\varphi = \widetilde{r}$, due to (3) and (5), (10) we can write $\widetilde{r}(t) = c_0(t)\overline{r(t)}$ with coefficient

$$c_0(t) = \frac{\overline{a(t)}}{a(t)} = \prod_{j=1}^{n} \left(\frac{t + i æ_j}{t - i æ_j}\right)^2 e^{-2i \arg a_1(t)}.$$

From (10) and the Sokhotski–Plemelj formulas it follows that $2 \arg a_1 = i S[\ln(1 - |r|^2)]$. Therefore, due to (22) and Theorem 1, we get

$$\arg a_1(t) \in \widetilde{C}^{2,\nu}_{\delta}(\mathbb{R}, \infty) \subseteq C^{2,\nu}_{0}(\mathbb{R}, \infty).$$

According to Lemma 1, it yields that $c_0 \in C^{2,\nu}_0(\mathbb{R}, \infty)$ and, thus $\widetilde{r} = c_0\overline{r} \in C^{2,\nu}_{\delta}$ $(\mathbb{R}, \infty)$. Therefore, (23) hold and for $\varphi = \widetilde{r}$.

(ii) In this case, for sufficiently small $\varepsilon > 0$, we introduce the function

$$g(t) = \ln[1 - |r(t)|^2] - 2\ln|t| \in C^{2,\nu}[-2\varepsilon, 2\varepsilon]. \tag{24}$$

Inside the semicircle $B_\varepsilon = \{|z| \le \varepsilon, \ \mathrm{Re}\,\zeta \ge 0\}$, we consider analytic functions

$$\psi(\zeta) = \ln\zeta - \frac{1}{\pi i}\int_{-2\varepsilon}^{2\varepsilon} \frac{\ln|t|dt}{t - \zeta}.$$

Due to the Sokhotski–Plemelj formulas, for its boundary value, we have the expression

$$\psi^+(t_0) = i \arg t_0 - \frac{1}{\pi i}\int_{-2\varepsilon}^{2\varepsilon} \frac{\ln|t|dt}{t - t_0}.$$

Thus, $\mathrm{Re}\,\psi^+ = 0$ and, consequently, the function ψ analytically continues inside the circle $\{|\zeta| < \varepsilon\}$. Together with (10), (24) it follows that

$$-2\ln a_1(\zeta) = \frac{1}{\pi i}\int_{|t|\ge 2\varepsilon} \frac{\ln[1 - |r(t)|^2]dt}{t - \zeta} + \frac{1}{\pi i}\int_{-2\varepsilon}^{2\varepsilon} \frac{g(t)dt}{t - \zeta} + 2\ln z - 2\psi(\zeta),$$

so that, taking into account Theorem 1, we have

$$\ln a_1(\zeta) + \ln\zeta \in C^{2,\nu}(B_\varepsilon). \tag{25}$$

In particular, it implies (4) with $\widetilde{a}(0) \ne 0$.

We write density $\ln(1-|r|^2)$ as sum $\varphi_0+\varphi_1$, where $\varphi_0(t)=0$ for $|t|\geq\varepsilon/2$ and $\varphi_1\in C^{\nu}_{\delta}(\mathbb{R},\infty)$. Then, on the basis of Theorem 1, in addition to (25), we obtain that

$$\ln a_1(\zeta)\in\widetilde{C}^{\nu}_{\delta}(B'_{\varepsilon},\infty) \tag{26}$$

outside of semi-circle $B'_{\varepsilon}=\{|\zeta|\geq\varepsilon,\ \operatorname{Im}\zeta\geq 0\}$. As a result, we establish validity of all statements of the first part of Theorem, including (26).

Conditions (7) are checked exactly by the same way as the case (i).

References

1. Levitan BM (1987) Inverse Sturm-Liouville problems. VNU Sci. Press, Utrecht
2. Faddeev LD (1967) Properties of the S-matrix of the one-dimensional Schrodinger equation. Am Math Soc Transl Ser 2(65):139–166
3. Marchenko VA (1986) Sturm–liouville operators and their applications. Birkhauser
4. Bitsadze AV (1984) Fundamentals of the theory of analytic functions. Nauka, Moscow (in Russian)
5. Levitan BM (1980) Sufficient conditions for the solvabilty of the inverse problem of scattering theory on the entire line. Sb Math 36(3):323–329
6. Soldatov AP, Zhura NA (2015) On a representation of the solution of the inverse Sturm–Liouville problem on the entire line. Differ Equ 51(8):1022–1032. https://doi.org/10.1134/S0012266115080066
7. Mushelishvili NI (1953) Singular integral equations. Noordhoff, Groningen
8. Soldatov AP (2020) Singular integral operators and elliptic boundary value problems. I. J Math Sci 245(6):695–891. https://doi.org/10.1007/s10958-020-04717-0

Variational Approach to Construction of Piecewise-Constant Approximations of the Solution of Dynamic Reconstruction Problem

Nina Subbotina and Evgenii Krupennikov

Abstract In the paper, the problem of dynamic reconstruction of controls and trajectories for deterministic control-affine systems is considered. The reconstruction is performed in real time using known discrete inaccurate measurements of the observed trajectory of the system. This trajectory is generated by an unknown measurable control that satisfies known geometric constraints. A well-posed statement of the problem is given. A solution is proposed using the variational approach developed by the authors. This approach uses auxiliary variational problem with regularized integral residual functional. The integrant of the functional is a d.c. function. The suggested algorithm reduces the reconstruction problem to integration of Hamiltonian systems of ordinary differential equations. This paper offers a method for construction of piecewise-constant approximations that satisfy the given geometric control constraints. The approximations converge almost everywhere to the desired control, and the reconstructed trajectories of the dynamical system converge uniformly to the observed trajectory.

Keywords Control theory · Inverse problems · Calculus of variations · Hamiltonian systems · D.C. functions

---

N. Subbotina (✉) · E. Krupennikov
N.N. Krasovskii Institute of Mathematics and Mechanics of the Ural Branch of the Russian Academy of Sciences (IMM UB RAS), 16 S.Kovalevskaya Str., Yekaterinburg 620108, Russia
e-mail: subb@uran.ru
URL: https://www.imm.uran.ru/eng

E. Krupennikov
e-mail: krupennikov@imm.uran.ru

Ural Federal University named after the first President of Russia B.N.Yeltsin, 19 Mira street, Ekaterinburg 620002, Russia

V. Vasilyev (ed.), *Differential Equations, Mathematical Modeling and Computational Algorithms*, Springer Proceedings in Mathematics & Statistics 423,
https://doi.org/10.1007/978-3-031-28505-9_16

1 Introduction

The theory of dynamic controlled processes has a lot of applications: in engineering, traffic control, economics, medicine, etc. The studying of such processes leads to the need to research and solve not only direct problems, aimed at constructing controls that optimize some quality criteria, but also to solve inverse problems of the control reconstruction based on data on the observed motion.

In this paper, control-affine deterministic systems are under consideration. The admissible controls are measurable functions with values in a known compact convex set. Discrete inaccurate measurements of the observed motion are known. The dynamic reconstruction problem is to find the whole trajectory and the unknown control that generated this motion. The reconstruction must be performed in real time synchronized with arrival of new measurements.

There are many different approaches to solving inverse problems (see, for example, [1–5]). Surveys of some of the approaches can be found in [6, 7].

The authors of this paper have suggested an original variational approach [8, 9]. It provides a method for construction of approximations of the desired controls with the use of constructions from auxiliary variational problems. The key feature of the method is that integrants of the functionals in the auxiliary problems are d.c. functions [10]. The method uses stationary points of the functionals.

This paper offers development of this method. In [8, 9], the authors developed an algorithm for construction of approximations of the desired control in the form of oscillatory high-frequency functions. These approximations are bounded, but not necessary satisfy the admissible controls' restrictions. In this paper, a modification of the previously developed method is suggested and justified. The new modification of the method allows to construct piecewise-constant approximations of the control, which satisfy the geometrical restrictions on the admissible controls. It is proved that they converge almost everywhere to the desired control, while the previously suggested approximations converge weakly*. The conditions on the approximation parameters are obtained that provide the convergence of the approximations.

The suggested algorithm reduces the dynamic reconstruction problem to solving systems of linear ordinary differential equations and numerical integration.

Results of numerical simulation are exposed on the example of a dynamical model from the area of medicine.

2 Previously Obtained Results

In [8, 9], a dynamic control reconstruction problem was stated. A new method for solving this problem, based on auxiliary constructions from variational problems, was suggested and justified. This section offers a brief review of these results.

2.1 Dynamics

Dynamic controlled systems of the following form are considered:

$$\frac{dx(t)}{dt} = G(t, x(t))u(t) + f(t, x(t)),$$
$$x \in \mathbb{R}^n,\ u \in \mathbb{R}^m,\quad G(\cdot) : [0, T] \times \mathbb{R}^n \to \mathbb{R}^{n\times m},\ f(\cdot) : [0, T] \times \mathbb{R}^n \to \mathbb{R}^n, \tag{1}$$
$$m \geq n,\quad t \in [0, T],\quad T < \infty,$$

where $x(\cdot)$ is the state variables vector and $u(\cdot)$ is the vector of the control parameters.

The admissible controls are measurable functions satisfying the restrictions

$$u(t) \in \mathbf{U} \subset \mathbb{R}^m \quad \text{a. e. on } [0, T], \tag{2}$$

where $\mathbf{U}$ is a convex compact set.

2.2 Input Data

Some trajectory of system (1), generated by an unknown admissible control, is being observed. It is called the basic trajectory $x^*(\cdot) : [0, T] \to \mathbb{R}^n$. Discrete inaccurate measurements of the basic trajectory arrive in real time. The measurements have error $\delta > 0$ and arrive with regular time step $h^\delta > 0$:

$$\{y_k^\delta : \|y_k^\delta - x^*(t_k)\| \leq \delta,\quad t_k = kh^\delta,\quad k = 0, \ldots, N,\quad N = \lceil T/h^\delta \rceil\}. \tag{3}$$

The notation $\|\cdot\|$ means the Euclidean norm.

The problem is to reconstruct the unknown control, generating $x^*(\cdot)$, by known data (3).

2.3 Assumptions

We assume that the input data (1)–(3) satisfy the following assumptions.

A.1 There exist constants $d_0 > 0$, $\delta_0 > 0$, $h_0 > 0$ and a compact $\Psi \subset \mathbb{R}^n$ such that for any accuracy $\delta \in (0, \delta_0]$ and any measurement step $h^\delta \in (0, h_0]$

$$\bigcup_{k=0,\ldots,N} B_{d_0}[y_k^\delta] \subset \Psi,$$

where $B_r[x]$ is the closed ball of the radius r with the center in x.

A.2 The matrix function $G(\cdot)$ and the vector function $f(\cdot)$ from dynamics (1) are Lipschitz continuous on $D_0 \triangleq [0, T] \times \Psi$ with the Lipschitz constant $L_{D_0} > 0$:

$$\forall (t_1, x_1), (t_2, x_2) \in D_0 \quad \|G(t_2, x_2) - G(t_1, x_1)\|_2 \le L_{D_0} \|(t_2, x_2) - (t_1, x_1)\|,$$
$$\|f(t_2, x_2) - f(t_1, x_1)\| \le L_{D_0} \|(t_2, x_2) - (t_1, x_1)\|.$$

The notation $\| \cdot \|_2$ means the spectral Matrix norm induced by the Euclidean norm.

A.3 Rang of $G(t, x)$ equals n for $(t, x) \in D_0$.

2.4 Dynamic Reconstruction Problem

It was shown in [9] that the problem of reconstruction of the unknown control, generating the basic trajectory, is ill-posed since such control may be not unique. To regularize this problem, a notation of the normal control was introduced.

Definition 1 Normal control $u^*(\cdot) : [0, T] \to \mathbb{R}^m$ is the measurable control, generating the basic trajectory $x^*(\cdot)$, that has the minimal norm in $L^2([0, T], \mathbb{R}^m)$ space.

It was proved [9] that for a basic trajectory $x^*(\cdot)$, satisfying assumptions **A.1–A.3**, there exists a unique normal control.

An additional assumption was introduced:

A.4 The normal control $u^*(\cdot)$ satisfies restrictions (2). Thus, it is an admissible control.

The following **dynamic reconstruction problem (the DRP)** was stated:

For any $\delta \in (0, \delta_0]$, $h^\delta \in (0, h_0]$ and the corresponding set of measurements $\{y_k^\delta\}$ (3) at time instant t_k $(t = 1, \ldots, N)$ to construct a measurable control $u^\delta(\cdot) : [0, t_k] \to \mathbb{R}^m$ such that at the terminal instant $t_N = T$ the following conditions are fulfilled:

B.1 The functions $u^\delta(\cdot) : [0, T] \to \mathbb{R}^m$ are bounded substantially and uniformly with respect to the parameter δ.

B.2 Each control $u^\delta(\cdot)$ generates a trajectory $x^\delta(\cdot) : [0, T] \to \mathbb{R}^n$ of system (1) with the boundary conditions $x^\delta(0) = y_0^\delta$ such that

$$\lim_{\delta \to 0} \|x^\delta(\cdot) - x^*(\cdot)\|_{C([0,T],\mathbb{R}^n)} = 0.$$

B.3 The functions $u^\delta(\cdot)$ weakly* converge to the normal control:

$$u^\delta(\cdot) \xrightarrow[\delta \to 0]{w^*} u^*(\cdot).$$

Here

$$\|f(\cdot)\|_{C([0,T],\mathbb{R}^n)} = \max_{t \in [0,T]} \|f(t)\|$$

is the norm in the $C([0, T], \mathbb{R}^n)$ space. The notation $\xrightarrow{w^*}$ stands for weak* convergence in the $L^1([0, T], \mathbb{R}^m)$ space:

$$u^\delta(\cdot) \xrightarrow[\delta\to 0]{w^*} u^*(\cdot) \Leftrightarrow \int_0^T \langle g(\tau), u^\delta(\tau) - u^*(\tau)\rangle d\tau \xrightarrow{\delta\to 0} 0 \quad \forall g(\cdot) \in C([0, T], \mathbb{R}^m). \tag{4}$$

The notation $\langle\cdot, \cdot\rangle$ means the scalar product.

2.5 Algorithm

A step-by-step algorithm for solving the DRP was described and justified in papers [8, 9].

On each step of this algorithm (that is, for $t \in [t_{k-1}, t_k]$, $k = 1, \dots, N$) three procedures are performed.

First, a third-order spline interpolation $y^\delta(\cdot) : [0, t_k] \to \mathbb{R}^n$ of the discrete measurements (3) is constructed. The interpolation is constructed on each step on the corresponding interval $[t_{k-1}, t_k]$ and the function $y^\delta(\cdot)$ is continuously differentiable on $[0, t_k]$.

Then, the function $y^\delta(\cdot)$ is used to state an auxiliary variational problem. It consists of find a pair of functions $x_k(\cdot) : [t_{k-1}, t_k] \to \mathbb{R}^n$, $u_k(\cdot) : [t_{k-1}, t_k] \to \mathbb{R}^m$ such that:

D.1 They are continuously differentiable functions that satisfy the dynamics equation (1) and there exist such function $s_k(\cdot) \in \mathbb{C}^1([t_{k-1}, t_k], \mathbb{R}^n)$ that $u_k(\cdot)$ has the structure

$$u_k(t) = -\frac{1}{\alpha^2} G^\top(t_{k-1}, y^\delta_{k-1}) s_k(t), \quad t \in [t_{k-1}, t_k].$$

D.2 They satisfy the boundary conditions

$$k = 1 : \quad x_1(0) = y^\delta_0, \quad s_1(0) = 0,$$
$$k = 2, \dots, N : \quad x_k(t_{k-1}) = y^\delta_{k-1}, \quad s_k(t_{k-1}) = s_{k-1}(t_{k-1}).$$

D.3 They provide a stationary point of the functional

$$I(x(\cdot), u(\cdot)) = \int_{t_{k-1}}^{t_k} -\frac{\|x(t) - y^\delta_k(t)\|^2}{2} + \frac{\alpha^2 \|u(t)\|^2}{2} dt,$$

where $\alpha > 0$ is a small regularizing [11] parameter.

Remark 1 In the suggested method just stationary points of this functional are used. So, there is no need to find the extremum of the functional. It is an original feature of the method.

The conditions for the stationary point can be written in the form of a Hamiltonian system of non-linear ODEs. This system is linearized and after linearization has the form

$$\begin{gathered}
\frac{dx_k(t)}{dt} = -\frac{1}{\alpha^2} Q_k s_k(t) + f_k, \\
\frac{ds_k(t)}{dt} = x_k(t) - y^\delta(t), \\
t \in [t_{k-1}, t_k], \\
k = 1: \quad x_1(0) = y_0^\delta, \quad s(0) = 0, \\
k = 2, \ldots, N: \quad x_k(t_{k-1}) = y_{k-1}^\delta, \quad s_k(t_{k-1}) = s_{k-1}(t_{k-1}),
\end{gathered} \tag{5}$$

where

$$Q_k \triangleq G_k G_k^\top, \quad G_k \triangleq G(t_{k-1}, y_{k-1}^\delta), \quad f_k \triangleq f(t_{k-1}, y_{k-1}^\delta).$$

In system (5), $s_k(\cdot)$ are adjoint variables. Finally, the solution $s_k^{\alpha,\delta}(\cdot) : [t_{k-1}, t_k] \to \mathbb{R}^n$ of system (5) is used to construct the DRP solution as the piecewise-defined functions

$$u^\delta(t) = \{u_k^{\alpha,\delta}(t), \ t \in [t_{k-1}, t_k]\}, \quad u_k^{\alpha,\delta}(t) = -\frac{1}{\alpha^2} G_k s_k^{\alpha,\delta}(t). \tag{6}$$

The algorithm is described in details in [9].

2.6 The Main Result

The following theorem is the main result of [8, 9].

Theorem 1 *If assumptions* **A.1–A.4** *hold for the input data (1)–(3), then the constructed functions $u^\delta(\cdot)$ (6) satisfy conditions* **B.1–B.3** *if the following agreement of the approximation parameters holds:*

$$\begin{gathered}
h^\delta = h^\delta(\delta), \quad \alpha = \alpha(\delta), \\
\lim_{\delta \to 0} h^\delta = 0, \quad \lim_{\delta \to 0} \alpha = 0, \quad \lim_{\delta \to 0} \frac{\delta}{h^\delta} = 0, \quad \lim_{\delta \to 0} \frac{\alpha}{(h^\delta)^2} = K_0 > 0.
\end{gathered} \tag{7}$$

3 The New Results

3.1 New DRP Statement

Let us now reformulate the **DRP** in another way (the differences are marked wits bold font):

For any $\delta \in (0, \delta_0]$, $h^\delta \in (0, h_0]$ and the corresponding set of measurements $\{y_k^\delta\}$ (3) at time instant t_k ($t = 1, \ldots, N$) to construct a **piecewise-constant** control $u^\delta(\cdot) : [0, t_k] \to \mathbb{R}^m$ such that at the terminal instant $t_N = T$ the following conditions are fulfilled:

$\mathcal{B}$**.1** The functions $u^\delta(\cdot) : [0, T] \to \mathbb{R}^m$ satisfy the restriction (2). In other words, they **are admissible controls.**

$\mathcal{B}$**.2** Each control $u^\delta(\cdot)$ generates a trajectory $x^\delta(\cdot) : [0, T] \to \mathbb{R}^n$ of system (1) with the boundary conditions $x^\delta(0) = y_0^\delta$ such that

$$\lim_{\delta \to 0} \|x^\delta(\cdot) - x^*(\cdot)\|_{C(\mathbb{R}^n;[0,T])} = 0.$$

$\mathcal{B}$**.3** The functions $u^\delta(\cdot)$ **converge almost everywhere** to the normal control:

$$u^\delta(t) \xrightarrow{\delta \to 0} u^*(t) \text{ for a. e. } t \in [0, T].$$

3.2 Algorithm for Solving the New DRP

In this paper a modification of the algorithm, described in Sect. 2.5, is suggested that allows to construct the solution of the DRP $\mathcal{B}$**.1**–$\mathcal{B}$**.3**.

The new algorithm is the same with the exception of adding two additional procedures on each step.

First, the constructed function $u_k^{\alpha,\delta}(\cdot)$ (6) is “averaged” by the formula

$$\bar{u}_k^{\alpha,\delta} \triangleq \frac{1}{h^\delta} \int\limits_{t_{k-1}}^{t_k} u_k^{\alpha,\delta}(\tau) d\tau.$$

Then, the following piecewise-constant “cut-off” functions are constructed:

$$\hat{u}^\delta(t) \triangleq \begin{cases} \bar{u}_k^{\alpha,\delta}, & \bar{u}_k^{\alpha,\delta} \in \mathbf{U} \\ \hat{u} \in \mathbf{U} : \ \|\hat{u} - \bar{u}_k^{\alpha,\delta}\| = \min\limits_{u \in \mathbf{U}} \|u - \bar{u}_k^{\alpha,\delta}\|, & \bar{u}_k^{\alpha,\delta} \notin \mathbf{U} \end{cases}, \quad (8)$$
$$t \in [t_{k-1}, t_k], \quad k = 1, \ldots, N.$$

We consider the functions $\hat{u}^\delta(\cdot)$ as the DRP $\mathcal{B}$**.1**–$\mathcal{B}$**.3** solution.

3.3 The Main Result

The following theorem is true.

Theorem 2 *If assumptions* **A.1–A.4** *hold for the input data (1)–(3), then the constructed functions $\hat{u}^{\delta}(\cdot)$ (8) satisfy conditions* $\mathcal{B}$**.1**–$\mathcal{B}$**.3** *if the agreement of the approximation parameters (7) holds.*

Proof Condition $\mathcal{B}$**.1** is fulfilled by the construction (8).

First, we prove that condition $\mathcal{B}$**.3** holds. Consider the first equation of (5). The matrix $Q_k = G_k G_k^{\top}$ is positively semi-definite since the rows of G_k are linearly independent [13, Chap. 1, p. 6]. So, we can substitute the solution $x_k^{\alpha,\delta}(t), s^{\alpha,\delta}(t)$ of (5) into this equation and multiply it by the inverse matrix Q_k^{-1}:

$$Q_k^{-1}\left(\frac{dx_k^{\alpha,\delta}(t)}{dt} - f_k\right) = -\frac{1}{\alpha^2} s^{\alpha,\delta}{}_k(t)$$

$$\Rightarrow$$

$$G_k^{\top} Q_k^{-1}\left(\frac{dx_k^{\alpha,\delta}(t)}{dt} - f_k\right) = -\frac{1}{\alpha^2} G_k^{\top} s^{\alpha,\delta}{}_k(t) \overset{def}{=} u_k^{\alpha,\delta}(t).$$

Apply the "averaging" procedure (8) to the latter expression:

$$\bar{u}_k^{\alpha,\delta} \overset{def}{=} \frac{1}{h^{\delta}} \int\limits_{t_{k-1}}^{t^k} u_k^{\alpha,\delta}(t)dt = \frac{1}{h^{\delta}} \int\limits_{t_{k-1}}^{t^k} G_k^{\top} Q_k^{-1}\left(\frac{dx_k^{\alpha,\delta}(t)}{dt} - f_k\right) dt$$

$$\Rightarrow \tag{9}$$

$$\bar{u}_k^{\alpha,\delta} = G_k^{\top} Q_k^{-1}\left(\frac{x_k^{\alpha,\delta}(t_k) - x_k^{\alpha,\delta}(t_{k-1})}{h^{\delta}} - f_k\right), \quad k = 1, \ldots, N.$$

Now, we will compare for each segment $[t_{k-1}, t_k]$ the "averaged" values $\bar{u}_k^{\alpha,\delta}$ with the values of the "averaged" normal control

$$\bar{u}^*(t) \triangleq \{\bar{u}_k^*,\ t \in [t_{k-1}, t_k]\}, \quad \bar{u}_k^* \triangleq \frac{1}{h^{\delta}} \int\limits_{t_{k-1}}^{t_k} u^*(t)dt. \tag{10}$$

It was proved in [9, Sect. 2.4] that

$$u^*(t) = G^{\top}(t, x^*(t)) Q^{-1}(t, x^*(t)) \left(\frac{dx^*(t)}{dt} - f(t, x^*(t))\right) \quad \text{a. e. on } [0, T],$$

$$Q(t, x^*(t)) \triangleq G(t, x^*(t)) G^{\top}(t, x^*(t)).$$

Substitute this expression into (10):

$$\bar{u}_k^* = \frac{1}{h^\delta} \int\limits_{tk-1}^{t_k} G^\top(t, x^*(t)) Q^{-1}(t, x^*(t)) \left(\frac{dx^*(t)}{dt} - f(t, x^*(t)) \right) dt$$

$$= \frac{1}{h^\delta} \int\limits_{tk-1}^{t_k} (G^\top(t, x^*(t)) Q^{-1}(t, x^*(t)) \pm G_k^\top Q_k^{-1}) \left(\frac{dx^*(t)}{dt} - f(t, x^*(t)) \pm f_k \right) dt$$

$$= \frac{1}{h^\delta} \int\limits_{tk-1}^{t_k} (G^\top(t, x^*(t)) Q^{-1}(t, x^*(t)) - G_k^\top Q_k^{-1}) \left(\frac{dx^*(t)}{dt} - f_k \right)$$

$$+ G_k^\top Q_k^{-1} (f_k - f(t, x^*(t)))dt + \frac{1}{h^\delta} \int\limits_{tk-1}^{t_k} G_k^\top Q_k^{-1} (\frac{dx^*(t)}{dt} - f_k) dt$$

$$= r_k(\delta, h^\delta) + G_k^\top Q_k^{-1} \left(\frac{x^*(t_k) - x^*(t_{k-1})}{h^\delta} - f_k \right), \tag{11}$$

where

$$r_k(\delta, h^\delta) = \frac{1}{h^\delta} \int\limits_{tk-1}^{t_k} (G^\top(t, x^*(t)) Q^{-1}(t, x^*(t)) - G_k^\top Q_k^{-1}) \left(\frac{dx^*(t)}{dt} - f_k \right) \tag{12}$$
$$+ G_k^\top Q_k^{-1} (f_k - f(t, x^*(t)))dt.$$

Estimate the norms of the following expressions from (12):

$$\begin{aligned} &\|G_k^\top Q_k^{-1} - G^\top(t, x^*(t)) Q^{-1}(t, x^*(t))\|_2 \\ &\stackrel{def}{=} \|G^\top(t_{k-1}, y_{k-1}^\delta) Q^{-1}(t_{k-1}, y_{k-1}^\delta) - G^\top(t, x^*(t)) Q^{-1}(t, x^*(t)) \\ &\pm G^\top(t_{k-1}, x^*(t_{k-1})) Q^{-1}(t_{k-1}, x^*(t_{k-1}))\|_2 \\ &\leq L_{G^\top Q^{-1}} (\delta + h^\delta (K + 1)), \end{aligned} \tag{13}$$

where

$$K = \max_{u \in \mathbf{U}, (t,x) \in D_0} \|G(t, x)u + f(t, x)\|,$$

and $L_{G^\top Q^{-1}}$ is the Lipschitz constant of the matrix function $G^\top(\cdot) Q^{-1}(\cdot) : [0, T] \to \mathbb{R}^{n \times n}$. This matrix function is Lipschitz continuous since assumption **A.2**. Indeed, consider arbitrary (t_1, x_1) and (t_2, x_2) from D_0.

$$
\begin{aligned}
G_i &\triangleq G(t_i, x_i), \quad Q_i \triangleq Q(t_i, x_i), \quad i = 1, 2, \\
\|G_2 Q_2^{-1} - G_1 Q_1^{-1} \pm G_2 Q_1^{-1}\|_2 &= \|G_2(Q_2^{-1} - Q_1^{-1}) + (G_2 - G_1) Q_1^{-1}\|_2 \\
&= \|G_2 Q_2^{-1}(Q_1 - Q_2) Q_1^{-1} + (G_2 - G_1) Q_1^{-1}\|_2 \\
&\leq L_{D_0} R_{Q^{-1}} (R_G R_{Q^{-1}} + 1) \|(t_2, x_2) - (t_1, x_1)\|,
\end{aligned}
$$

where

$$
R_G = \max_{(t,x)\in D_0} \|G(t,x)\|_2, \quad R_{Q^{-1}} \max_{(t,x)\in D_0} \|Q^{-1}(t,x)\|_2.
$$

Note that $R_{Q^{-1}} < \infty$ since $Q^{-1}(\cdot)$ is continuous [13, Chap. 8, p. 4].

Then,

$$
\begin{aligned}
\|f(t, x^*(t)) - f_k\| &\overset{def}{=} \|f(t, x^*(t)) - f(t_{k-1}, y_{k-1}^\delta) \pm f(t_{k-1}, x^*(t_{k-1}))\| \\
&\leq L_{D_0}(\delta + h^\delta(K+1))
\end{aligned} \tag{14}
$$

And finally,

$$
\begin{aligned}
\|f(t, x^*(t))\| &\leq R_f = \max_{(t,x)\in D_0} \|f(t,x)\|, \\
\|G^\top(t, x^*(t)) Q^{-1}(t, x^*(t))\|_2 &\leq R_{G^\top Q^{-1}} = \max_{(t,x)\in D_0} \|G^\top(t,x) Q^{-1}(t,x)\|_2.
\end{aligned} \tag{15}
$$

Applying (13)–(15) to (12), we get

$$
\|r_k(\delta, h^\delta)\| \leq (\delta + h^\delta(K+1)) \left(L_{G^\top Q^{-1}}(K + R_f) + R_{G^\top Q^{-1}} L_{D_0} \right). \tag{16}
$$

Now, we compare the expressions for $\bar{u}_k^{\alpha,\delta}$ (9) and $\bar{u}_k^*$ (11):

$$
\begin{aligned}
&\|\bar{u}_k^* - \bar{u}_k^{\alpha,\delta}\| \\
&\leq \|r_k(\delta, h^\delta)\| + \left\| G_k^\top Q_k^{-1} \frac{\left(x^*(t_k) - x_k^{\alpha,\delta}(t_k)\right) - \left(x^*(t_{k-1}) - x_k^{\alpha,\delta}(t_{k-1})\right)}{h^\delta} \right\|.
\end{aligned} \tag{17}
$$

It was proved in [9, p. 3.4] that

$$
\begin{aligned}
\|x_k^{\alpha,\delta}(t) - y^\delta(t)\| &\leq \frac{T}{h^\delta} \alpha (\lambda^*)^{0.5} r_s(\delta, h^\delta, \alpha) + r_z(\delta, h^\delta, \alpha) \triangleq r_x(\delta, h^\delta, \alpha), \\
& t \in [t_{k-1}, t_k], \quad k = 1, \ldots, N,
\end{aligned} \tag{18}
$$

were

$$r_s(\delta, h^\delta, \alpha) = n\left(\frac{L_{D_0}}{\lambda_{1,i}}(2\delta + h^\delta(K+1)) + 12\frac{\alpha}{\lambda_*^{1.5}}\frac{(2\delta + h^\delta K)}{(h^\delta)^2} + 48\frac{\alpha^3}{\lambda_*^2}\frac{(2\delta + h^\delta K)}{(h^\delta)^3}\right),$$

$$r_z(\delta, h^\delta, \alpha) = n\left(48\frac{\alpha^2}{\lambda_*}\frac{(2\delta + h^\delta K)}{(h^\delta)^2} + 24\frac{\alpha^3}{\lambda_*^{1.5}}\frac{(2\delta + h^\delta K)}{(h^\delta)^3}\right).$$

The parameters λ_* and λ^* are respectively the minimum and the maximum values of the eigenvalues of the symmetric [13, Chap. 1, p. 6] matrix function $Q(t, x) = G(t, x)G^\top(t, x)$, $(t, x) \in D_0$. Note that the matrix function $Q(\cdot)$ is continuous on D_0 (see Assumption **A.2**). Therefore, its eigenvalues $\{\lambda_1(\cdot), \ldots, \lambda_n(\cdot)\}$ are also continuous on D_0 [13, Chap. 8, p. 8]. Therefore, λ_* and λ^* exist.

So, we use (18) to get that

$$\|x_k^{\alpha,\delta}(t_k) - x^*(t_k)\| \le \|x_k^{\alpha,\delta}(t) - y_k^\delta\| + \|y_k^\delta - x^*(t_k)\| \le r_x(\delta, h^\delta, \alpha) + \delta. \quad (19)$$

Also, note the boundary conditions from (12):

$$x_k^{\alpha,\delta}(t_{k-1}) = y_{k-1}^\delta. \quad (20)$$

Applying (19) and substituting (20) into (17), we get that

$$\|\bar{u}_k^* - \bar{u}_k^{\alpha,\delta}\| \le r_k(\delta, h^\delta) + R_{G^\top Q^{-1}}\frac{r_x(\delta, h^\delta, \alpha) + 2\delta}{h^\delta} \triangleq r_{\hat{u}}(\delta, h^\delta, \alpha), \quad k = 1, \ldots, N. \quad (21)$$

It follows from the definitions (16), (18) that

$$r_{\hat{u}}(\delta, h^\delta, \alpha) \xrightarrow{\delta \to 0} 0, \quad (22)$$

if the agreement conditions (7) hold.

We will now show that the piecewise-constant functions $\hat{u}^\delta(\cdot)$ converge pointwise almost everywhere on $[0, T]$ to $u^*(\cdot)$. First, consider the function

$$U(t) = \int_0^t u^*(\tau)d\tau, \quad t \in [0, T].$$

Since $u^*(\cdot)$ is measurable, $U(\cdot) : [0, T] \to \mathbb{R}^m$ is differentiable almost everywhere on $[0, T]$ [12]. Fix such an arbitrary $t \in [0, T]$, where $\dot{U}(t) = u^*(t)$ exists. For any $h^\delta > 0$ there exists a unique number $k_{t,\delta} \in \{1, \ldots, N\}$ such that $t \in [t_{k_{t,\delta}-1}, t_{k_{t,\delta}})$.

Consider the following discrepancy for $t \in [t_{k_{t,\delta}-1}, t_{k_{t,\delta}})$:

$$\|\hat{u}^\delta(t) - u^*(t)\| \le \|\hat{u}^\delta(t) - \bar{u}_{k_{t,\delta}}^{\alpha,\delta}\| + \|\bar{u}_{k_{t,\delta}}^{\alpha,\delta} - \bar{u}_{k_{t,\delta}}^*\| + \|\bar{u}_{k_{t,\delta}}^* - u^*(t)\|. \quad (23)$$

Since $\dot{U}(t)$ exists, we use the fundamental increment lemma on the following expressions:

$$\bar{u}^*_{k_{t,\delta}} \overset{def}{=} \frac{1}{h^\delta}\left(\int_{t_{k_{t,\delta}-1}}^{t} u^*(\tau)d\tau + \int_{t}^{t_{k_{t,\delta}}} u^*(\tau)d\tau\right)$$
$$= \frac{1}{h^\delta}\big((U(t_{k_{t,\delta}}) - U(t)) + (U(t) - U(t_{k_{t,\delta}-1}))\big)$$
$$= \frac{1}{h^\delta}\big(u^*(t)(t_{k_{t,\delta}} - t) + o(t_{k_{t,\delta}} - t) + u^*(t)(t - t_{k_{t,\delta}-1}) + o(t - t_{k_{t,\delta}-1})\big) \quad (24)$$
$$= u^*(t) + \frac{o(t_{k_{t,\delta}} - t) + o(t - t_{k_{t,\delta}-1})}{h^\delta}$$
$$\Rightarrow$$
$$\|\bar{u}^*_{k_{t,\delta}} - u^*(t)\| = \frac{o(t_{k_{t,\delta}} - t) + o(t - t_{k_{t,\delta}-1})}{h^\delta} \xrightarrow{h^\delta \to 0} 0$$

Applying (21), (8) and (24) to (23), we obtain that the functions $\hat{u}^\delta(\cdot)$ converge pointwise almost everywhere on $[0, T]$ to $u^*(\cdot)$ as $\delta \to 0$ provided the agreement conditions (7) hold. So, condition $\mathcal{B}.3$ is fulfilled.

Let us now check that condition $\mathcal{B}.2$ holds. In other words, that the trajectories $\hat{x}^\delta(\cdot) : [0, T] \to \mathbb{R}^n$ of system (1), generated by the controls $\hat{u}^\delta(\cdot)$, uniformly converge to $x^*(\cdot)$.

By definition,

$$\|\hat{x}^\delta(t) - x^*(t)\|$$
$$= \Big\| y_0^\delta - x^*(0) + \int_0^t G(\tau, \hat{x}^\delta(\tau))\hat{u}^\delta(\tau) + f(\tau, \hat{x}^\delta(\tau))d\tau$$
$$- \int_0^t G(\tau, x^*(\tau))u^*(\tau) - f(\tau, x^*(\tau))d\tau$$
$$+ \int_0^t \pm G(\tau, x^*(\tau))\hat{u}^\delta(\tau) \pm G_k(\hat{u}^\delta(\tau) - u^*(\tau))d\tau \Big\|$$
$$\leq \delta + R_G \left\| \int_0^t \hat{u}^\delta(\tau) - u^*(\tau)d\tau \right\| + \int_0^t \|G(\tau, x^*(\tau)) - G_k\|_2 \|\hat{u}^\delta(\tau) - u^*(\tau)\|d\tau$$
$$+ \int_0^t \|f(\tau, \hat{x}^\delta(\tau)) - f(\tau, x^*(\tau))\| + R_u \|G(\tau, \hat{x}^\delta(\tau)) - G(\tau, x^*(\tau))\|_2 d\tau, \quad (25)$$

where

$$R_G = \max_{(t,x)\in D_0} \|G(t,x)\|_2, \quad R_u = \max_{u\in \mathbf{U}} \|u\|.$$

In (25), considering (8) and (21),

$$\begin{aligned}
&\left\| \int_0^t \hat{u}^\delta(\tau) - u^*(\tau)d\tau \right\| \\
&= \left\| \sum_{j=1}^{k_{t,\delta}-1} \left[\int_{t_{j-1}}^{t_j} \hat{u}^\delta(\tau)d\tau - \int_{t_{j-1}}^{t_j} u^*(\tau)d\tau \right] + \int_{k_{t,\delta-1}}^{t} \hat{u}^\delta(\tau) - u^*(\tau)d\tau \right\| \qquad (26)\\
&\leq \left\| \sum_{j=1}^{k_{t,\delta}-1} \left[\int_{t_{j-1}}^{t_j} \hat{u}^\delta(\tau) - \bar{u}_k^{\alpha,\delta} d\tau + h^\delta(\bar{u}_k^{\alpha,\delta} - \bar{u}_k^*) \right] \right\| + h^\delta 2R_u.
\end{aligned}$$

Note that since $u^*(t) \in \mathbf{U}$ a. e. on $[0, T]$ (see Assumption **A.4**), it is true that $\bar{u}_k^* \in \mathbf{U}$, $k = 1, \ldots, N$. Therefore, it follows form (21), (22), (8) that

$$\begin{aligned}
&\hat{u}^\delta(t) \in \mathbf{U}, \quad t \in [0, T], \\
&\|\hat{u}^\delta(t) - \bar{u}^{\alpha,\delta}(t)\| \leq r_{\hat{u}}(\delta, h^\delta, \alpha), \quad t \in [0, T].
\end{aligned} \qquad (27)$$

Applying (21), (27) to (26), we get

$$\left\| \int_0^t \hat{u}^\delta(\tau) - u^*(\tau)d\tau \right\| \leq h^\delta \frac{t - t_0}{h^\delta} 2r_{\hat{u}}(\delta, h^\delta, \alpha) + 2h^\delta R_u. \qquad (28)$$

Applying estimates (13), (14) and (28) to (25), we get

$$\begin{aligned}
&\|\hat{x}^\delta(t) - x^*(t)\| \leq \delta + 2h^\delta R_G R_u + 2TL_{D_0} R_u(\delta + h^\delta(K+1)) \\
&+(t - t_0)2R_G r_{\hat{u}}(\delta, h^\delta, \alpha) + L_{D_0}(R_u + 1) \int_0^t \|\hat{x}^\delta(\tau) - x^*(\tau)\| d\tau.
\end{aligned}$$

Since the function $\|\hat{x}^\delta(\cdot) - x^*(\cdot)\|$ is continuous, we can apply the generalized Grönwall's inequality [14, Chap.1, p. 1]:

$$\begin{aligned}
&\|\hat{x}^\delta(t) - x^*(t)\| \leq \big(\delta + 2h^\delta R_G R_u + 2TL_{D_0} R_u(\delta + h^\delta(K+1))\big) e^{L_{D_0}(R_u+1)T} \\
&+2\frac{R_G r_{\hat{u}}(\delta, h^\delta, \alpha)}{L_{D_0}(R_u+1)}(e^{L_{D_0}(R_u+1)T} - 1) \xrightarrow{\delta\to 0} 0, \quad t \in [0, T].
\end{aligned}$$

Thus, condition $\mathcal{B}$.**2** holds and the theorem is proved. □

4 Example

Consider as an example a dynamical model from the area of medicine. It is a simplified model of the process of penicillin fermentation [15]. The dynamics are

$$\begin{pmatrix} \dot{X}(t) \\ \dot{S}(t) \end{pmatrix} = \begin{pmatrix} X(t) & -\dfrac{X(t)}{S_F V} \\ -\dfrac{X(t)}{Y_{X/S}} & \dfrac{S_F - S(t)}{S_F} \end{pmatrix} \begin{pmatrix} \mu(t) \\ U(t) \end{pmatrix} + \begin{pmatrix} 0 \\ -\rho \frac{X(t)}{Y_{P/S}} - \frac{S(t)X(t)}{K_m + S(t)} \end{pmatrix},$$

$$U(t) \in [0, 30], \quad \mu(t) \in [0, 0.3], \quad t \in [0, 10],$$

$$X(0) = 1.5, \quad S(0) = 0.01,$$

$$Y_{X/S} = 0.47, \quad Y_{P/S} = 1.2, \quad S_F = 500, \quad K_m = 0.0001, \quad \rho = 0.0055.$$

Here the state variables $X(\cdot)$ and $S(\cdot)$ are the concentrations of biomass and substrate in the organism. The control $U(\cdot)$ is the substrate feeding profile, and the unknown parameter $\mu(\cdot)$, which is the specific biomass growth rate, is considered as the second control.

To simulate the process of measuring the basic trajectory $X^*(\cdot)$, $S^*(\cdot)$, it was numerically constructed for the controls

$$\mu^*(t) \equiv 0.11 \sin(t) + 0.03 \sin t, \quad U^*(t) = \begin{cases} 15 + t/T, & t \in [0, 0.5T], \\ 0, & t \in (0.5T, T] \end{cases}.$$

Then, the constructed basic trajectory was randomly perturbed to simulate arrival of the measurements (3). Upon this data, both the old algorithm (Sect. 2.5) and the new algorithm (Sect. 3.2) were applied to obtain approximations of the normal control that generates the basic trajectory.

The results of numerical simulations are the graphs of the controls $U^\delta(\cdot)$, $\mu^\delta(\cdot)$, constructed by the formula (6) (the old algorithm) with the trajectory $X^\delta(\cdot)$, $S^\delta(\cdot)$, generated by these controls, and graphs of the controls $\hat{U}^\delta(\cdot)$, $\hat{\mu}^\delta(\cdot)$, constructed by the formula (8) (the new algorithm) with the trajectory $\hat{X}^\delta(\cdot)$, $\hat{S}^\delta(\cdot)$, generated by these controls. The approximation parameters are

$$\delta = 10^{-4}, \quad \alpha = 5 \cdot 10^{-4}, \quad N = 50, \quad h^\delta = 0.2.$$

The results are presented on the Fig. 1—$U^\delta(\cdot)$, $\mu^\delta(\cdot)$, Fig. 2—$\hat{U}^\delta(\cdot)$, $\hat{\mu}^\delta(\cdot)$ and Fig. 3—$\hat{X}^\delta(\cdot)$, $\hat{S}^\delta(\cdot)$.

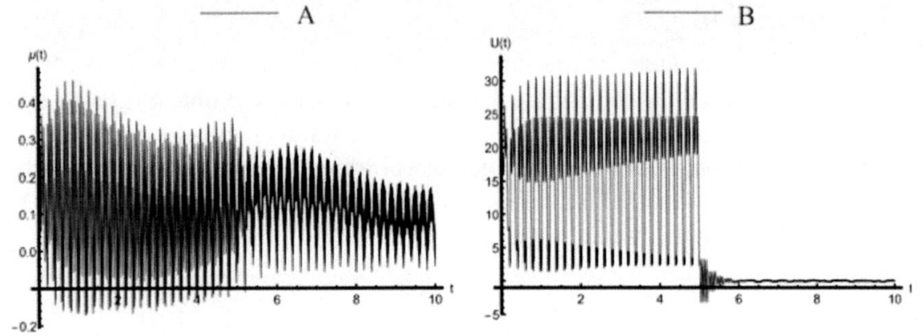

Fig. 1 Reconstructed controls (by the old algorithm). Legend: A is $U^{\delta}(t)$, B is $\mu^{\delta}(t)$

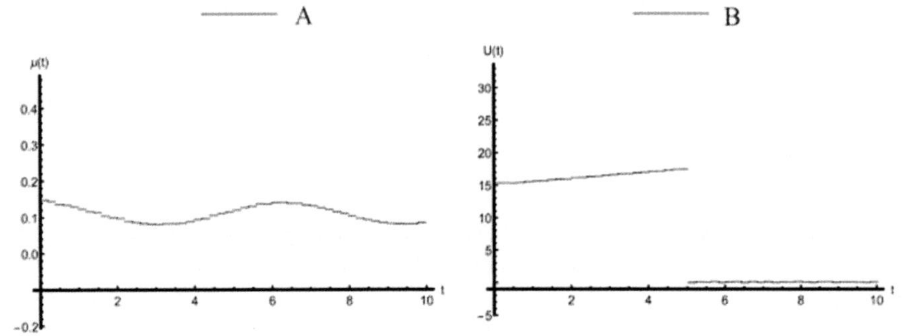

Fig. 2 Reconstructed controls (by the new algorithm). Legend: A is $\hat{U}^{\delta}(t)$, B is $\hat{\mu}^{\delta}(t)$

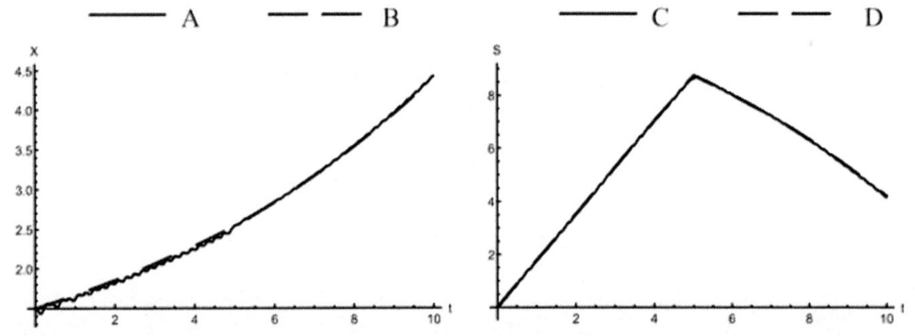

Fig. 3 Trajectories, generated by $\hat{U}^{\delta}(\cdot)$, $\hat{\mu}^{\delta}(\cdot)$. Legend: A is $\hat{X}^{\delta}(t)$, B is $X^{*}(t)$, C is $\hat{S}^{\delta}(t)$, D is $S^{*}(t)$

5 Conclusion

An original method for solving the dynamic reconstruction problem is being developed. Namely, it's modification is justified, which allows to construct piecewise-constant approximations of the desired control which satisfy the given geometrical restrictions on the admissible controls. These approximations converge almost everywhere to the desired control.

References

1. Kryazhimskij AV, Osipov YS (1983) Modelling of a control in a dynamical system. In: Ehg Cybern **21**(2):38–47. Jzv. Akad. Nauk SSSR, Moscow
2. Maksimov VI (2021) Reconstruction of an unbounded input of a system of differential equations. In.: Proceedings of Steklov institute of mathematics, vol 315. Steklov Institute, Moscow, pp 149–160. https://doi.org/10.1134/S0081543821050114
3. Kabanikhin SI, Krivorotko OI (2015) Identification of biological models described by systems of nonlinear differential equations. J Inverse and Ill-posed Prob 23(5):519–527. De Gruyter, Berlin. https://doi.org/10.1515/jiip-2015-0072
4. Ren C, Wang N, Liu Q, Liu Ch (2019) Dynamic force identification problem based on a novel improved Tikhonov regularization. Math Prob Eng 2019:6095184. Hindawi. https://doi.org/10.1155/2019/6095184
5. Schmitt U, Louis AK, Wolters C, Vauhkonen M (2002) Efficient algorithms for the regularization of dynamic inverse problems. Inverse Theory. Inverse Prob 18(3):659–676. IOP Publishing. https://doi.org/10.1088/0266-5611/18/3/308
6. Schuster N, Burger M, Hahn B (2018) Dynamic inverse problems: modelling-regularization-numerics. Preface. Inverse Prob 34:040301. IOP Publishing, Bristol. https://doi.org/10.1088/1361-6420/aab0f5
7. Sabatier PC (2000) Past and future of inverse problems. J Math Phys 41(6):4082–4124. AIP Publishing. https://doi.org/10.1063/1.533336
8. Subbotina NN, Krupennikov EA (2021) Weak* approximations for the solution of a dynamic reconstruction problem. Tr Inst Math Mech 27(2):208–220. IMM UrO RAN, Ekaterinburg. https://doi.org/10.21538/0134-4889-2021-27-2-208-220 (In Russian)
9. Subbotina NN, Krupennikov EA (2021) Weak* solution to a dynamic reconstruction problem. Proc Steklov Inst Math 315:233–246. Pleiades Publishing. https://doi.org/10.1134/S0081543821050187
10. Strekalovsky AS (2003) Elements of nonconvex optimization. Nauka, Novosibirsk (in Russian)
11. Tikhonov AN (1963) Solution of incorrectly formulated problems and the regularization method. Soviet Math Dokl 4(5):1035–1038. RAS, Moscow
12. Kolmogorov AN, Fomin SV (1957) Elements of the theory of functions and functional analysis. Graylock Press, Rochester, N. Y.
13. Neudecker H, Magnus JR (2019) Matrix differential calculus with applications in statistics and econometrics. Wiley, New Jersey. https://doi.org/10.1002/9781119541219
14. Barbashin EA (1970) Introduction to the theory of stability. Wolters-Noordhoff Publishing, Groningen
15. Zhai C, Qiu T, Palazoglu A, Sun W (2018) The emergence of feedforward periodicity for the fed-batch penicillin fermentation process. In: IFAC-PapersOnLine, vol 51, Issue 32. Elsevier, Amsterdam, pp 130–135. https://doi.org/10.1016/j.ifacol.2018.11.367

Discrete Operators and Equations: Analysis and Comparison

Alexander Vasilyev, Vladimir Vasilyev, and Asad Esmatullah

Abstract We develop a discrete variant of a theory of pseudo-differential equations and boundary value problems in canonical domains which are model situations for manifolds with non-smooth boundaries. Using digitization process for ordinary functional spaces we construct certain discrete functional spaces or spaces of functions of a discrete variable and define discrete pseudo-differential operators acting in such spaces. A main problem in which we are interested is to establish a correspondence between continual and discrete solutions of considered continual and discrete equations and in future boundary value problems. We have illustrated our considerations by certain examples of Calderon–Zygmund operators for which we have some interesting conclusions.

Keywords Discrete operator · Solvability

1 Introduction

We deal with some special operators namely pseudo-differential operators. Our global main goal is to construct a theory of discrete pseudo-differential operators and corresponding boundary value problems on smooth manifolds with a boundary which may be non-smooth.

A basic equation in an operator form is the following

$$(Au)(x) = v(x), \quad x \in D, \tag{1}$$

where $D \subset \mathbf{R}^m$ is a some domain, A is a pseudo-differential operator which is defined by the formula

A. Vasilyev (✉) · V. Vasilyev · A. Esmatullah
Belgorod State National Research University, Belgorod 308015, Russia
e-mail: alexvassel@gmail.com

V. Vasilyev (ed.), *Differential Equations, Mathematical Modeling and Computational Algorithms*, Springer Proceedings in Mathematics & Statistics 423,
https://doi.org/10.1007/978-3-031-28505-9_17

$$(Au)(x) = (2\pi)^{-m} \int\limits_D \int\limits_{\mathbf{R}^m} e^{i(y-x)\cdot\xi} \tilde{A}(x, \xi)\tilde{u}(\xi)dyd\xi, \quad x \in D, \tag{2}$$

and a sign $\sim$ over the function u denotes its Fourier transform

$$\tilde{u}(\xi) = \int\limits_{\mathbf{R}^m} e^{-ix\cdot\xi} u(x)dx.$$

Definition 1 The function $\tilde{A}(x, \xi)$ is called a symbol of a pseudo-differential operator A. A symbol $\tilde{A}(x, \xi)$ is called an elliptic symbol if $ess \inf\limits_{(x,\xi)\in\mathbf{R}^m\times\mathbf{R}^m} |\tilde{A}(x, \xi)| > 0$.

As far as I know it is impossible to find an exact solution of the equation (1) for an arbitrary domain D. Therefore all researches are interested in describing Fredholm properties of the equation at least. But for simplest cases it can very easy by the Fourier transform.

Example 1 Let K(x) be a Calderon–Zygmund kernel and the operator A is defined by the formula [4]

$$(Ku)(x) = v.p. \int\limits_{\mathbf{R}^m} K(x - y)u(y)dy, \tag{3}$$

so that it can represented in the form (6)

$$(Ku)(x) = (2\pi)^{-m} \int\limits_{\mathbf{R}^m} \int\limits_{\mathbf{R}^m} e^{i(y-x)\cdot\xi} \sigma(\xi)\tilde{u}(\xi)dyd\xi,$$

and the function $\sigma(\xi)$ is called a symbol of the operator A. It is well known that for the operator A to be invertible in the space $L_2(\mathbf{R}^m)$ necessary and sufficient its symbol $\sigma(\xi)$ should be an elliptic [4].

Let $D_d = D \cap h\mathbf{Z}^m$, $h > 0$. We are interested in studying some discrete equations which we call discrete pseudo-differential equations and which are related to the Eq. (1). Let us define a discrete pseudo-differential operator by the formula

$$(A_d u_d)(\tilde{x}) = \sum_{\tilde{y}\in D_d} \int\limits_{\hbar\mathbf{T}^m} e^{i(\tilde{y}-\tilde{x})\cdot\xi} A_d(\tilde{x}, \xi)\tilde{u}_d(\xi)d\xi, \quad \tilde{x} \in D_d,$$

where $u_d(\tilde{x})$ is a function of a discrete variable $\tilde{x} \in h\mathbf{Z}^m$, $\tilde{u}_d(\xi)$ denotes its discrete Fourier transform

$$\tilde{u}_d(\xi) \equiv (F_d u_d)(\xi) = \sum_{\tilde{y}\in h\mathbf{Z}^m} e^{i\tilde{y}\cdot\xi} \tilde{u}_d(y), \quad \xi \in \hbar\mathbf{T}^m, \tag{4}$$

$\mathbf{Z}^m$ is an integer lattice in $\mathbf{R}^m$, $\mathbf{T}^m$ is m-dimensional cube $[-\pi, \pi]^m$, $\hbar = \frac{h^{-1}}{2\pi}$, and given function $A_d(\tilde{x}, \xi), \tilde{x} \in h\mathbf{Z}^m, \xi \in \hbar\mathbf{T}^m$, is called a symbol of the discrete pseudo-differential operator A_d.

We would like to study the equation

$$(A_d u_d)(\tilde{x}) = v_d(\tilde{x}), \quad \tilde{x} \in D_d, \tag{5}$$

in some discrete functional spaces. Since it is difficult to study such general operators (as it was said above) for discrete cases also we'll consider certain model situations.

2 The Concept of the Research

We'll present here main ideas for studying this large problem. In contrast of algebraic approaches [2, 3, 5] we use analytical methods based on properties of the Fourier transform and considered operators. A plan of the studying is the following:

- infinite discrete and finite discrete Fourier transform
- discrete functional spaces
- solvability of infinite discrete equation
- solvability of finite discrete equation
- comparison of continual and infinite discrete solution
- comparison of infinite and finite discrete solution.

2.1 *Local Discrete Operators*

We'll illustrated the above scheme with very simple model pseudo-differential operator namely operator A from example 1 because many our results are related to this operator. In addition we assume that kernel $K(x)$ of the operator A is differentiable on $\mathbf{R}^m \setminus \{0\}$.

2.2 *Discrete and Continual*

Discrete Fourier Transform To obtain a good approximation for the integral equation (1) we will use the following reduction. First instead of the integral in (1) we introduce the series

$$\sum_{\tilde{y} \in h\mathbf{Z}^m} K(\tilde{x} - \tilde{y}) u_d(\tilde{y}) h^m, \tag{6}$$

which generates a discrete operator

$$(K_d u_d)(\tilde{x}) = \sum_{\tilde{y} \in h\mathbf{Z}^m} K(\tilde{x} - \tilde{y}) u_d(\tilde{y}) h^m, \quad \tilde{x} \in h\mathbf{Z}^m, \tag{7}$$

defined on functions u_d of discrete variable $\tilde{x} \in h\mathbf{Z}^m$. Since the Calderon–Zygmund kernel has a strong singularity at the origin we mean $K(0) = 0$. Convergence for the series (6) means that the following limit

$$\lim_{N \to +\infty} \sum_{\tilde{y} \in h\mathbf{Z}^m \cap Q_N} K(\tilde{x} - \tilde{y}) u_d(\tilde{y}) h^m$$

exists, where $Q_N = \{x \in \mathbf{R}^m : \max_{1 \le k \le m} |x_k| < N\}$. It was shown earlier that a norm of the operator $K_d : L_2(h\mathbf{Z}^m) \to L_2(h\mathbf{Z}^m)$ does not depend on h [11]. But although the operator is a discrete object it is an infinite one.

Let us define the infinite discrete Fourier transform for functions u_d of a discrete variable $\tilde{x} \in h\mathbf{Z}^m$

$$(F_d u_d)(\xi) = \sum_{\tilde{x} \in h\mathbf{Z}^m} u_d(\tilde{x}) e^{i\tilde{x} \cdot \xi} h^m, \quad \xi \in \hbar\mathbf{T}^m.$$

Such discrete Fourier transform preserves all basic properties of the classical Fourier transform, particularly for a discrete convolution of two discrete functions u_d, v_d

$$(u_d * v_d)(\tilde{x}) \equiv \sum_{\tilde{y} \in h\mathbf{Z}^m} u_d(\tilde{x} - \tilde{y}) v_d(\tilde{y}) h^m$$

we have the well known multiplication property

$$(F_d(u_d * v_d))(\xi) = (F_d u_d)(\xi) \cdot (F_d v_d)(\xi).$$

If we apply this property to the operator K_d we obtain

$$(F_d(K_d u_d))(\xi) = (F_d K_d)(\xi) \cdot (F_d u_d)(\xi).$$

Let us denote $(F_d K_d)(\xi) \equiv \sigma_d(\xi)$ and give the following

Definition 2 The function $\sigma_d(\xi), \xi \in \hbar\mathbf{T}^m$, is called a symbol of the discrete operator K_d.

We will assume below that the symbol $\sigma_d(\xi) \in C(\hbar\mathbf{T}^m)$ therefore we have immediately the following

Property 1 The operator K_d is invertible in the space $L_2(h\mathbf{Z}^m)$ iff $\sigma_d(\xi) \neq 0, \forall \xi \in \hbar\mathbf{T}^m$.

We say that a continuous symbol is called an **elliptic symbol** if $\sigma_d(\xi) \neq 0, \forall \xi \in \hbar\mathbf{T}^m$.

So we see that an arbitrary elliptic symbol $\sigma_d(\xi)$ corresponds to an invertible operator K_d in the space $L_2(h\mathbf{Z}^m)$.

A very interesting fact was proved in [8, 9].

Theorem 1 *Operators (3) and (7) are invertible or non-invertible in spaces* $L_2(\mathbf{R}^m)$ *and* $L_2(h\mathbf{Z}^m)$ *simultaneously* $\forall h > 0$.

If we consider the equation

$$(K_d u_d)(\tilde{x}) = v_d(\tilde{x}), \quad \tilde{x} \in h\mathbf{Z}^m,$$

in the space $L_2(h\mathbf{Z}^m)$ then we solve the equation by the discrete Fourier transform F_d. Indeed after applying the Fourier transform we have the trivial equation

$$\sigma_d(\xi)\tilde{u}_d(\xi) = \tilde{v}_d(\xi), \quad \xi \in \hbar\mathbf{T}^m,$$

in the dual space $L_2(\hbar\mathbf{T}^m)$.

We have first difficulties when consider this equation in the space $L_2(h\mathbf{Z}^m_+)$, where $\mathbf{Z}^m_+ = \{\tilde{x} \in \mathbf{Z}^m : \tilde{x} = (\tilde{x}_1, \ldots, \tilde{x}_m), \tilde{x}_m > 0\}$. We can not apply the Fourier transform directly as above because the functions under consideration are defined not on a whole space. Thus we need to describe images of such function after the discrete Fourier transform, ant it leads us to the next extensions.

A Half-Space Case If we consider Eqs. (3) and (7) in spaces $L_2(\mathbf{R}^m_+)$ and $L_2(h\mathbf{Z}^m_+)$ or in other words operators $K : L_2(\mathbf{R}^m_+) \to L_2(\mathbf{R}^m_+)$ and $K_d : L_2(h\mathbf{Z}^m_+) \to L_2(h\mathbf{Z}^m_+)$ then for studying invertibility of the operator K_d one has constructed a special periodic Riemann boundary value problem [10]. A solvability of mentioned Riemann problem depends on a certain topological invariant æ related to a symbol of an elliptic operator. This number æ is called an index of periodic Riemann boundary value problem. It was shown these topological numbers for elliptic operators K and K_d are the same and it implies the following [8, 9]

Theorem 2 *Operators (3) and (7) are invertible or non-invertible in spaces* $L_2(\mathbf{R}^m_+)$ *and* $L_2(h\mathbf{Z}^m_+)$ *simultaneously* $\forall h > 0$.

Studying more complicated situations related to cones [6] was started in [14], first steps were done.

Discrete Boundary Value Problems These arise first in the case $h\mathbf{Z}^m_+$ then we have a boundary, and it is possible the mentioned index æ is not a zero. To exclude a non-uniqueness of solution one needs some boundary conditions [1, 6]. Some similar situations were considered for difference equations in papers [12, 13, 15].

2.3 Infinite and Finite

Finite Discrete Fourier Transform Here we will introduce a special discrete periodic kernel $K_{d,N}(\tilde{x})$ which is defined in the following way. We take a restriction of the discrete kernel $K_d(\tilde{x})$ on the set $Q_N \cap h\mathbf{Z}^m \equiv Q_N^d$ and periodically continue it to a whole $h\mathbf{Z}^m$. Further we consider discrete periodic functions $u_{d,N}$ with discrete cube of periods Q_N^d. We can define a cyclic convolution for a pair of such functions $u_{d,N}, v_{d,N}$ by the formula

$$(u_{d,N} * v_{d,N})(\tilde{x}) = \sum_{\tilde{y} \in Q_N^d} u_{d,N}(\tilde{x} - \tilde{y}) v_{d,N}(\tilde{y}) h^m. \tag{8}$$

Further we introduce finite discrete Fourier transform by the formula

$$(F_{d,N} u_{d,N})(\tilde{\xi}) = \sum_{\tilde{x} \in Q_N^d} u_{d,N}(\tilde{x}) e^{i\tilde{x}\cdot\tilde{\xi}} h^m, \quad \tilde{\xi} \in R_N^d,$$

where $R_N^d = \hbar\mathbf{T}^m \cap \hbar\mathbf{Z}^m$. Let us note that here $\tilde{\xi}$ is a discrete variable.

Finite Discrete Operator According to the formula (8) one can introduce the operator

$$K_{d,N} u_{d,N}(\tilde{x}) = \sum_{\tilde{y} \in Q_N^d} K_{d,N}(\tilde{x} - \tilde{y}) u_{d,N}(\tilde{y}) h^m$$

on periodic discrete functions $u_{d,N}$ and a finite discrete Fourier transform for its kernel

$$\sigma_{d,N}(\tilde{\xi}) = \sum_{\tilde{x} \in Q_N^d} K_{d,N}(\tilde{x}) e^{i\tilde{x}\cdot\tilde{\xi}} h^m, \quad \tilde{\xi} \in R_N^d.$$

Definition 3 A function $\sigma_{d,N}(\tilde{\xi})$, $\tilde{\xi} \in R_N^d$, is called s symbol of the operator $K_{d,N}$. This symbol is called an elliptic symbol if $\sigma_{d,N}(\tilde{\xi}) \neq 0$, $\forall \tilde{\xi} \in R_N^d$.

Theorem 3 *Let $\sigma_d(\xi)$ be an elliptic symbol. Then for enough large N the symbol $\sigma_{d,N}(\tilde{\xi})$ is elliptic symbol also.*

A proof of the theorem follows immediately.

As before an elliptic symbol $\sigma_{d,N}(\tilde{\xi})$ corresponds to the invertible operator $K_{d,N}$ in the space $L_2(Q_N^d)$.

3 Discrete Functional Spaces

Since we'll use projectors on points of lattice we need subspaces of continuous functions instead of Lebesgue spaces. We introduce the space C_h which is the space of functions u_d of discrete variable $\tilde{x} \in h\mathbf{Z}^m$ with the norm

$$||u_d||_{C_h} = \max_{\tilde{x} \in h\mathbf{Z}^m} |u_d(\tilde{x})|.$$

In other words, the space C_h is the space of functions $u \in C(\mathbf{R}^m)$ restricted on lattice points $\mathbf{Z}_h^m$. Here we remind, that the operator K isn't bounded in the space $C(\mathbf{R}^m)$, but it is bounded in the space $L_2(\mathbf{R}^m)$, and it is well-known, that if the right hand side of the equation

$$(Ku)(x) = v(x)$$

has some smoothness properties (for example, it satisfies the Hölder condition), then the solution of this (if it exists in the space $L_2(\mathbf{R}^m)$) has the same smoothness property [4].

Further we define the discrete space $C_h(\alpha, \beta)$ as a functional space of discrete variable $\tilde{x} \in h\mathbf{Z}^m$ with finite norm

$$||u_d||_{C_h(\alpha,\beta)} = ||u_d||_{C_h} + \sup_{\tilde{x},\tilde{y} \in h\mathbf{Z}^m} |u_d(\tilde{x}) - u_d(\tilde{y})|,$$

and additional assumptions

$$|u_d(\tilde{x}) - u_d(\tilde{y})| \le c \, \frac{|\tilde{x} - \tilde{y}|^\alpha}{(\max\{1 + |\tilde{x}|,\ 1 + |\tilde{y}|\})^\beta},$$

$$|u_d(\tilde{x})| \le \frac{c}{(1 + |\tilde{x}|)^{\beta-\alpha}}, \qquad \forall \tilde{x}, \tilde{y} \in h\mathbf{Z}^m,\ \alpha, \beta > 0,\ 0 < \alpha < 1.$$

4 Approximate Solutions

4.1 Infinite Discrete Solutions

Let's denote P_h the restriction operator on the lattice $h\mathbf{Z}^m$, i.e. the operator, which an arbitrary function, defined on $\mathbf{R}^m$, maps to the set of its discrete values in lattice points $h\mathbf{Z}^m$.

Definition 4 The approximation rate for the operators K and K_d in vector normed space X of functions defined on $\mathbf{R}^m$, is called the operator norm

$$||P_h K - K_d P_h||_{X \to X_d},$$

where X_d is the normed space of functions defined on the lattice $h\mathbf{Z}^m$ with norm, which is induced by the norm of the space X.

For the space $C_h(\alpha, \beta)$ we have

Theorem 4 *If $m < \beta < \alpha + m$, then the estimate*

$$||K_d u_d||_{C_h(\alpha,\beta)} \leq c||u_d||_{C_h(\alpha,\beta)},$$

is valid, and c doesn't depend on h.

The continual analogue of such spaces is the space $H^{\alpha}_{\beta}(\mathbf{R}^m)$ of functions, which are continuous in $\mathbf{R}^m$ and satisfy the Hölder condition of order $0 < \alpha < 1$ and with weight $(1+|x|)^{\beta}$. It is well known from results of S.K. Abdullaev (Sov. Math., Dokl. 40, No.2, 417-421,1990) that the operator K is a linear bounded operator $K : H^{\alpha}_{\beta}(\mathbf{R}^m) \to H^{\alpha}_{\beta}(\mathbf{R}^m)$ under the condition $m < \beta < \alpha + m$.

We will give the approximation rate for the operators K and K_d in the space $C_h(\alpha,\beta)$. It will permit to obtain the error estimate for approximate solution, if we will change the continual operator K by its discrete analogue K_d.

Theorem 5 *The approximate rate for the operators K and K_d is the following*

$$||P_h K - K_d P_h||_{C_h(\alpha,\beta)} \leq c h^{\tilde{\alpha}},$$

where c doesn't depend on h, $\tilde{\alpha} < \alpha$, $\tilde{\beta} > \beta$.

Some of these results were obtained in [7].

4.2 Finite Discrete Solutions

Let us denote P_N the projector $L_2(h\mathbf{Z}^m) \to L_2(Q^d_N)$.

Theorem 6 *For operators K_d and $K_{d,N}$ we have the following estimate*

$$||(P_N K_d - K_{d,N} P_N)u_d||_{L_2(Q^d_N)} \leq C N^{m+2(\alpha-\beta)}$$

for arbitrary $u_d \in C_h(\alpha,\beta)$, $\beta > \alpha + m/2$.

Now we consider the equation

$$K_{d,N} u_{d,N} = P_N v_d \tag{9}$$

instead of the equation

$$K_d u_d = v_d \tag{10}$$

and give a comparison for these two solutions assuming that operator K_d is invertible in $L_2(h\mathbf{Z}^m)$.

Theorem 7 *If $v_d \in C_h(\alpha,\beta)$, $\beta > \alpha + m/2$, u_d is a solution of the Eq. (10), $u_{d,N}$ is a solution of (9) then the estimate*

$$||u_d - u_{d,N}||_{L_2(h\mathbf{Z}^m)} \leq CN^{m+2(\alpha-\beta)}$$

is valid, and C is a constant non-depending on N.

Conclusion

These considerations are first steps to realize the declared programm. We hope that obtained results will help us to study more general discrete operators and equations and to describe a correspondence between discrete and continual objects, and also between finite and infinite discrete objects.

References

1. Eskin G (1981) Boundary value problems for elliptic pseudodifferential equations. AMS, Providence
2. Didenko V, Silbermann B (2008) Approximation of additive convolution-like operators. Real C^*-algebra approach. Birkhäuser, Basel
3. Hagen R, Roch S, Silbermann B (2001) C^*-algebras and numerical analysis. Marcel Dekker, New York
4. Mikhlin SG, Prössdorf S (1986) Singular integral operators. Akademie-Verlag, Berlin
5. Prössdorf S, Silbermann B (1991) Numerical analysis for integral and related operator equations. Birkhäuser, Basel
6. Vasil'ev VB (2000) Wave factorization of elliptic symbols: theory and applications. Kluwer, Dordrecht
7. Vasilyev AV, Vasilyev VB (2012) Numerical analysis for some singular integral equations. Neural Parall Sci Comput 20:313–326
8. Vasilyev AV, Vasilyev VB (2013) Discrete singular operators and equations in a half-space. Azerb J Math 3:84–93
9. Vasilyev AV, Vasilyev VB (2015) Discrete singular integrals in a half-space. In: Mityushev V, Ruzhansky M (eds) Current trends in analysis and its applications. Proceedings of 9th ISAAC congress, Kraków, Poland, 2013. Birkhäuser, Basel, pp 663–670
10. Vasilyev AV, Vasilyev VB (2015) Periodic Riemann problem and discrete convolution equations. Differ Equ 51:652–660
11. Vasil'ev AV, Vasil'ev VB (2015) On the solvability of certain discrete equations and related estimates of discrete operators. Dokl Math 92:585–589
12. Vasilyev AV, Vasilyev VB (2016) On solvability of some difference-discrete equations. Opusc Math 36:525–539
13. Vasilyev AV, Vasilyev VB (2016) Difference equations in a multidimensional space. Math Model Anal 21:336–349
14. Vasilyev VB (2016) Discrete equations and periodic wave factorization. AIP Conf Proc 1759:0200126
15. Vasilyev AV, Vasilyev VB (2016) Difference equations and boundary value problems. In: Pinelas S et al (eds) Differential and difference equations and applications. Springer proceeding mathematics & statistics vol 164. Birkhäuser, Basel, pp 421–432

Pseudo-Differential Equations in Spaces of Different Smoothness Exponents on Variables

Vladimir Vasilyev, Victor Polunin, and Igor Shmal

Abstract We study a model elliptic pseudo-differential equation and simplest boundary value problems for a half-space and a special cone in Sobolev–Slobodetskii spaces which have different smoothness with respect to separate variables. Sufficient conditions for a unique solvability for such boundary value problems are described

Keywords Sobolev–Slobodetskii space · Pseudo-differential equation · Cone · Boundary value problem

1 Introduction

The theory of pseudo-differential operators was appeared near a half-century ago, and it has taken attention of mathematicians for a long time [1–3]. More general Fourier integral operators and new functional spaces were studied in this context. As a rule the theory means constructing a symbolic calculus and an index formula. Such a theory is very convenient for generalization on smooth compact manifolds without a boundary, but for more complicated situations new constructions and new approaches were needed. More complicated situations mean presence of a smooth boundary, or more generally a non-smooth boundary. For manifolds with a smooth boundary a certain approach was suggested in [4], and it was based the factorization principle for an elliptic symbol at boundary points. This method is not applicable for manifolds with a non-smooth boundary, and it has initiated a lot of approaches for "non-smooth" situations [5, 6, 10–14].

This paper presents a future development of the second author's approach [17–20] to the theory of pseudo-differential equations and related boundary value problems in non-smooth domains.

V. Vasilyev (✉) · V. Polunin · I. Shmal
Belgorod State National Research University, Belgorod 308015, Russia
e-mail: vbv57@inbox.ru

V. Vasilyev (ed.), *Differential Equations, Mathematical Modeling and Computational Algorithms*, Springer Proceedings in Mathematics & Statistics 423,
https://doi.org/10.1007/978-3-031-28505-9_18

2 Elliptic Pseudo-differential Operators

2.1 Sobolev–Slobodetskii Spaces of Different Smoothness

Following to [15] (see also [16]) we introduce useful notations. A multidimensional Euclidean space $\mathbf{R}^M$ is represented as an orthogonal sum of subspaces in which only some of coordinates $x_1, x_2, \dots, x_M$ are nor vanishing. Namely, if $K \subset 1, ..., M$ is not empty set we put

$$\mathbf{R}^K = \{x \in \mathbf{R}^M : x = (x_1, \dots, x_M), x_j = 0, \forall j \notin K\} \subset \mathbf{R}^M.$$

Let $K_1, K_2, \dots, K_n \subset \{1, 2, \dots, M\}$ be a nonempty set so that

$$\bigcup_{j=1}^{n} K_j = \{1, 2, \dots, M\}, \quad K_i \cap K_j = \emptyset, i \neq j, \quad card\ K_j = k_j.$$

Thus, we obtain the representation

$$\mathbf{R}^M = \mathbf{R}^{K_1} \oplus \mathbf{R}^{K_2} \oplus \dots \oplus \mathbf{R}^{K_n},$$

where x_{K_j} is an element of the space $\mathbf{R}^{K_j}$. For functions defined in $\mathbf{R}^M$ we use the standard Fourier transform

$$\tilde{u}(\xi) = \int_{\mathbf{R}^M} e^{ix\cdot\xi} u(x)dx, \quad \xi = (\xi_1, \dots, \xi_M).$$

Let $S = (s_1, \dots, s_n)$. Now we introduce the Sobolev–Slobodetskii space $H^S(\mathbf{R}^M)$ as a Hilbert space with the inner product

$$(f, g) = \int_{\mathbf{R}^M} f(x)\overline{g(x)}dx$$

and the norm

$$||f||_S = \left(\int_{\mathbf{R}^M} (1 + |\xi_{K_1}|)^{2s_1} (1 + |\xi_{K_2}|)^{2s_2} \cdots (1 + |\xi_{K_n}|)^{2s_n} |\tilde{f}(\xi)|^2 d\xi \right)^{1/2}.$$

Such H^S-spaces have the same properties similar to usual Sobolev–Slobodetskii spaces [16]. Particularly, the usual space $H^s(\mathbf{R}^M)$ is obtained under the following choice of subsets K_j and parameters s_j:

$$K_1 = K_2 = \dots = K_{n-1} = \emptyset, \quad K_n = \{1, 2, \dots, M\}, \quad S = (0, 0, \dots, 0, s).$$

2.2 Model Operators and Equations

According to the local principle we will concentrate on studying a model pseudo-differential equation with operator with a symbol non-depending on a spatial variable.

Model pseudo-differential operators Let $\tilde{A}(\xi)$, $\xi \in \mathbf{R}^M$ be a measurable function. A model pseudo-differential operator A is defined as follows

$$(Au)(x0 = \frac{1}{(2\pi)^M} \int\limits_{\mathbf{R}^M} \int\limits_{\mathbf{R}^M} e^{i(x-y)\cdot\xi} \tilde{A}(\xi) u(y) dy d\xi,$$

and the function $\tilde{A}(\xi)$ is called a symbol of the pseudo-differential operator A.

We consider here the following class of symbols $A(\xi)$ satisfying the condition

$$c_1 \prod_{j=1}^{n} (1 + |\xi_{K_j}|)^{\alpha_j} \leq |A(\xi)| \leq c_2 \prod_{j=1}^{n} (1 + |\xi_{K_j}|)^{\alpha_j}, \quad \alpha_j \in \mathbf{R},\ j = 1, 2, \ldots, n, \tag{$*$}$$

with positive constants c_1, c_2.

Let $\alpha = (\alpha_1, \ldots, \alpha_n)$

Lemma 1 *Let A be a pseudo-differential operator with the symbol $\tilde{A}(\xi)$ satisfying the condition $(*)$. Then $A : H^S(\mathbf{R}^M) \to H^{S-\alpha}(\mathbf{R}^M)$ is a linear bounded operator.*

Proof Indeed, we have

$$||Au||^2_{S-\alpha} = \int\limits_{\mathbf{R}^M} \prod_{j=1}^{n} (1 + |\xi_{K_j}|)^{2(s_j - \alpha_j)} \widetilde{Au}(\xi)|^2 d\xi =$$

$$= \int\limits_{\mathbf{R}^M} (1 + |\xi_{K_1}|)^{2(s_1-\alpha_1)} (1 + |\xi_{K_2}|)^{2(s_2-\alpha_2)} \cdots (1 + |\xi_{K_n}|)^{2(s_n-\alpha_n)} |\tilde{A}(\xi)\tilde{u}(\xi)|^2 d\xi \leq$$

$$\leq c_2 \int\limits_{\mathbf{R}^M} (1 + |\xi_{K_1}|)^{2s_1} (1 + |\xi_{K_2}|)^{2s_2} \cdots (1 + |\xi_{K_n}|)^{2s_n} |\tilde{u}(\xi)|^2 d\xi = c_2 ||u||^2_S,$$

and the proof is completed. □

Thus, we can start studying a solvability for the equation

$$(Au)(x) = v(x), \quad x \in \mathbf{R}^M, \tag{1}$$

where A is a pseudo-differential operator with the symbol $\tilde{A}(\xi)$ satisfying the condition $(*)$, and the right hand side $v \in H^{S-\alpha}(\mathbf{R}^M)$.

Corollary 1 *If A is a pseudo-differential operator with the symbol $\tilde{A}(\xi)$ satisfying the condition $(*)$ then the Eq. (1) with an arbitrary right hand side $v \in H^{S-\alpha}(\mathbf{R}^M)$ has unique solution $u \in H^S(\mathbf{R}^M)$. The a priori estimate*

$$||u||_S \leq C||v||_{S-\alpha}$$

holds.

Proof The operator A^{-1} with the symbol $\tilde{A}^{-1}(\xi)$ is a pseudo-differential operator. Its symbol satisfies the condition $(*)$ with order $-\alpha$ instead of α. Then we have

$$\tilde{u} = \tilde{A}^{-1}\tilde{v},$$

and therefore

$$||u||_S^2 = ||A^{-1}v||_S^2 =$$

$$\int_{\mathbf{R}^M} (1 + |\xi_{K_1}|)^{2s_1}(1 + |\xi_{K_2}|)^{2s_2} \cdots (1 + |\xi_{K_n}|)^{2s_n} |\tilde{A}^{-1}(\xi)\tilde{v}(\xi)|^2 d\xi \leq c_1^{-2}||v||_{S-\alpha}^2,$$

and the sentence is proved. □

Unfortunately, such a simple conclusion is possible for the space $\mathbf{R}^M$. If we will take a domain $D \subset \mathbf{R}^M$ and will try to study a solvability for similar equation then we will obtain a lot of difficulties related to invertibility of operators.

We extract special *canonical* domains D in Euclidean space $\mathbf{R}^M$. Such domains are conical domains and we will start from a standard convex cone in Euclidean space non-including a whole straight line. Let $C_{K_j} \subset \mathbf{R}^{K_j}$ and we would like to consider the equation

$$(Au)(x) = v(x), \quad x \in C_{K_j}. \tag{2}$$

Direct applying the Fourier transform does not give the required answer since we have no the convolution theorem. The Eq. (2) can be rewritten in the form

$$(P_{K_j}Au)(x) = v(x), \quad x \in C_{K_j},$$

where P_{K_j} is the restriction on C_{K_j},

$$(P_{K_j}u)(x) = \begin{cases} u(x), \ x \in C_{K_j}; \\ \quad 0, \ x \notin \overline{C}_{K_j}. \end{cases}$$

and to use the Fourier transform we need to know what is the Fourier image of the operator P_{K_j}.

Structure of projectors For a general convex cone $C^m \subset \mathbf{R}^m$ one can define the Bochner kernel [7–9]

$$B_m(z) = \int\limits_C e^{ix\cdot z}dx, \quad z = (z_1, \ldots, z_m),$$

and the following representation in Fourier imaged

$$(FP_+u)(\xi) = \lim_{\tau\to 0+} \int\limits_{\mathbf{R}^m} B_m(\xi' - \eta', \xi_m - \eta_m + i\tau)\tilde{u}(\eta', \eta_m)d\eta,$$

here P_+ is the projector on the cone C^m [14, 17]. There are certain concrete realizations in the latter formula.

Example 1 We consider here one-dimensional case in which we have only one cone, and this cone is $\mathbf{R}_+$ [4]. For this case it was proved for a function $u(x)$, $x \in \mathbf{R}$, that

$$(FP_+u)(\xi) = \frac{1}{2}\tilde{u}(\xi) + \frac{i}{2\pi}p.v. \int\limits_{-\infty}^{+\infty} \frac{\tilde{u}(\eta)d\eta}{\xi - \eta}.$$

As a consequence we have for a function $u(x)$, $x \in \mathbf{R}^m$ and the cone $\mathbf{R}^m_+ = \{x \in \mathbf{R}^m : x = (x', x_m), x_m > 0\}$ the following result

$$(FP_+u)(\xi) = \frac{1}{2}\tilde{u}(\xi) + \frac{i}{2\pi}p.v. \int\limits_{-\infty}^{+\infty} \frac{\tilde{u}(\xi', \eta_m)d\eta_m}{\xi_m - \eta_m}, \quad \xi = (\xi', \xi_m).$$

Example 2 Let $m = 2$, and

$$C^a_+ = \{x \in \mathbf{R}^2 : x = (x_1, x_2), x_2 > a|x_1|, a > 0\}.$$

Then we have [21]

$$(FP_{C^a_+}u)(\xi) = \frac{\tilde{u}(\xi_1 + a\xi_2, \xi_2) + \tilde{u}(\xi_1 - a\xi_2, \xi_2)}{2} +$$

$$+v.p.\frac{i}{2\pi} \int\limits_{-\infty}^{+\infty} \frac{\tilde{u}(\eta, \xi_2)d\eta}{\xi_1 + a\xi_2 - \eta} - v.p.\frac{i}{2\pi} \int\limits_{-\infty}^{+\infty} \frac{\tilde{u}(\eta, \xi_2)d\eta}{\xi_1 - a\xi_2 - \eta}.$$

Example 3 Let $m = 3$, and $C^{a_1a_2}_+ = \{x \in \mathbf{R}^3 : x_2 > a_1|x_1| + a_2|x_2|\}$. Then

$$(FP_{C_+^{a_1a_2}}u)(\xi_1,\xi_2,\xi_3) =$$

$$= \frac{\tilde{u}(\xi_1 - a_1\xi_3, \xi_2 - a_2\xi_3, \xi_3) + \tilde{u}(\xi_1 + a_1\xi_3, \xi_2 - a_2\xi_3, \xi_3)}{4} +$$

$$+\frac{1}{2}(S_1\tilde{u})(\xi_1 + a_1\xi_3, \xi_2 - a_2\xi_3, \xi_3) - \frac{1}{2}(S_1\tilde{u})(\xi_1 - a_1\xi_3, \xi_2 - a_2\xi_3, \xi_3) +$$

$$+\frac{\tilde{u}(\xi_1 - a_1\xi_3, \xi_2 + a_2\xi_3, \xi_3) + \tilde{u}(\xi_1 + a_1\xi_3, \xi_2 + a_2\xi_3, \xi_3)}{4} +$$

$$+\frac{1}{2}(S_1\tilde{u})(\xi_1 + a_1\xi_3, \xi_2 + a_2\xi_3, \xi_3) - \frac{1}{2}(S_1\tilde{u})(\xi_1 - a_1\xi_3, \xi_2 + a_2\xi_3, \xi_3) +$$

$$+\frac{(S_2\tilde{u})(\xi_1 - a_1\xi_3, \xi_2 + a_2\xi_3, \xi_3) + (S_2\tilde{u})(\xi_1 + a_1\xi_3, \xi_2 + a_2\xi_3, \xi_3)}{2} +$$

$$+(S_1S_2\tilde{u})(\xi_1 + a_1\xi_3, \xi_2 + a_2\xi_3, \xi_3) - (S_1S_2\tilde{u})(\xi_1 - a_1\xi_3, \xi_2 + a_2\xi_3, \xi_3) -$$

$$-\frac{(S_2\tilde{u})(\xi_1 - a_1\xi_3, \xi_2 - a_2\xi_3, \xi_3) - (S_2\tilde{u})(\xi_1 + a_1\xi_3, \xi_2 - a_2\xi_3, \xi_3)}{2} -$$

$$-(S_1S_2\tilde{u})(\xi_1 + a_1\xi_3, \xi_2 - a_2\xi_3, \xi_3) + (S_1S_2\tilde{u})(\xi_1 - a_1\xi_3, \xi_2 - a_2\xi_3, \xi_3).$$

where

$$(S_1u)(\xi_1,\xi_2,\xi_3) = v.p.\frac{i}{2\pi}\int_{-\infty}^{+\infty} \frac{u(\tau,\xi_2,\xi_3)d\tau}{\xi_1 - \tau},$$

$$(S_2u)(\xi_1,\xi_2,\xi_3) = v.p.\frac{i}{2\pi}\int_{-\infty}^{+\infty} \frac{u(\xi_1,\eta,\xi_3)d\eta}{\xi_2 - \eta}.$$

This case was studied in [22]

Elliptic equations and complex variables This approach is related to the function theory of many complex variables, namely to functions which are holomorphic in radial tube domains [7–9].

Let $C_{K_j} \subset \mathbf{R}^{K_j}$, $j = 1, \ldots, n$, be convex cones non-including a whole straight line in $\mathbf{R}^{K_j}$. Let us compose the set $C = C_{K_1} \times \ldots C_{K_n}$.

Lemma 2 *The set C is a cone in $\mathbf{R}^M$ non-including a whole straight line in $\mathbf{R}^M$.*

Proof Indeed, C is a cone since each C_j is a cone. If we will assume that C includes a certain line in $\mathbf{R}^M$ then we will conclude that each cone C_j includes a certain straight line. □

Now we will start studying a solvability of the equation

$$(Au)(x) = v(x), \quad x \in C, \tag{3}$$

and the solution is sought in the space $H^S(C)$.

Definition 1 The space $H^S(C)$ consists of functions (distributions) from $H^S(\mathbf{R}^M)$ with supports in $\overline{C}$.

The right-hand side v is chosen from the space $H_0^{S-\alpha}(C)$; by definition the space $H_0^S(C)$ is a space of distributions on C, admitting a continuation on $H^S(\mathbf{R}^M)$. The norm in the space $H_0^S(C)$ is defined as

$$||v||_S^+ = \inf ||\ell f||_S,$$

where the *infimum* is taken over all continuations $\ell l f$ on the whole $\mathbf{R}^M$.

Fourier image of the space $H^S(C)$ will be denoted by $\tilde{H}^S(C)$

Definition 2 A radial tube domain over the cone C is called a domain in M-dimensional complex space $\mathbf{C}^M$ of the following type

$$T(C) \equiv \{z \in \mathbf{C}^M : z = x + iy, x \in \mathbf{R}^M, y \in C\}.$$

A conjugate cone $\overset{*}{C}$ is called such a cone in which for all points the condition

$$x \cdot y > 0, \quad \forall y \in C,$$

holds; $x \cdot y$ means inner product for x and y.

Definition 3 The wave factorization of an elliptic symbol $A(\xi)$ with respect to the cone C is called its representation in the form

$$A(\xi) = A_{\neq}(\xi)A_{=}(\xi),$$

where factors $A_{\neq}(\xi)$, $A_{=}(\xi)$ must satisfy the following conditions:

(1) $A_{\neq}(\xi)$, $A_{=}(\xi)$ are defined for all $\xi \in \mathbf{R}^M$ may be except the points $\xi \in \partial \overset{*}{C}$;

(2) $A_{\neq}(\xi)$, $A_{=}(\xi)$ admit an analytic continuation into radial tube domains $T(\overset{*}{C})$, $T(-\overset{*}{C})$ respectively with estimates

$$|A_{\neq}^{\pm 1}(\xi + i\tau)| \leq c_1 \prod_{j=1}^{n} (1 + |\xi_{K_j}| + |\tau_{K_j}|)^{\pm æ_j},$$

$$|A_{=}^{\pm 1}(\xi - i\tau)| \leq c_2 \prod_{j=1}^{n} (1 + |\xi_{K_j}| + |\tau_{K_j}|)^{\pm(\alpha_j - æ_j)}, \quad \forall \tau \in \overset{*}{C}, \quad æ_j \in \mathbf{R}.$$

The vector $æ = (æ_1, \dots, æ_n)$ is called an index of the wave factorization.

To apply the Fourier transform to the Eq. (3) we need to know what is FP_C; here F denotes the Fourier transform in M-dimensional space. Let us introduce the following notations. For every C_{K_j} we consider corresponding radial tube domain $T(\overset{*}{C}_{K_j})$ over the conjugate cone and an element of $T(\overset{*}{C}_{K_j})$ will be denoted by $\xi_{K_j} + i\tau_{K_j}$. Moreover, for ξ_{K_j} we will use the notation $\xi_{K_j} = (\xi'_{K_j}, \xi_{k_j})$, where ξ_{k_j} is the k_jth coordinate, and ξ'_{K_j} denotes left other coordinates. The same notations will be used for $x \in \mathbf{R}^{K_j}$, $x_{K_j} = (x'_{K_j}, x_{k_j})$.

As before we denote by P_C the restriction operator on C. Obviously,

$$P_C = \prod_{j=1}^{n} P_{K_j}.$$

and then

$$B_M(z) = \prod_{j=1}^{n} B_{k_j}(z_{K_j}), \quad z = (z_{K_1}, \dots, z_{K_n}).$$

The last our observation is the following:

$$T(\overset{*}{C}) = \prod_{j=1}^{n} T(\overset{*}{C}_{K_j}),$$

and the Bochner kernel $B_M(z)$ will be a holomorphic function in $T(\overset{*}{C})$.

Theorem 1 *If the symbol $A(\xi)$ admits the wave factorization with respect to the cone C with the index æ such that $|æ_j - s_j| < 1/2$, $j = 1, \dots, n$, then the Eq. (3) has unique solution in the space $H^s(C)$ for arbitrary right hand side $v \in H_0^{s-\alpha}(C)$.*

The a priori estimate

$$||u||_s \leq const\ ||v||^+_{s-\alpha}$$

holds.

Proof We use the Wiener–Hopf method [4, 14]. Let ℓv be an arbitrary continuation of v onto $\mathbf{R}^M$ then we put

$$u_-(x) = (\ell v)(x) - (Au)(x),$$

so that $v_-(x) = 0$ for $x \in C$. Further,

$$(Au)(x) + u_-(x) = (\ell v)(x),$$

and after applying the Fourier transform and the wave factorization we obtain

$$A_{\neq}(\xi)\tilde{u}(\xi) + A_{=}^{-1}(\xi)\tilde{u}_-(\xi) = A_{=}^{-1}(\xi)\widetilde{(\ell v)}(\xi) \tag{4}$$

Now we can use the following result (see [14]).

Property 1 *If* $\tilde{u}_- \in \widetilde{H}^S(\mathbf{R}^M \setminus C)$, $A_{=}^{-1}$ *is a factor of the wave factorization then* $A_{=}^{-1}\tilde{u}_- \in \widetilde{H}^{S+\alpha-æ}(\mathbf{R}^M \setminus C)$.

Obviously, the summand $A_{\neq}(\xi)\tilde{u}(\xi)$ belongs to $\widetilde{H}^{S-æ}(C)$ according to Lemma 1 and holomorphic properties, and $A_{=}^{-1}(\xi)\tilde{u}_-(\xi)$ belongs to $\widetilde{H}^{S-æ}(\mathbf{R}^M \setminus C)$ according to Property 1.

The right hand side $A_{=}^{-1}(\xi)\widetilde{(\ell v)}(\xi)$ belongs to the space $\widetilde{H}^{S-æ}(\mathbf{R}^M)$ (Lemma 1), and since $|æ_j - s_j| < 1/2$, $j = 1, \dots, n$, it can be uniquely represented as

$$A_{=}^{-1}(\xi)\widetilde{(\ell v)}(\xi) = \tilde{v}_+(\xi) + \tilde{v}_-(\xi), \tag{5}$$

where

$$\tilde{v}_+(\xi) = B_M\left(A_{=}^{-1}(\xi)\widetilde{(\ell v)}(\xi)\right), \quad \tilde{v}_-(\xi) = (I - B_M)\left(A_{=}^{-1}(\xi)\widetilde{(\ell v)}(\xi)\right).$$

The representation (5) is true since the operator $B_M : \widetilde{H}^\delta(\mathbf{R}^M) \to \widetilde{H}^\delta(\mathbf{R}^M)$ for $|\delta_J| < 1/2$, $j = 1, \dots, n$, and we remind that $|æ_j - s_j| < 1/2$, $j = 1, \dots, n$,.

Further, we rewrite the equality (4) in the form

$$A_{\neq}(\xi)\tilde{u}(\xi) - \tilde{v}_+(\xi) = \tilde{v}_-(\xi) - A_{=}^{-1}(\xi)\tilde{u}_-(\xi),$$

and we obtain that a distribution from $H^\delta(C)$ equals to a distribution from $H^\delta(\mathbf{R}^M \setminus \overline{C})$. But for such small δ this common distribution should be zero only [14]. Thus,

$$A_{\neq}(\xi)\tilde{u}(\xi) - \tilde{v}_+(\xi) = 0,$$

or in other words

$$\tilde{u}(\xi) = A_{\neq}^{-1}(\xi)B_M\left(A_{=}^{-1}(\xi)\widetilde{(\ell v)}(\xi)\right).$$

A priori estimate is based on Lemma 1 and boundedness property of the operator $B_M : \widetilde{H}^\delta(\mathbf{R}^M) \to \widetilde{H}^\delta(\mathbf{R}^M)$. Indeed,

$$||u||_S = ||\tilde{u}||_S = ||A_{\neq}^{-1}(\xi)B_M\left(A_{=}^{-1}(\xi)\widetilde{(\ell v)}(\xi)\right)||_S \le$$

$$\le const\ ||B_M\left(A_{=}^{-1}(\xi)\widetilde{(\ell v)}(\xi)\right)||_{S-æ} \le const\ ||A_{=}^{-1}(\xi)\widetilde{(\ell v)}(\xi)||_{S-æ} \le$$

$$\le const\ ||\widetilde{(\ell v)}(\xi)||_{S-\alpha} = const\ ||\ell v||_{S-\alpha} \le const\ ||v||^+_{S-\alpha},$$

and Theorem 1 is proved. □

Multiply solutions For the cone C_{K_j}, $j = 1, \ldots, n$, we suppose that a surface of this cone is given by the equation $x_{k_j} = \varphi_j(x'_{K_j})$, where $\varphi_j : \mathbf{R}^{k_j-1} \to \mathbf{R}$ is a smooth function in $\mathbf{R}^{k_j-1} \setminus \{0\}$, and $\varphi_j(0) = 0$.

Let us introduce the following change of variables

$$\begin{cases} t'_{K_j} = x'_{K_j} \\ t_{k_j} = x_{k_j} - \varphi_j(x'_{K_j}) \end{cases}$$

and we denote this operator by $T_{\varphi_j} : \mathbf{R}^{K_j} \to \mathbf{R}^{K_j}$. Since the cone is in one part of a half-space then points of the second part of a half-space will be fixed. Such change of variables can be defined for distributions also [22].

Below we will use notation F_m for the Fourier transform in m-dimensional space, so that the notation F_{K_j} will be the Fourier transform in $\mathbf{R}^{K_j}$.

Following to [22] we conclude

$$F_{K_j} T_{\varphi_j} = V_{\varphi_j} F_{K_j}.$$

Further, we introduce $T_\varphi : \mathbf{R}^M \to \mathbf{R}^M$ by the formula

$$T_\varphi = \prod_{j=1}^{n} T_{\varphi_j}$$

and construct the operator

$$V_\varphi = \prod_{j=1}^{n} V_{\varphi_j},$$

for which we have

$$F_M T_\varphi = V_\varphi F_M.$$

Let us introduce vectors $N = (n_1, \ldots, n_n)$, $L = (l_1, \ldots, l_n)$, $\delta=(\delta_1, \ldots, \delta_n)$, n_j, $l_j \in \mathbf{N}$, $|\delta_j| < 1/2$, $j = 1, \ldots, n$, and a polynomial $Q_N(\xi)$, $\xi \in \mathbf{R}^M$ satisfying the condition

$$|Q_N(\xi)| \sim \prod_{j=1}^{n} (1 + |\xi_{K_j}|)^{n_j}, \tag{6}$$

Theorem 2 *If the symbol $A(\xi)$ admits the wave factorization with the index æ, æ $-$ $S = N + \delta$, then a general solution of the Eq. (3) in Fourier images is given by the formula*

$$\tilde{u}(\xi) = A_{\neq}^{-1}(\xi) Q_N(\xi) B_M Q_N^{-1}(\xi) A_{=}^{-1}(\xi) \widetilde{(\ell v)}(\xi) +$$

$$+ A_{\neq}^{-1}(\xi) V_\varphi^{-1} \left(\sum_{l_1=1}^{n_1} \sum_{l_2=1}^{n_2} \cdots \sum_{l_n=1}^{n_n} \tilde{c}_L(\xi'_K) \xi_{k_1}^{l_1-1} \xi_{k_2}^{l_2-1} \cdots \xi_{k_n}^{l_n-1} \right),$$

where $c_L(x'_K) \in H^{S_L}(\mathbf{R}^{M-n})$ *are arbitrary functions,* $S_L=(s_1-æ_1+l_1-1/2,\ldots, s_n-æ_n+l_n-1/2)$, $l_j=1,2,\ldots,n_j)$, $j=1,2,...,n$, ℓv *is an arbitrary continuation of* v *onto* $H^{S-\alpha}(\mathbf{R}^M)$.

The a priori estimate

$$||u||_S \leq const \left(||v||^+_{S-\alpha} + \sum_{l_1=1}^{n_1}\sum_{l_2=1}^{n_2}\cdots\sum_{l_n=1}^{n_n} ||c_L||_{S_L} \right)$$

holds.

Proof Similar to the proof of Theorem 1 we obtain the equality (4). Further, let us note that the function $A_{=}^{-1}(\xi)\widetilde{(\ell v)}(\xi)$ belongs to the space $\tilde{H}^{S-æ}(\mathbf{R}^M)$. So, if take an arbitrary polynomial $Q_N(\xi)$ satisfying the condition (6) then the function $Q_N^{-1}(\xi)A_{=}^{-1}(\xi)\widetilde{(\ell v)}(\xi)$ will belong to the space $\tilde{H}^{-\delta}(\mathbf{R}^M)$.

Further, according to the theory of multidimensional Riemann problem [14] we can represent the latter function as a sum of two summands, this is so called a jump problem which can be solved by the operator B_M:

$$Q_N^{-1}A_{=}^{-1}\widetilde{(\ell v)} = f_+ + f_-,$$

where $f_+ \in \tilde{H}^{-\delta}(C)$, $f_- \in \tilde{H}^{-\delta}(\mathbf{R}^M \setminus C)$,

$$f_+ = B_M(A_{=}^{-1}\widetilde{(\ell v)}), \quad f_- = (I - B_M)(A_{=}^{-1}\widetilde{(\ell v)}).$$

Multiplying the equality (4) by $Q_N^{-1}(\xi)$ we rewrite it in the form

$$Q_N^{-1}A_{\neq}\tilde{u} + Q_N^{-1}A_{=}^{-1}\tilde{u}_- = f_+ + f_-,$$

or

$$Q_N^{-1}A_{\neq}\tilde{u} - f_+ = f_- - Q_N^{-1}A_{=}^{-1}\tilde{u}_-$$

In other words

$$A_{\neq}\tilde{u} - Q_N f_+ = Q_N f_- - A_{=}^{-1}\tilde{u}_-. \tag{7}$$

The left hand side of the equality (7) belongs to the space $\tilde{H}^{-N-\delta}(C)$, bur the right hand side belongs to the space $\tilde{H}^{-N-\delta}(\mathbf{R}^M \setminus C)$. Therefore, we have

$$F_M^{-1}(A_{\neq}\tilde{u} - Q_N f_+) = F_M^{-1}(Q_N f_- - A_{=}^{-1}\tilde{u}_-),$$

where the left hand side belongs to the space $H^{-N-\delta}(C)$, but right hand side belongs to the space $H^{-N-\delta}(\mathbf{R}^M \setminus C)$, from which we conclude immediately that this is a distribution supported on the surface ∂C.

The form for such a distribution is given in [22] for the cone C_{K_j} with help of the operator V_{φ_j}. Thus, we apply the operator T_φ to the latter equality and obtain

$$T_\varphi F_M^{-1}(A_{\neq}\tilde{u} - Q_N f_+) = T_\varphi F_M^{-1}(Q_N f_- - A_=^{-1}\tilde{u}_-),$$

so that both left hand side and right hand side is a distribution supported on the hyper-plane $x_{k_1} = 0, x_{k_2} = 0, \ldots, x_{k_n} = 0$. Then

$$T_\varphi F_M^{-1}(A_{\neq}\tilde{u} - Q_N f_+) =$$

$$= \sum_{l_1=1}^{n_1}\sum_{l_2=1}^{n_2}\cdots\sum_{l_n=1}^{n_n} c_L(x'_K)\delta^{(l_1-1)}(x_{k_1})\delta^{(l_2-1)}(x_{k_2})\cdots\delta^{(l_n-1)}(x_{k_n}),$$

where $L = l_1, \ldots, l_n)$, $x'_K = (x'_{K_1}, \ldots, x'_{K_n}) \in \mathbf{R}^{M-n}$, δ is the Dirac mass-function.

Applying the Fourier transform we obtain

$$F_M T_\varphi F_M^{-1}(A_{\neq}\tilde{u} - Q_N f_+) =$$

$$= \sum_{l_1=1}^{n_1}\sum_{l_2=1}^{n_2}\cdots\sum_{l_n=1}^{n_n} \tilde{c}_L(\xi'_K)\xi_{k_1}^{l_1-1}\xi_{k_2}^{l_2-1}\cdots\xi_{k_n}^{l_n-1}, \tag{8}$$

Taking into account that $F_M T_\varphi F_M^{-1}$ we can write

$$A_{\neq}\tilde{u} - Q_N f_+ = V_\varphi^{-1}\left(\sum_{l_1=1}^{n_1}\sum_{l_2=1}^{n_2}\cdots\sum_{l_n=1}^{n_n} \tilde{c}_L(\xi'_K)\xi_{k_1}^{l_1-1}\xi_{k_2}^{l_2-1}\cdots\xi_{k_n}^{l_n-1}\right),$$

or finally

$$\begin{aligned}\tilde{u}(\xi) = {} & A_{\neq}^{-1}(\xi)Q_N(\xi)B_M Q_N^{-1}(\xi)A_=^{-1}(\xi)\widetilde{(\ell v)}(\xi) + \\ & + A_{\neq}^{-1}(\xi)V_\varphi^{-1}\left(\sum_{l_1=1}^{n_1}\sum_{l_2=1}^{n_2}\cdots\sum_{l_n=1}^{n_n} \tilde{c}_L(\xi'_K)\xi_{k_1}^{l_1-1}\xi_{k_2}^{l_2-1}\cdots\xi_{k_n}^{l_n-1}\right).\end{aligned} \tag{9}$$

To obtain a priori estimates let us note that all summands in the formula (8) should belong to the space $\widetilde{H}^{S-æ}(\mathbf{R}^M)$. We take one of summands and estimate corresponding integral.

$$||\tilde{c}_L(\xi'_K)\xi_{k_1}^{l_1-1}\xi_{k_2}^{l_2-1}\cdots\xi_{k_n}^{l_n-1}||_{S-æ}^2 \leq$$

$$\leq \int_{\mathbf{R}^M} |\tilde{c}_L(\xi'_K)|^2 \prod_{j=1}^{n}(1+|\xi_{K_j}|)^{2(s_j-æ_j)}\prod_{j=1}^{n}|\xi_{k_j}|^{2(l_j-1)}d\xi'_K\prod_{j=1}^{n}d\xi_{k_j} \leq$$

$$\leq \int_{\mathbf{R}^M} |\tilde{c}_L(\xi'_K)|^2 \prod_{j=1}^{n}(1+|\xi_{K_j}|^{2(s_j-æ_j+l_j-1)}d\xi'_K\prod_{j=1}^{n}d\xi_{k_j},$$

and for existence of each integral of the type

$$\int_{-\infty}^{+\infty}(1+|\xi'_{K_j}|+|\xi_{k_j}|)^{2(s_j-æ_j+l_j-1)}d\xi_{k_j}$$

the condition

$$2(s_j-æ_j+l_j-1)<-1 \tag{10}$$

is necessary. It is equivalent to the following condition

$$s_j-æ_j+l_j<1.$$

Since we have $s_j-æ_j+l_j=-n_j-\delta_j+l_j$ then we see that the condition (10) is satisfied for all

$$l_j=1,2,\ldots,n_j,$$

but it is not satisfied for $l_j=n_j+1$. After integration on all ξ_{k_j} we will find that $\tilde{c}_L(\xi'_K)\in\widetilde{H}^{S_L}(\mathbf{R}^{M-n})$, where $S_L=(s_1-æ_1+l_1-1/2,\ldots,s_n-æ_n+l_n-1/2)$, and $l_j=1,2,\ldots,n_j$, $j=1,2.\ldots,n$.

For a priori estimates we have

$$||A_{\neq}^{-1}(\xi)Q_N(\xi)B_MQ_N^{-1}(\xi)A_{=}^{-1}(\xi)\widetilde{(\ell v)}(\xi)||_S\leq$$

$$\leq const\ ||B_MQ_N^{-1}(\xi)A_{=}^{-1}(\xi)\widetilde{(\ell v)}(\xi)||_{S-æ+N}\leq$$

$$\leq const\ ||Q_N^{-1}(\xi)A_{=}^{-1}(\xi)\widetilde{(\ell v)}(\xi)||_{S-æ+N}\leq$$

$$\leq const\ ||\widetilde{(\ell v)}(\xi)||_{S-æ+N-N+æ-\alpha}=const\ ||\widetilde{(\ell v)}(\xi)||_{S-\alpha}\leq const\ ||v||^{+}_{S-\alpha}$$

according to Lemma 1 and the fact that $S-æ+N=\delta.|\delta_j|<1/2$, $j=1,\ldots,n$.

To estimate other summands in the formula (9) we use above considerations. Really, if $\tilde{c}_L(\xi'_K)\in\widetilde{H}^{S_L}(\mathbf{R}^{M-n})$ then each summand $\tilde{c}_L(\xi'_K)\xi_{k_1}^{l_1-1}\xi_{k_2}^{l_2-1}\cdots\xi_{k_n}^{l_n-1}$ in the formula (9) belongs to the space $\widetilde{H}^{S-æ}(\mathbf{R}^M)$. Thus, we have

$$||A_{\neq}^{-1}(\xi)V_{\varphi}^{-1}\tilde{c}_L(\xi'_K)\xi_{k_1}^{l_1-1}\xi_{k_2}^{l_2-1}\cdots\xi_{k_n}^{l_n-1}||_S\leq$$

$$\leq const\ ||V_{\varphi}^{-1}\tilde{c}_L(\xi'_K)\xi_{k_1}^{l_1-1}\xi_{k_2}^{l_2-1}\cdots\xi_{k_n}^{l_n-1}||_{S-æ}\leq$$

$$\leq const\ ||\tilde{c}_L(\xi'_K)\xi_{k_1}^{l_1-1}\xi_{k_2}^{l_2-1}\cdots\xi_{k_n}^{l_n-1}||_{S-æ}\leq const\ ||\tilde{c}_L||_{S_L}.$$

The latter estimate was obtained above. The Theorem 2 is proved. □

Remark 1 This formula includes the operator V_φ. Examples 2 and 3 give exact representation for this operator for certain concrete cones.

Conclusion

These studies led to different boundary value problems for such elliptic pseudo-differential equations in cones similar to [14, 17, 18]. Particularly, for the case of Theorem 2 a general solution of the Eq. (3) includes a lot of arbitrary functions from corresponding Sobolev–Slobodetskii spaces. To determine these functions uniquely one needs some additional conditions (not necessary boundary conditions). We will try to describe certain statements of boundary value problems in forthcoming papers.

References

1. Treves F (1980) Introduction to pseudodifferential operators and Fourier integral operators. Springer, New York
2. Taylor M (1981) Pseudodifferential operators. Princeton University Press, Princeton
3. Hörmander L (1983) Analysis of partial differential operators, Vol I–IV. Springer, Berlin
4. Eskin G (1981) Boundary value problems for elliptic pseudodifferential equations. AMS, Providence
5. Schulze B-W, Sternin B, Shatalov V (1998) Differential equations on singular manifolds: semiclassical theory and operator algebras. Wiley-VCH, Berlin
6. Schulze B-W (1998) Boundary value problems and singular pseudo-differential operators. Wiley, Chichester
7. Bochner S, Martin WT (1948) Several complex variables. Princeton University Press, Princeton
8. Vladimirov VS (2007) Methods of the theory of functions of many complex variables. Dover Publications, Mineola
9. Vladimirov VS (1979) Generalized functions in mathematical physics. Mir Publishers, Moscow
10. Carvalho C, Nistor V, Qiao Y (2017) Fredholm criteria for pseudodifferential operators and induced representations of groupoid algebras. Electron Res Announc Math Sci 24:68–77. https://doi.org/10.3934/era.2017.24.008
11. Ruzhansky M, Tokmagambetov N (2016) Nonharmonic analysis of boundary value problems. Int Math Res Notices 2016:3548–3615. https://doi.org/10.1093/imrn/rnv243
12. Plamenevskii B (2010) Solvability of algebras of pseudodifferential operators with piecewise smooth coefficients on smooth manifolds. St Petersbg Math J 21:317–351. https://doi.org/10.1090/S1061-0022-10-01097-6
13. Rabinovich V (2015) Boundary problems for domains with conical exits at infinity and limit operators. Complex Var Elliptic Equ 60:293–309. https://doi.org/10.1080/17476933.2014.926339
14. Vasil'ev VB (2000) Wave factorization of elliptic symbols: theory and applications. Kluwer Academic Publishers, Dordrecht-Boston-London
15. Nagel A, Ricci F, Stein EM, Wainger S (2018) Algebras of singular integral operators with kernels controlled by multiple norms. Memoirs of AMS. 2018. Vol 256, Issue 1230. vii, 141 pp. https://doi.org/10.1090/memo/1230
16. Gindikin SG, Volevich LR (1992) Distributions and convolution equations. Gordon and Breach Science Publishers, Philadelphia

17. Vasilyev VB (2018) Pseudodifferential equations, wave factorization, and related problems. Math Methods Appl Sci 41:9252–9263. https://doi.org/10.1002/mma.5212
18. Vasilyev VB (2020) On certain 3-dimensional limit boundary value problems. Lobachevskii J Math 41:917–925. https://doi.org/10.1134/S1995080220050133
19. Vasilyev V (2019) Elliptic operators and their symbols. Demonstr Math 52:361–369. https://doi.org/10.1515/dema-2019-0025
20. Vasilyev V (2020) Operator symbols and operator indices. Symmetry 12:1–12. https://doi.org/10.3390/sym12010064
21. Vasilyev VB (2019) On some distributions associated to boundary value problems. Complex Var Elliptic Equ 64(5):888–898. https://doi.org/10.1080/17476933.2019.1571048
22. Vasilyev VB (2018) Pseudo-differential equations, wave factorization, and related problems. Math Meth Appl Sci 41:9252–9263. https://doi.org/10.1002/mma.5212

Thermodynamic Limit in Vector Lattice Models

Yuri P. Virchenko

Abstract Classes of Gibbs random fields $\mathbf{u}(x)$, $x \in \mathcal{Z}^d$ on finite sets $\Lambda \subset \mathcal{Z}^d$, $d \in \mathcal{N}$ with values in the space $\mathcal{R}^n$, $n \in \mathcal{N}$ are studied. Each class is connected with the sequence $\langle \Lambda;\ \Lambda \subset \mathcal{Z}^d \rangle$ unboundedly expanding according to the definite rule when $\Lambda \to \mathcal{Z}^d$. Each random field is generated by the Hamiltonian $\mathsf{H}_\Lambda[\mathbf{u}(z)]$. Classes of all functionals $\mathsf{H}_\Lambda[\mathbf{u}(z)]$ corresponding to sequence $\langle \Lambda;\ \Lambda \subset \mathcal{Z}^d \rangle$ form the Banach space H_ν. It is proved the existence of the limit statistical characteristic $\ln Z_\Lambda/|\Lambda|$ in each class when $\Lambda \to \mathcal{Z}^d$ which is the continuous functional in H_ν.

Keywords Vector models · Hamiltonian · Gibbs' random field · Free energy · Phase space · Thermo-dynamic limit

1 Introduction

The object of study in this paper is Gibbs random fields on the integer lattice $\mathcal{Z}^d$, $d \in \mathcal{N}$. The importance of studying such mathematical objects is due to the fact that models of statistical mathematical physics are constructed on their basis (about the subject of the study and the terminology used, see, for example, [1–6]). We will call such models as *vector lattice systems*. From the point of view of theoretical physics, these models describe, within the microscopic approach and with appropriate interpretation of the parameters defining theirs, the thermodynamic behavior of single-crystal solid-state structures in a wide temperature range. Despite the fact that a considerable amount of literature is devoted to the mathematical analysis of such theoretical models, in most mathematical works related to their study within the framework of the formalism of statistical mechanics of classical (non-quantum) systems, the greatest attention is paid to such of them which are called *lattice gases*. For

Y. P. Virchenko (✉)
Belgorod State Technological University, Kostyukova st. 46, Belgorod 308012, Russia
e-mail: virch@bsu.edu.ru

Belgorod State University, Pobedy st., 85, Belgorod 308015, Russia

V. Vasilyev (ed.), *Differential Equations, Mathematical Modeling and Computational Algorithms*, Springer Proceedings in Mathematics & Statistics 423,
https://doi.org/10.1007/978-3-031-28505-9_19

such mathematical objects, the terminology has been developed that unites them. In terms of this terminology their properties are established at the level of those requirements that are imposed on mathematical texts. The purpose of this work is to extend these basic concepts to a much wider class of models of statistical mechanics of classical systems which we call, as mentioned above, the vector lattice models. For such systems, we will prove, within the framework of accepted general restrictions, the validity of one of the basic provisions of statistical mechanics, namely, we establish the presence of *extensive* asymptotic $F_\Lambda \sim |\Lambda|$ of the *free energy* F_Λ if the sets Λ tends to $\mathcal{Z}^d$ according to a certain principle dictated by physical considerations.

2 The Gibbs Random Fields

Consider a random field $\tilde{\mathbf{u}}(\Lambda) = \{\tilde{\mathbf{u}}(x)\,;\, x \in \Lambda\}$ on an arbitrary finite subset Λ of the integer lattice $\mathcal{Z}^d$, $d \in \mathcal{N}$, with elements $x = \langle l_1, ..., l_d\rangle, l_j \in \mathcal{Z}, j = 1 \div d$. We will call the lattice elements as *vertexes*.[1] This means that corresponding probability space $\mathbf{P}_\Lambda = \langle \Omega_\Lambda, \mathbf{B}_\Lambda, \mathsf{P}_\Lambda\rangle$ consists of Ω_Λ elementary random events (random configurations), σ-algebra $\mathbf{B}_\Lambda$ of measurable subsets of Ω_Λ, each element of which is considered as the random event, and the probability distribution P_Λ on $\mathbf{B}_\Lambda$.

For the Gibbs random fields of vector lattice models considered in this paper, the listed components of the probability space $\mathbf{P}_\Lambda$ are defined as follows. Denote the set $\Omega \equiv \mathcal{R}^n$, which we will call the phase space of each vertex in $\mathcal{Z}^d$. The number $n \in \mathcal{N}$ is the dimension of the vector field fixed during the work.

For any subset of $\Lambda \subset \mathcal{Z}^d$, we define the space $\Omega_\Lambda = \Omega^\Lambda$. This means that each vertex $x = \langle l_1, ..., l_d\rangle \in \Lambda$ is mapped to a point of the Ω space which is assigned the label x and, as a result of such an operation, the phase space Ω_x is obtained. Then, for any $\Lambda \subset \mathcal{Z}^d$, the space of elementary events, which we will call *the space of states* (configurations), is represented by the formula

$$\Omega_\Lambda = \bigotimes_{x\in\Lambda} \Omega_x\,. \tag{1}$$

On the space Ω, there is a natural measurability structure defined by the σ-algebra $\mathbf{B}$ of Borel sets in $\mathcal{R}^n$. Then, similarly to the formula (1), by assigning labels, σ-algebras $\mathbf{B}_x$, $x \in \mathcal{Z}$ are introduced on each of the spaces Ω_x and, on the bases of them, the σ-algebra $\mathbf{B}_\Lambda$ is constructed on Ω_Λ

$$\mathbf{B}_\Lambda = \bigotimes_{x\in\Lambda} \mathbf{B}_x\,. \tag{2}$$

In accordance with this structure of measurability, we will also assume that the measure M is defined on the σ-algebra $\mathbf{B}$. For simplicity of further constructions, we

[1] Here and further throughout the text, random variables are marked with the "tilde" sign.

will assume with respect to this measure that it does not contain a *singular* component, that is, it has a derivative $d\mathsf{M}/d\mathbf{u} = D(\mathbf{u}) \geq 0$ of the Lebesgue measure in $\mathcal{R}^n$ with the differential $d\mathbf{u} = du_1...du_n$. This derivative is expressed as a generalized function with respect to the countably normalized space of locally continuous functions on $\mathcal{R}^n$. In particular, in the case of $n = 1$, this means that in the Lebesgue decomposition of the measure M on $\mathcal{R}$, there are only absolutely continuous and discrete components.

On the basis of the measure M, by assigning labels $x \in \mathcal{Z}^d$, we introduce measures M_x on σ-algebras $\mathbf{B}_x$ and, as a result, the measure $\mathbf{B}_\Lambda$ is defined as a product of measures

$$\mathsf{M}_\Lambda = \prod_{x\in\Lambda} \mathsf{M}_x\,, \quad d\mathsf{M}_\Lambda = \prod_{x\in\Lambda} D(\mathbf{u}(x))d\mathbf{u}(x)\,. \tag{3}$$

Each Gibbs random field is $\tilde{\mathbf{u}}(\Lambda)$ is constructed by the definition of the probability distribution P_Λ on a measurable space $\langle \Omega_\Lambda, \mathbf{B}_\Lambda, \mathsf{M}_\Lambda\rangle$. Its random realizations $\tilde{\mathbf{u}}(\Lambda) \in \Omega_\Lambda$ are represented by mappings $\tilde{\mathbf{u}}(\Lambda) : \Lambda \mapsto \mathcal{R}^n$. Due to the finiteness of the set Λ, each such mapping can be considered as a collection of $\{\tilde{\mathbf{u}}(x)\,;\, x \in \Lambda\}$ of $|\,\Lambda\,|$ (number of vertexes in Λ) random variables taking the value in $\mathcal{R}^n$. The fact that we consider further this set of random variables as a *Gibbs random field* means that the probability distribution P_Λ has a non-negative density on the measure M_Λ defined by the formula

$$d\mathsf{P}_\Lambda = \frac{1}{Z_\Lambda} \exp\big(- \mathsf{H}_\Lambda[\mathbf{u}(z)]\big) d\mathsf{M}_\Lambda\,, \tag{4}$$

where each of the functionals $\mathsf{H}_\Lambda[\mathbf{u}(z)]$, $\Lambda \subset \mathcal{Z}^d$ is called the *Hamiltonian* of the Gibbs random field.

Statistical characteristic Z_Λ of the probability distribution (4) called the *partition function*, is determined on the basis of the normalization condition $\mathsf{P}_\Lambda(\Omega_\Lambda) = 1$ of the distribution P_Λ

$$Z_\Lambda = \int_{\Omega_\Lambda} \exp\big(- \mathsf{H}_\Lambda[\mathbf{u}(z)]\big) d\mathsf{M}_\Lambda\,. \tag{5}$$

Thus, for a fixed measure M, we assume the choice only of such functionals $\mathsf{H}_\Lambda[\mathbf{u}(z)]$ for which this integral is finite.

In order to connect probability spaces $\{\mathbf{P}_\Lambda\,;\, \Lambda \subset \mathcal{Z}^d\}$ defined at various $\Lambda \subset \mathcal{Z}^d$ by the fixed phase the space Ω and the fixed measure M on it with statistical mechanics models, it is necessary to distribute these spaces by equivalence classes such that one may take into account the property of physical uniformity.

This is done, firstly, taking into account the fact that the translation of the set Λ should not change the physical predictions, that is, it should not change values of statistical averages obtained as a result of calculations on the basis of a mathematical model.

Secondly, it should be taken into account that the sets Λ for statistical mechanics systems consist of an indefinitely large number of vertexes so that each *intensive thermodynamic characteristic*, related to one vertex of the lattice, is practically independent on $|\Lambda|$.

The first of these requirements can be satisfied by assuming that the collection of all Hamiltonians $\mathsf{H}_\Lambda[\cdot]$, $\Lambda \subset \mathcal{Z}^d$ describing the same physical system, subject to a condition that reflects independence of all statistical averages on the location of the set Λ in $\mathcal{Z}^d$. This is expressed by the property of the *translational invariance*. Let us formulate the simplest version of such a condition. Let z be an arbitrary vertex of $\mathcal{Z}^d$. Then the space Ω_Λ and the measure M_Λ on it have the property

$$\Omega_{\Lambda+z} = \Omega_\Lambda|_{\mathsf{u}(x)\to\mathsf{u}(x+z)}, \quad \mathsf{M}_{\Lambda+z} = \mathsf{M}_\Lambda|_{\mathsf{u}(x)\to\mathsf{u}(x+z)}\,. \tag{6}$$

Hamiltonian $\mathsf{H}_\Lambda[\mathbf{u}(x)]$ is called the translationally invariant one if it has the following property

$$\mathsf{H}_{\Lambda+z}[\mathbf{u}(x)] = \mathsf{H}_\Lambda[\mathbf{u}(x+z)]\,. \tag{7}$$

Each Hamiltonian $\mathsf{H}_\Lambda[\mathbf{u}(x)]$ is defined as a function on vector variables $\{x \in \Lambda\}$ for each set Λ. We denote this function as $\mathbf{u}(\Lambda) = \{\mathbf{u}(x)\,;\, x \in \Lambda\}$. Then, the property (7) means that all these functions are the same for all sets $\Lambda + z$, $z \in \mathcal{Z}^d$.

Theorem 1 *If Hamiltonian* $\mathsf{H}[\mathbf{u}(z)]$ *is translationally invariant, then the probability distributions* P_Λ *and* $\mathsf{P}_{\Lambda+z}$ *are equivalent in the sense that*

$$d\mathsf{P}_{\Lambda+z}[\mathbf{u}(y)] = d\mathsf{P}_\Lambda[\mathbf{u}(y+z)]\,. \tag{8}$$

Proof Statement directly follows from (5)–(7).

Let us now proceed to the discussion of the second requirement for a Gibbs random field with Hamiltonians $\mathsf{H}_\Lambda[\mathbf{u}(z)]$ which allows distribute them into equivalence classes. Let us fix some lattice vertex $\mathcal{Z}^d$ which we will call the zero one. We will consider only Gibbs fields on sets Λ that contain this vertex. Due to the necessity to use a large number of vertexes $|\Lambda|$ (even for the smallest experimentally studied nanoparticles of a solid state substance $|\Lambda| \approx 10^6$ and more), it does not make sense to accurately calculate the expectations of $\mathsf{E}_\Lambda(\cdot)$ on the basis of the probability measure P_Λ.

On the contrary, in the practice of using of probability theory methods in theoretical statistical physics, it is necessary only to have confidence the fact that the calculated thermodynamic characteristics have a quite definite asymptotic behavior at unlimited increase of the set Λ occupied by the thermodynamically homogeneous medium under study. In this case, only the main asymptotic terms of expectations $\mathsf{E}_\Lambda(\cdot)$ on the probability measure P_Λ are of interest when Λ is expanded to $\mathcal{Z}^d$ according to a definite rule. Transition to the limit at $\Lambda \to \mathcal{Z}^d$ according to corresponding expanding sequences of statistical characteristics of Gibbs random fields is called the transition to *thermodynamic limit* in statistical mechanics.

In this paper we study so-called the *extensive systems* which are traditional to statistical mechanics when the function $F_\Lambda = \ln Z_\Lambda$, which is named theirs free energy, has the asymptotic

$$F_\Lambda[\mathsf{H}_\Lambda] = |\,\Lambda\,|\big(f(\mathsf{M}, \mathsf{H}_\Lambda) + o(1)\big)\,, \tag{9}$$

that is, this thermodynamic characteristic has the certain density $f(\mathsf{M}, \mathsf{H}_\Lambda)$ which is the functional on the measure M and on the Hamiltonian $\mathsf{H}_\Lambda[\mathbf{u}(z)]$.

The concept of the thermodynamic limit transition needs the serious clarification, since there are some different ways to construct expanding sequences $\langle\Lambda;\, \Lambda \subset \mathcal{Z}^d\rangle$ which are associated with fundamentally different physical situations, and which, generally speaking, should not lead to the same result.

The simplest type of sequences $\langle\Lambda;\, \Lambda \subset \mathcal{Z}^d\rangle$ used in statistical mechanics, whose components serve as geometric models of crystals and which we will consider further is represented by the sets $\Lambda = \{0, 1, ..., L\}^d$ where $L \in \mathcal{N}$ is the size of the "crystal". The number of vertexes in each of such sets is equal $|\,\Lambda\,| = (L+1)^d < \infty$.

Let us consider the equality

$$\mathsf{H}_\Lambda[\mathbf{u}(z)] = \sum_{\Gamma \subset \Lambda : |\Gamma| > 1} V_\Gamma\big(\mathbf{u}(\Gamma)\big) \tag{10}$$

where each function $V_\Gamma\big(\mathbf{u}(\Gamma)\big)$ at fixed set $\Gamma \subset \Lambda$ of vertexes depends on corresponding collection $\mathbf{u}(\Gamma) = \{\mathbf{u}(x)\,;\, x \in \Gamma\}$. One may consider this equality as the functional equation defining functions $V_\Gamma(\cdot)$. These functions, which we further call *potentials*, are defined by recursively as the solution of this equation, using the induction on the number $|\,\Lambda\,|$ and putting $V_\Gamma\big(\mathbf{u}(\Gamma)\big) = 0$ with $|\Gamma| = 1$. By induction, it is also established that the potentials $V_\Gamma(\cdot)$ have a property similar to (7). Namely, since

$$\sum_{\Gamma \subset \Lambda} V_\Gamma(\mathbf{u}(\Gamma + z)) = \mathsf{H}_\Lambda[\mathbf{u}(x+z)]\,, \qquad \sum_{\Gamma \subset \Lambda + z} V_\Gamma\big(\mathbf{u}(\Gamma)\big) = \mathsf{H}_{\Lambda+z}[\mathbf{u}(x)]\,,$$

then the potentials $V_\Gamma(\mathbf{u}(\Gamma))$ for all sets, which are differed from each other only by shifts with arbitrary vector $z \in \mathcal{Z}^d$, coincides, $V_\Gamma(\mathbf{u}(\Gamma)) = V_{\Gamma+z}(\mathbf{u}(\Gamma))$.

We do not include terms $\Gamma = \{x\}$ with $|\Gamma| = 1$ in $\mathsf{H}_\Lambda[\mathbf{u}(x)]$ and refer them to the definition of M_x, $x \in \Lambda$. At the same time, as already mentioned above, we restrict ourselves to the case when all measures M_x are isomorphic between themselves, that is, they are instances of the same measure M.

Definition 1 The class of Gibbs random fields whose probability spaces $\langle \mathbf{P}_\Lambda$, $\Lambda \subset \mathcal{Z}^d\rangle$ are constructed on the basis of the same measurable phase space $\langle \Omega, \mathbf{B}, \mathsf{M}\rangle$, whose Hamiltonians are determined by the same set of potentials $V_\Gamma(\mathbf{u}(\Gamma))\,;\ |\Gamma| \in \mathcal{N} \setminus \{1\}\rangle$ so that corresponding partition functions are finite when the sequence

$\langle \Lambda\ \Lambda \subset \mathcal{Z}^d \rangle$ of sets coincides with $\langle \Lambda(L) = \{0, 1, ..., L\}^d\, ;\, L \in \mathcal{N} \rangle$ with their suitable translation, we will call the limit Gibbs random field on $\mathcal{Z}^d$.

Thus, the limit Gibbs random field is determined by the sequence of Hamiltonians $\langle \mathsf{H}_{\Lambda(L)}[\cdot]\, ;\, \Lambda(L) \subset \mathcal{Z}^d \rangle$ which is constructed on the basis of potentials by decomposition (10) where the sets $\Lambda(L)$ are defined by $L \in \mathcal{N}$.

Note that, accepting this definition, we adhere to a more traditional view about the thermodynamic limit for statistical characteristics of Gibbs random fields within the framework of statistical mechanics, in contrast to the approach known in statistical mathematical physics. It consists of determination of the Gibbs random field on the entire lattice $\mathcal{Z}^d$ by means of a set of conditional probabilities allowed by the fixed set of *relative Hamiltonians* (see, [7]).

3 The Hamiltonians Space $\mathbf{H}_\nu$ of Limit Gibbs Fields

Note that the study of limit Gibbs random fields is sufficient to carry out fixing only their generating family of sets $\{\Lambda(L)\, ;\, L \in \mathcal{N}\}$ without the account of translations, on which the further presentation in this paper is based. Moreover, we will study a family of Gibbs random fields with the fixed measure M. With the account of these remarks, every limit Gibbs random field uniquely characterized by the class of Hamiltonians $\mathbf{H} = \{\mathsf{H}_{\Lambda(L)}[\cdot]\, ;\, L \in \mathcal{N}\}$ which is defined by the fixed set of potentials $\{V_\Gamma(\mathbf{u}(\Gamma))\, ;\, |\Gamma| \in \mathcal{N} \setminus \{1\}\}$. It is obvious that all such classes form a linear manifold with natural linear operations.

Let us further assume, throughout the work, that there is a monotone function $\nu(s) > 0$, $s \in (0, \infty)$ such that the integral $\int_{\mathcal{R}^n} \exp\big(a\nu(|\mathbf{u}|)\big) d\mathsf{M}(\mathbf{u}) < \infty$ defined by the density $D(\mathbf{u})$ of the measure M, converges for any $a > 0$. In particular, this takes place if the support of the measure M is compact, that is, it is concentrated on the interval $[0, s_*]$, $s_* < \infty$ and its density $D(\mathbf{u})$ is zero at $|\mathbf{u}| > s_*$. It takes place in the case for the standard vector model (see, for example, [8]). In this case one may consider $\nu = 1$.

We connect the study of Gibbs random fields when their measures M have non-compact supports in order to apply our results for such objects of statistical mathematical physics as, for example, the Berlin-Katz spherical model [9, 10], the Gaussian model and the φ^4 model which play an important role in the fluctuation theory of phase transitions (see, [11]). One may note that the above described Gibbs random field on $\mathcal{Z}^d$ include, in particular, all classical lattice models at $n = 1$ specified in [2]. To see this fact it is sufficient to introduce the measure M with the density $D(u) = \sum_{l=1}^{N} \delta(u - l) e^{\mu(u)}$ on the space $\Omega = \mathcal{R}$.

Further, we fix the function $\nu(\cdot)$ connected with the measure M. Let the potentials $V_\Gamma(\mathbf{u}(\Gamma))$ depend continuously on the values of the field $\mathbf{u}(x)$, $x \in \Lambda$. Then, there exists the function

$$\mathsf{G}(\Gamma) \equiv \sup_{\mathbf{u}(\Gamma)} \frac{\left|V_\Gamma(\mathbf{u}(\Gamma))\right|}{\sum_{x\in\Gamma} \nu(|\mathbf{u}(x)|)} < \infty \tag{11}$$

for each set $\Gamma \subset \mathcal{Z}^d$. Let us additionally assume that the Hamiltonians $\mathsf{H}_{\Lambda(L)}[\mathbf{u}(x)]$ included in each fixed class $\mathbf{H}$ have such a property that for any vertex $z \in \Lambda$ it takes place

$$\mathsf{N}\big[\mathsf{H}_{\Lambda(L)}\big] \equiv \sum_{\Gamma\subset\mathcal{Z}^d :\, z\in\Gamma,\, |\Gamma|>1} \mathsf{G}(\Gamma) < \infty\,. \tag{12}$$

Due to the translational invariance of potentials, the values of the functional $\mathsf{N}[\cdot]$ on classes $\mathbf{H}$ of Hamiltonians $\{\mathsf{H}_{\Lambda(L)}\,;\ L \in \mathcal{N}\}$ does not depend on the choice of the vertex $z \in \mathcal{Z}^d$. Then, on the linear manifold of all such classes of Hamiltonians, it is possible to introduce the norm $\mathsf{N}[\cdot]$ that turns this manifold into the Banach space H_ν. In order to simplify the presentation, we omit the proof of the completeness of this space. It is very important that this norm allows also the following definition

$$\|\mathsf{H}_{\Lambda(L)}\| \equiv \sup_{L\in\mathcal{N}} \sup_{\mathbf{u}(\Lambda(L))} \frac{\mathsf{W}_{\Lambda(L)}\big[\mathsf{H}_{\Lambda(L)}\big]}{\sum_{x\in\Lambda(L)} \nu(|\mathbf{u}(x)|)}\,,$$

$$\mathsf{W}_{\Lambda(L)}\big[\mathsf{H}_{\Lambda(L)}\big] = \sum_{\Gamma\subset\Lambda(L):\, |\Gamma|>1} |V_\Gamma(\mathbf{u}(\Gamma))|\,. \tag{13}$$

It is valid the following statement.

Theorem 2 *It takes place the equality*

$$\|\mathsf{H}_{\Lambda(L)}\| = \mathsf{N}\big[\mathsf{H}_{\Lambda(L)}\big]\,. \tag{14}$$

Proof Let us consider the inequalities

$$|V_\Gamma(\mathbf{u}(\Gamma))| \le \mathsf{G}(\Gamma)\sum_{x\in\Gamma} \nu(|\mathbf{u}(x)|)\,, \quad \Gamma \subset \Lambda(L)\,.$$

Summing them on all $\Gamma \subset \Lambda(L)$ at $|\Gamma| > 1$, we obtain

$$\mathsf{W}_{\Lambda(L)}\big[\mathsf{H}_{\Lambda(L)}\big] = \sum_{\Gamma\subset\Lambda(L):\, |\Gamma|>1} |V_\Gamma(\mathbf{u}(\Gamma))| \le \sum_{\Gamma\subset\Lambda(L):\, |\Gamma|>1} \mathsf{G}(\Gamma) \sum_{x\in\Lambda(L)} \nu(|\mathbf{u}(x)|) \le$$

$$\le \sum_{x\in\Lambda(L)} \nu(|\mathbf{u}(x)|) \sum_{x\in\Gamma\subset\Lambda(L)} \mathsf{G}(\Gamma) \le \mathsf{N}\big[\mathsf{H}_{\Lambda(L)}\big]\cdot \sum_{x\in\Lambda(L)} \nu(|\mathbf{u}(x)|)$$

and, therefore,

$$\|\mathsf{H}_{\Lambda(L)}\| = \sup_{\Lambda(L)\subset \mathcal{Z}^d} \frac{\mathsf{W}_{\Lambda(L)}\big[\mathsf{H}_{\Lambda(L)}\big]}{\sum\limits_{x\in\Lambda(L)} \nu(|\mathbf{u}(x)|)} \le \mathsf{N}\big[\mathsf{H}_{\Lambda(L)}\big]. \tag{15}$$

Let us establish the inverse inequality. Choose a value $\varepsilon > 0$. Then, there will be such $L \in \mathcal{N}$ and the field $\mathbf{u}(x)$, $x \in \Lambda(L)$ for which the following inequality

$$\mathsf{W}_{\Lambda(L)}\big[\mathsf{H}_{\Lambda(L)}\big] \ge \big(\|\mathsf{H}_{\Lambda(L)}\| - \varepsilon\big) \sum_{x\in\Lambda(L)} \nu(|\mathbf{u}(x)|)$$

takes place. On the other hand, we have

$$\mathsf{W}_{\Lambda(L)}\big[\mathsf{H}_{\Lambda(L)}\big] = \sum_{\Gamma\subset\Lambda(L):\,|\Gamma|>1} \mathsf{D}(\Gamma) \sum_{x\in\Gamma} \nu(|\mathbf{u}(x)|) \le \mathsf{N}\big[\mathsf{H}_{\Lambda(L)}\big] \sum_{x\in\Lambda(L)} \nu(|\mathbf{u}(x)|)$$

and, therefore, $\|\mathsf{H}_{\Lambda(L)}\| - \varepsilon \le \mathsf{N}\big[\mathsf{H}_{\Lambda(L)}\big]$. Due to the arbitrariness of the value $\varepsilon > 0$, there is an inequality $\|\mathsf{H}_{\Lambda(L)}\| \le \mathsf{N}\big[\mathsf{H}_{\Lambda(L)}\big]$. The validity of (14) follows from it and from the inequality (15).

We show that if the limit random field defined by the class of Hamiltonians $\{\mathsf{H}_{\Lambda(L)};\ L \in \mathcal{N}\}$ which belongs to the space H_ν with a function $\nu(\cdot)$, then it is correctly defined. Namely, it is valid

Theorem 3 *If the integral $\int_{\mathcal{R}^n} \exp\big(a\nu(|\mathbf{u}|)\big)d\mathsf{M}(\mathbf{u}) < \infty$ converges for a monotone function $\nu(s) > 0$, $s \in (0,\infty)$ and for any $a > 0$ and if the class $\mathbf{H}$ of Hamiltonians defined by the set of potentials $\{V_\Gamma(\mathbf{u}(\Gamma))\,;\,|\Gamma| \in \mathcal{N}\setminus\{1\}\}$ belongs to H_ν, then the partition function Z_Λ, defined by (5), is finite and, therefore, the corresponding Gibbs random the field is defined for all $\Lambda \subset \mathcal{Z}^d$.*

Proof Let Hamiltonians $\mathsf{H}_{\Lambda(L)}$ be satisfied the condition (12). Then, on the basis of definition (5) and according to (14), the following estimates are valid

$$\mathrm{Z}_\Lambda \le \int\limits_{\Omega_\Lambda} \exp\big(|\mathsf{H}_\Lambda[\mathbf{u}(z)]|\big)d\mathsf{M}_\Lambda \le \int\limits_{\Omega_\Lambda} \exp\Big(\|\mathsf{H}_{\Lambda(L)}[\mathbf{u}(z)]\| \cdot \sum_{x\in\Lambda} \nu(|\mathbf{u}(x)|)\Big)d\mathsf{M}_\Lambda \le$$

$$\le \prod_{x\in\Lambda} \int\limits_{\Omega_x} \exp\Big(\|\mathsf{H}_{\Lambda(L)}\| \cdot \nu(|\mathbf{u}(x)|)\Big)d\mathsf{M}_x(\mathbf{u}(x)) =$$

$$= \Big[\int\limits_{\Omega} \exp\Big(\|\mathsf{H}_{\Lambda(L)}\| \cdot \nu(|\mathbf{u}|)\Big)d\mathsf{M}(\mathbf{u})\Big]^{|\Lambda|}. \tag{16}$$

We give the following

Definition 2 Let the class $\{\mathsf{H}_{\Lambda(L)} : \Lambda(L) = \{0, 1, ..., L\}^d; L \in \mathcal{N}\}$ of Hamiltonians determines the limit Gibbs random field with fixed measure M on the phase space Ω. The set of limit Gibbs random fields defined by the set $\beta\mathbf{H} = \{\beta\mathsf{H}_{\Lambda(L)}[\cdot] : L \in \mathcal{N}\}$ of classes of Hamiltonians contained in H_ν where each set is parameterized by $\beta > 0$, is called the lattice classical model of statistical mechanics corresponding to $\beta\mathbf{H}$.

Introduction of the set of classes of Hamiltonians which is represented as a rectilinear ray in the space H_ν, is connected with the fact that the model of equilibrium statistical mechanics is defined by the thermodynamic interpretation of measurable parameters of corresponding limit Gibbs field. First of all, it refers to the main thermodynamic parameter, that is the temperature. According to the canons of statistical mechanics, it is proportional to β^{-1}.

4 The Extensive Asymptotics of Free Energy

Our aim is the proof the asymptotic formula (9) at $\Lambda(L) \to \mathcal{Z}^d$ for each lattice system of statistical mechanics.

Definition 3 The Hamiltonian (10) has the finite range of action if there exists such a finite set $\Delta \subset \mathcal{Z}^d$, $\mathbf{0} \in \Delta$ of vertexes for which $V_\Gamma(\mathbf{u}(\Gamma)) \neq 0$ only in the case when there is such a vertex $z \in \Gamma$ that $\Gamma - z \subset \Delta$.

If the Hamiltonian $\mathsf{H}_{\Lambda(L)}$ has a finite range of action, the pointed out set Δ is named its *support*. It is obvious that all such Hamiltonians form the linear manifold $\mathrm{H}^{(0)}$ in the Banach space H_ν. We begin the proof of the extensiveness of the free energy from the proof of the following statement.

Theorem 4 *For the fixed measure* M *and any finite set* $\Lambda \subset \mathcal{Z}^d$, *the corresponding manifold* $\mathrm{H}^{(0)}$ *of Hamiltonians* H_Λ *is dense in the space* H_ν.

Proof Let us fix the value $\varepsilon > 0$. Since the sum in (12) is finite for the fixed Hamiltonian $\mathsf{H}_\Lambda[\cdot]$, one may choose the finite family Σ of finite subsets $\Gamma \subset \mathcal{Z}^d$ such that each of them contains the vertex $\mathbf{0}$ and it takes place the inequality

$$\sum_{\Gamma \subset \mathcal{Z}^d : \mathbf{0} \in \Gamma, \Gamma \notin \Sigma} \mathsf{G}(\Gamma) < \varepsilon . \tag{17}$$

Let us introduce the set

$$\Delta = \bigcup_{\Gamma \in \Sigma} \Gamma .$$

We add the family Σ such that it should contain all sets $\Gamma \subset \Delta$. The inequality (17) is strengthened only at such an expansion. After that, we define $\widehat{V}_\Gamma(\mathbf{u}(\Gamma)) = V_\Gamma(\mathbf{u}(\Gamma))$, if one may find such a vertex $z \in \mathcal{Z}^d$ for which the inclusion $(\Gamma + z) \in \Sigma$ takes place. In opposite case, we define $\widehat{V}_\Gamma(\mathbf{u}(\Gamma)) = 0$. The latter means that $\widehat{V}_{\Gamma'}(\mathbf{u}(\Gamma')) = 0$ every time when the set $\Gamma' \subset \mathcal{Z}^d$ is such that $\Gamma' + z \not\subset \Delta$ takes place for any vertex $z \in \mathcal{Z}^d$.

Further, we define the Hamiltonian

$$\widehat{\mathsf{H}}_\Lambda[\mathbf{u}(z)] = \sum_{\Gamma \subset \Lambda : |\Gamma| > 1} \widehat{V}_\Gamma(\mathbf{u}(\Gamma)) \,. \tag{18}$$

It belongs to the linear manifold $\mathrm{H}^{(0)}$. Then, using the determination of potentials $\widehat{V}_\Gamma(\mathbf{u}(\Gamma))$, due to the Theorem 1 the following equality

$$\|\mathsf{H}_\Lambda - \widehat{\mathsf{H}}_\Lambda\| = \mathsf{N}\big[\mathsf{H}_\Lambda - \widehat{\mathsf{H}}_\Lambda\big] = \sum_{\Gamma \subset \mathcal{Z}^d : \, \mathbf{0} \in \Gamma, \, \Gamma \notin \Sigma} \mathsf{G}(\Gamma) < \varepsilon$$

takes place that is any Hamiltonian H_Λ may be approximate arbitrarily accurate in the space H_ν by the Hamiltonian $\widehat{\mathsf{H}}_\Lambda$ with finite range of action.

To solve the problem which is set at the beginning of the section, some following supplementary properties of density $D(\cdot)$ should be used. According to the basic supposition, the measure M has the density $D(\cdot)$ which is a generalized function relative to the space of continuous functions. It consists of two summands $D(\cdot) = D_c(\cdot) + D_d(\cdot)$ where $D_c(\cdot)$ is measurable bounded nonnegative function on $\mathcal{R}^n$ and $D_d(\mathbf{u}) = \sum_k \mu_k \delta(\mathbf{u} - \mathbf{v}_k)$; $\mu_k > 0$, $\mathbf{v}_k \in \mathcal{R}^n$. Denote $D^\alpha(\mathbf{u}) = D_c^\alpha(\mathbf{u}) + \sum_k \mu_k^\alpha \delta(\mathbf{u} - \mathbf{v}_k)$ at $0 < \alpha < 1$. We will say that such a density $D(\cdot)$ is bounded by the value K if $\max D_c(\mathbf{u}) \le K$ and $\mu_l \le K, l \in \mathcal{N}$.

In addition to the existence of positive monotone function $\nu(s)$ on $(0, \infty)$ such that the density $D(\mathbf{u})$ possesses the property $\int_\Omega \exp(a\nu(|\mathbf{u}|)) D(\mathbf{u}) d\mathbf{u} < \infty$ at any $a > 0$, we will suppose also the availability of some supplementary more strong restrictions for the density when the basic result of the paper will be obtained in this section.

Lemma 1 *Let the Hamiltonians class* $\{\mathsf{H}_{\Lambda(L)}\,;\, L \in \mathcal{N}\}$ *belongs to the space* H_ν. *Let also the density* $D(\cdot)$ *of measure* M *defines the limit Gibbs random field together with this class. If* $D(\cdot)$ *is bounded by the value K and there exists such a nonnegative function* $\nu(s)$, *the value* $\alpha \in (0, 1)$ *for which the integral* $\int_\Omega D^\alpha(\mathbf{u}) e^{a\nu(\mathbf{u})} d\mathbf{u} < \infty$ *is finite and also the function* $\nu(\mathbf{u}) D^{1-\alpha}(\mathbf{u})$ *is bounded by the value* $K_\nu > 0$, *then the following inequality is valid for expectation* $\mathsf{E}_{\Lambda(L)} \nu(|\tilde{\mathbf{u}}(x)|) < K_\nu K^{1-\alpha}$ *and for any vertex* $x \in \mathcal{Z}^d$.

Proof Since the function $\nu(\mathbf{u}) D^{1-\alpha}(\mathbf{u})$ is bounded by the value K_ν, then, for the following integral with any nonnegative weight function $W(\cdot)$ on $\mathcal{R}^n$, the estimate

$$\int_{\mathcal{R}^n} \nu(\mathbf{u}) D(\mathbf{u}) W(\mathbf{u}) d\mathbf{u} < K_\nu \int_{\mathcal{R}^n} D^\alpha(\mathbf{u}) W(\mathbf{u}) d\mathbf{u} \tag{19}$$

takes place. By the same way, since the density $D(\mathbf{u})$ is bounded by the value K, the inequality

$$\int_{\mathcal{R}^n} D(\mathbf{u})W(\mathbf{u})d\mathbf{u} < K^{1-\alpha} \int_{\mathcal{R}^n} D^{\alpha}(\mathbf{u})W(\mathbf{u})d\mathbf{u} \tag{20}$$

is valid.

Now, we note that, due to the lemma conditions relative the integral with the density $D(\cdot)$, the following partition function is finite (see the proof of Theorem 2),

$$Z_{\Lambda(L),\alpha} = \int_{\Omega_x} D^{\alpha}(\mathbf{u}(|x|))d\mathbf{u}(x) \int_{\Omega_{\Lambda(L)\setminus x}} \exp(-\mathsf{H}_{\Lambda(L)}[\mathbf{u}(z)])d\mathsf{M}_{\Lambda(L)\setminus x} <$$

$$< \int_{\Omega} \exp\Big(\|\mathsf{H}_{\Lambda(L)}\| \cdot \nu(|\mathbf{u}|)\Big) D^{\alpha}(\mathbf{u})d\mathbf{u} \cdot \Big[\int_{\Omega} \exp\Big(\|\mathsf{H}_{\Lambda(L)}\| \cdot \nu(|\mathbf{u}|)\Big)d\mathsf{M}(\mathbf{u})\Big]^{|\Lambda|-1},$$

since $|\mathsf{H}_{\Lambda(L)}| \le \mathsf{W}[\mathsf{H}_{\Lambda(L)}] \le \|\mathsf{H}_{\Lambda(L)}\| \sum_{x\in\Lambda(L)} \nu(|\mathbf{u}|)$.

Then, on the basis of the identity $1 = Z_{\Lambda(L)}/Z_{\Lambda(L)}$, using the inequality (20) for the denominator, we find that

$$Z_{\Lambda(L),\,\alpha} \ge K^{\alpha-1} Z_{\Lambda(L)}\,. \tag{21}$$

By the same way, due to the condition for the integral pointed out and due to the inequality (19), we find the estimate

$$\int_{\Omega_x} [\nu(|u(x)|)D^{1-\alpha}(\mathbf{u}(x))]D^{\alpha}(\mathbf{u}(x))d\mathbf{u}(x) \int_{\Omega_{\Lambda(L)\setminus x}} \exp(-\mathsf{H}_{\Lambda(L)}[\mathbf{u}(z)])d\mathsf{M}_{\Lambda(L)\setminus x}$$

$$< K_{\nu} Z_{\Lambda(L),\,\alpha}\,. \tag{22}$$

The expression for the expectation $\mathrm{E}_{\Lambda(L)}\big[\,\nu(|\mathbf{u}(x)|)\big]$ is written in the following form

$$\mathrm{E}_{\Lambda(L)}\big[\,\nu(|\mathbf{u}(x)|)\big] = \frac{\int_{\Omega_x} \nu(|\mathbf{u}(x)|)d\mathsf{M}_x \int_{\Omega_{\Lambda(L)\setminus x}} \exp(-\mathsf{H}_{\Lambda(L)}[\mathbf{u}(z)])d\mathsf{M}_{\Lambda(L)\setminus x}}{\int_{\Omega_x} d\mathsf{M}_x \int_{\Omega_{\Lambda(L)\setminus x}} \exp(-\mathsf{H}_{\Lambda(L)}[\mathbf{u}(z)])d\mathsf{M}_{\Lambda(L)\setminus x}}\,.$$

We apply the estimate (22) for the nominator and the estimate (21) for the denominator. Then

$$\mathrm{E}_{\Lambda(L)}\big[\,\nu(|\mathbf{u}(x)|)\big] \le K_{\nu} K^{1-\alpha}\,.$$

Further, we suppose that always the measure M satisfies conditions of Lemma 1.

Let $\Lambda' \subset \Lambda$ and $\mathbf{u}(\Lambda')$ is the restriction of the field $\mathbf{u}(\Lambda)$ on the set Λ'. If the Hamiltonian $\mathsf{H}_\Lambda[\mathbf{u}(z)]$ has the property $\mathsf{H}_\Lambda[\mathbf{u}(\Lambda)] = \mathsf{H}_\Lambda[\mathbf{u}(\Lambda')]$, then we will say that $\mathsf{H}_\Lambda[\mathbf{u}(\Lambda')]$ is the *natural restriction* of the Hamiltonian $\mathsf{H}_\Lambda[\mathbf{u}(z)]$ on the linear manifold $\Omega_{\Lambda'}$ of vector fields $\mathbf{u}(\Lambda')$. We will denote this natural restriction by means of $\mathsf{H}_{\Lambda'}[\mathbf{u}(z)]$.

Lemma 2 *Let $\Lambda \subset \mathbb{Z}^d$. Then, for the partition functions*

$$\mathsf{Z}_\Lambda[\mathsf{H}_\Lambda^{(m)}] = \int_{\Omega_\Lambda} \exp\big(-\mathsf{H}_\Lambda^{(m)}[\mathbf{u}(z)]\big) d\mathsf{M}_\Lambda$$

which are defined by Hamiltonians $\mathsf{H}_\Lambda^{(m)}$, $m \in \{1, 2\}$ of the space H_ν such that the difference $\mathsf{H}_{\Lambda'}^{(1)} - \mathsf{H}_{\Lambda'}^{(2)}$ at $\Lambda' \subset \Lambda$ is the natural restriction of the Hamiltonian $\mathsf{H}_\Lambda^{(1)} - \mathsf{H}_\Lambda^{(2)}$ on $\Omega_{\Lambda'}$, the following inequality is valid

$$\big|\ln \mathsf{Z}_\Lambda[\mathsf{H}_\Lambda^{(1)}] - \ln \mathsf{Z}_\Lambda[\mathsf{H}_\Lambda^{(2)}]\big| \le \Big(\mathsf{E}_\Lambda \nu(|\tilde{\mathbf{u}}|)\Big) \cdot |\Lambda'| \cdot \|\mathsf{H}_{\Lambda'}^{(1)} - \mathsf{H}_{\Lambda'}^{(2)}\|\,. \tag{23}$$

Proof The Hamiltonian $\mathsf{H}_{\Lambda'}^{(1)} - \mathsf{H}_{\Lambda'}^{(2)}$ possesses the finite norm $\|\cdot\|$. We introduce the family of Hamiltonians $\mathsf{H}[\mathbf{u}(z); t] = \mathsf{H}_{\Lambda'}^{(2)}[\mathbf{u}(z)] + t\big(\mathsf{H}_{\Lambda'}^{(1)}[\mathbf{u}(z)] - \mathsf{H}_{\Lambda'}^{(2)}[\mathbf{u}(z)]\big)$, $t \in [0, 1]$ so that all belong to H_ν, and also we consider the family of corresponding partition functions

$$\mathsf{Z}_\Lambda(t) = \int_{\Omega_\Lambda} \exp\big(-\mathsf{H}[\mathbf{u}(z); t]\big) d\mathsf{M}_\Lambda\,.$$

These functions are finite due to Theorem 2.

Now, we note that the following estimates are valid

$$\left|\frac{d}{dt} \ln \mathsf{Z}_\Lambda(t)\right| \le \mathsf{Z}_\Lambda^{-1}(t) \int_{\Omega_\Lambda} \left|\frac{d}{dt} \mathsf{H}[\mathbf{u}(z); t]\right| \exp\big(-\mathsf{H}[\mathbf{u}(z); t]\big) d\mathsf{M}_\Lambda \le$$

$$\left\|\frac{d}{dt} \mathsf{H}[\mathbf{u}(z); t]\right\| \cdot Z^{-1}(t) \int_{\Omega_\Lambda} \sum_{x \in \Lambda} \nu(|\mathbf{u}(x)|) \exp\big(-\mathsf{H}[\mathbf{u}(z); t]\big) d\mathsf{M}_\Lambda =$$

$$= \Big(\mathsf{E}_\Lambda \nu(|\tilde{\mathbf{u}}|)\Big) |\Lambda'| \cdot \|\mathsf{H}_{\Lambda'}^{(1)} - \mathsf{H}_{\Lambda'}^{(2)}\|\,,$$

if we take into account the definition (13) of the norm and also that the difference $\mathsf{H}_{\Lambda'}^{(1)} - \mathsf{H}_{\Lambda'}^{(2)}$ is the natural restriction on $\Omega_{\Lambda'}$. Here, the expectation $\mathsf{E}\nu(|\tilde{\mathbf{u}}|)$ is finite. Due to Lemma 1, it does not exceed $K_\nu K^{1-\alpha}$. Integrating the obtained inequality from 0 up to 1 and taking into account that $\mathsf{Z}_\Lambda(0) = \mathsf{Z}_\Lambda[\mathsf{H}_\Lambda^{(2)}]$, $\mathsf{Z}_\Lambda(1) = \mathsf{Z}_\Lambda[\mathsf{H}_\Lambda^{(1)}]$, the inequality (23) follows.

Lemma 3 *Let the Hamiltonian* $\mathsf{H}_\Lambda \in \mathrm{H}^{(0)}$ *has the finite range of action and* Δ *is the finite subset in* $\mathbb{Z}^d$ *which is its support. Let also* Δ_1 *and* Δ_2 *be any nonintersecting finite subsets in* $\mathbb{Z}^d$, $\Delta_1 \cap \Delta_2 = \emptyset$ *and* $\Sigma_*(\Delta_1, \Delta_2; \Delta)$ *be the set of such vertexes* $z \in \mathbb{Z}^d$ *for which* $(\Delta + z) \cap \Delta_1 \neq \emptyset$, $(\Delta + z) \cap \Delta_2 \neq \emptyset$ *are fulfilled simultaneously.*

Let the Hamiltonians $\mathsf{H}_{\Delta_1 \cup \Delta_2}$, H_{Δ_1}, H_{Δ_2} *are natural restrictions of the Hamiltonian* H_Λ *on* $\Omega_{\Delta_1 \cup \Delta_2}$, Ω_{Δ_1}, Ω_{Δ_2}, *correspondingly. Then, the difference* $(H_{\Delta_1 \cup \Delta_2} - \mathsf{H}_{\Delta_1} - \mathsf{H}_{\Delta_2})$ *has the natural restriction on* $\Omega_{\Sigma_*(\Delta_1, \Delta_2; \Delta)}$ *and the following estimate*

$$|\mathsf{H}_{\Delta_1 \cup \Delta_2}[\mathbf{u}(z)] - \mathsf{H}_{\Delta_1}[\mathbf{u}(z)] - \mathsf{H}_{\Delta_2}[\mathbf{u}(z)]| \le \|\mathsf{H}_{\Delta_1 \cup \Delta_2}\| \sum_{x \in \Sigma_*(\Delta_1, \Delta_2; \Delta)} \nu(|x|) \quad (24)$$

is valid for it.

Proof Let us estimate the left-hand side of the inequality (24)

$$\left|\mathsf{H}_{\Delta_1 \cup \Delta_2}[\mathbf{u}(z)] - \mathsf{H}_{\Delta_1}[\mathbf{u}(z)] - \mathsf{H}_{\Delta_2}[\mathbf{u}(z)]\right| \le$$

$$\le \Big(\sum_{\substack{\Gamma \subset \Delta_1 \cup \Delta_2 : \\ |\Gamma| > 1}} - \sum_{\substack{\Gamma \subset \Delta_1 : \\ |\Gamma| > 1}} - \sum_{\substack{\Gamma \subset \Delta_2 : \\ |\Gamma| > 1}} \Big) \left|V_\Gamma(\mathbf{u}(\Gamma))\right| = \sum_{\substack{\Gamma \subset \Delta_1 \cup \Delta_2 : |\Gamma| > 1 \\ \Gamma \cap \Delta_1 \neq \emptyset,\ \Gamma \cap \Delta_2 \neq \emptyset}} \left|V_\Gamma(\mathbf{u}(\Gamma))\right| \le$$

$$\le \sum_{x \in \Delta_1 \cup \Delta_2} \sum_{\substack{\Gamma \subset \Delta_1 \cup \Delta_2 :\ x \in \Gamma,\ \Gamma - x \subset \Delta, \\ \Gamma \cap \Delta_1 \neq \emptyset,\ \Gamma \cap \Delta_2 \neq \emptyset,\ |\Gamma| > 1}} \left|V_\Gamma(\mathbf{u}(\Gamma))\right| \le$$

$$\le \sum_{\Gamma \subset \mathbb{Z}^d : \mathbf{0} \in \Gamma,\ |\Gamma| > 1} \mathsf{G}(\Gamma) \sum_{x \in \Sigma(\Delta_1, \Delta_2; \Delta)} \nu(|x|)\,.$$

Here, we take into account that $V_\Gamma(\mathbf{u}(\Gamma)) \neq 0$ only in the case when there exists such a vertex $x \in \Gamma$ for which the relation $\Gamma - x \subset \Delta$ is valid and, therefore, we introduce the set $\Sigma(\Delta_1, \Delta_2; \Delta)$ of vertexes $x \in \Delta_1 \cup \Delta_2$. For each vertex in this set there exists a subset Γ with the following properties $\Gamma \subset \Delta_1 \cup \Delta_2$, $x \in \Gamma$, $\Gamma - x \subset \Delta$, $\Gamma \cap \Delta_1 \neq \emptyset$, $\Gamma \cap \Delta_2 \neq \emptyset$.

Now, we show that the inclusion $\Sigma(\Delta_1, \Delta_2; \Delta) \subset \Sigma_*(\Delta_1, \Delta_2; \Delta)$ takes place. Indeed, from two last inclusions we conclude $(\Gamma - x) \cap (\Delta_1 - x) \neq \emptyset$ and $(\Gamma - x) \cap (\Delta_2 - x) \neq \emptyset$. Then, combining these inclusions with the following $\Gamma - x \subset \Delta$, we may assert that relationships $\Delta \cap (\Delta_1 - x) \neq \emptyset$ and $\Delta \cap (\Delta_2 - x) \neq \emptyset$ are realized. Thus, $(\Delta + x) \cap \Delta_1 \neq \emptyset$, $(\Delta + x) \cap \Delta_2 \neq \emptyset$ and, therefore, the last inequality leads to the inequality (24) if we take into account the statement of Theorem 2.

The following lemma is the consequence of Lemmas 2 and 3.

Lemma 4 *Let Hamiltonian* $\mathsf{H}_\Lambda \in \mathrm{H}^{(0)}$ *has the finite range of action and* Δ *is the finite subset in* $\mathbb{Z}^d$ *which is its support. If* Δ_1 *are* Δ_2 *are nonintersecting finite subsets in* $\mathbb{Z}^d$, $\Delta_1 \cap \Delta_2 = \emptyset$, *then the following estimate takes place*

$$\left|\ln Z_{\Delta_1\cup\Delta_2} - \ln Z_{\Delta_1} - \ln Z_{\Delta_2}\right| \le \left(\mathsf{E}_\Lambda \nu(|\tilde{\mathbf{u}}|)\right)\cdot \|\mathsf{H}_\Lambda\|\cdot|\Sigma_*(\Delta_1,\ \Delta_2;\ \Delta)|\,. \quad (25)$$

Proof We define the following Hamiltonians $\mathsf{H}^{(1)}_\Lambda = \mathsf{H}_{\Delta_1\cup\Delta_2}$ and $\mathsf{H}^{(2)}_\Lambda = \mathsf{H}_{\Delta_1} + \mathsf{H}_{\Delta_2}$. Then, using this definition, we have $Z_{\Delta_1\cup\Delta_2} = Z_\Lambda[\mathsf{H}_{\Delta_1\cup\Delta_2}]$, $Z_{\Delta_1} = Z_\Lambda[\mathsf{H}_{\Delta_1}]$, $Z_{\Delta_2} = Z_\Lambda[\mathsf{H}_{\Delta_2}]$. Due to $\Delta_1 \cap \Delta_2 = \emptyset$, the Hamiltonians H_{Δ_1}, H_{Δ_2} act in linear manifolds which have the empty intersection. Consequently,

$$\left|\ln Z_{\Delta_1\cup\Delta_2} - \ln Z_{\Delta_1} - \ln Z_{\Delta_2}\right| = \left|\ln Z_\Lambda[\mathsf{H}^{(1)}_\Lambda] - \ln Z_\Lambda[\mathsf{H}^{(2)}_\Lambda]\right|.$$

Further, we apply the inequality (24) to partition functions $Z_\Lambda[\mathsf{H}^{(m)}_\Lambda]$, $m = 1, 2$,

$$\left|\ln Z_\Lambda[\mathsf{H}^{(1)}_\Lambda] - \ln Z_\Lambda[\mathsf{H}^{(2)}_\Lambda]\right| \le \left(\mathsf{E}_\Lambda \nu(|\tilde{\mathbf{u}}|)\right)\cdot|\Sigma_*(\Delta_1,\ \Delta_2;\ \Delta)|\cdot\|\mathsf{H}^{(1)}_\Lambda - \mathsf{H}^{(2)}_\Lambda\|\,,$$

where we take into account that the difference $\mathsf{H}^{(1)}_\Lambda - \mathsf{H}^{(2)}_\Lambda$ has the natural restriction on $\Sigma_*(\Delta_1,\ \Delta_2;\ \Delta)$. Since $\mathsf{H}^{(1)}_\Lambda - \mathsf{H}^{(2)}_\Lambda = \mathsf{H}_{\Sigma_*(\Delta_1,\ \Delta_2;\ \Delta)}$, then $\|\mathsf{H}^{(1)}_\Lambda - \mathsf{H}^{(2)}_\Lambda\| = \|\mathsf{H}_{\Sigma_*(\Delta_1,\ \Delta_2;\ \Delta)}\| \le \|\mathsf{H}_\Lambda\|$ because of the nondecreasing of the Hamiltonian norm when the set Λ is expanded. From here, it follows the inequality (25).

Corollary 1 *Let Hamiltonian* $\mathsf{H}_\Lambda \in \mathrm{H}^{(0)}$ *has the finite rang of action and* Δ *is finite subset in* $\mathcal{Z}^d$ *which is its support. If* Δ_j, $j = 1 \div m$ *are finite subsets in* $\mathcal{Z}^d$ *such that* $\Delta_j \cap \Delta_k = \emptyset$ *at* $j \ne k$, *then the following estimate*

$$\left|\ln Z_{\Upsilon_m} - \sum_{j=1}^{m}\ln Z_{\Delta_j}\right| < \left(\mathsf{E}_{\Upsilon_m}\nu(|\tilde{\mathbf{u}}|)\right)\cdot\|\mathsf{H}_{\Upsilon_m}\|\cdot\sum_{j=2}^{m}|\Sigma_*(\Upsilon_{j-1},\ \Delta_j;\ \Delta)| \quad (26)$$

takes place where $\Upsilon_l = \bigcup_{j=1}^{l}\Delta_j$ *and* $\Sigma_*(\Upsilon_{j-1},\ \Delta_j;\ \Delta)$ *is the set of such vertexes* $z \in \mathcal{Z}^d$ *for which the relationships* $(\Delta + z) \cap \Upsilon_l \ne \emptyset$, $(\Delta + z) \cap \Delta_l \ne \emptyset$ *follow simultaneously for each* $l = 2 \div m$.

Proof The proof is carried out by induction according to $m \in \mathcal{N}$ with the use of the inequality (25), starting out $m = 2$.

Let us proceed to the proof of the main result of this work. It is carried out according to the same scheme that is proposed in [2], and it is based on the representation of a lattice model as the sum of a large number of isomorphic disjoint identical "weakly interacting" lattice models.

Theorem 5 *If* $\mathsf{H}_\Lambda \in \mathrm{H}^{(0)}$, *then there exists the finite limit*

$$f(\mathsf{M}, \mathsf{H}_\Lambda) = \lim_{L\to\infty}\frac{\ln Z_\Lambda}{|\Lambda(L)|}\,. \quad (27)$$

Proof On the basis of the set $\Lambda(a-1)$, $a \in \mathcal{N}$, $a \ge 2$ and vertexes $y \in \mathcal{Z}^d$, we define the sets $\Lambda_y = \Lambda(a-1) + ay$. Let $L = aN - 1$. Consider the set $\Lambda(aN - 1)$

which contains $a^d N^d$ vertexes. We represent it in the form

$$\Lambda(aN-1) = \bigcup_{y\in \Lambda(N-1)} \Lambda_y$$

where $\Lambda_{y_1} \cap \Lambda_{y_2} = \emptyset$ for any pair of vertexes $\{y_1, y_2\} \subset \Lambda(N-1)$.

Let us introduce lexicographical order of the set $\Lambda(N-1)$ containing N^d vertexes, we demote the fact that the vertex y_2 follows the vertex y_1 by $y_1 < y_2$. It means that for each pair of such vertexes $y_1 = \langle y_1^{(1)}, ..., y_d^{(1)}\rangle$, $y_2 = \langle y_1^{(2)}, ..., y_d^{(2)}\rangle$; $y_j^{(m)} = 0 \div N-1$; $j = 1 \div d$, $m \in \{1, 2\}$ there exists such a number $k = 1 \div d$ for which $y_j^{(1)} = y_j^{(2)}$, $j = 1 \div k-1$, $y_k^{(1)} < y_k^{(2)}$.

The values of functionals $\ln Z_{\Lambda_{y_j}}$ do not depend on $j = 1 \div N^d$ due to translational invariance. Then

$$\left|\ln Z_{\Lambda(L)} - N^d \ln Z_{\Lambda(a-1)}\right| = \left|\ln Z_{\Lambda(L)} - \sum_{j=1}^{N^d} \ln Z_{\Lambda_{y_j}}\right|. \tag{28}$$

To estimate the right-hand side of this equality we apply the inequality (26) connected with sets $\Delta_j = \Lambda_{y_j}$, $j = 1 \div N^d$, $\Upsilon_{N^2} = \Lambda(L)$ in the sense of the introduced order,

$$\left|\ln Z_{\Lambda(L)} - \sum_{j=1}^{N^d} \ln Z_{\Lambda_{y_j}}\right| < \Big(\mathsf{E}_{\Lambda(L)}\, \nu(|\tilde{\mathbf{u}}|)\Big) \cdot \|\mathsf{H}_{\Lambda(L)}\| \cdot \sum_{j=2}^{N^d} \left|\Sigma_*(\Upsilon_{j-1}, \Lambda_{y_j}; \Delta)\right| \tag{29}$$

where $\Upsilon_l = \bigcup_{j=1}^{l} \Lambda_{y_j}$, $\Upsilon_{y_1} = \Lambda(a-1)$. We choose the number $a \in \mathcal{N}$ so large that the inclusion $\Lambda(a-1) \supset (\Delta + z)$ is fulfilled for a vertex z.

Suppose there are two sets Λ_{y_j} and Λ_{y_k} such that there exists such a vertex $x \in \mathcal{Z}^d$ for them when the relationships $(\Delta + x) \cap \Lambda_{y_j} \neq \emptyset$ and $(\Delta + x) \cap \Lambda_{y_k} \neq \emptyset$ are valid. Then from the inclusion $\Lambda(a-1) \supset (\Delta + z)$ it follows that $(\Lambda(a-1) + x - z) \cap \Lambda_{y_j} \neq \emptyset$ and $(\Lambda(a-1) + x - z) \cap \Lambda_{y_k} \neq \emptyset$. Such a situation is possible only in the case when Λ_{y_j} and Λ_{y_k} are "neighboring" sets, namely, $y_j = \langle y_1^{(j)}, ..., y_d^{(j)}\rangle$, $y_k = \langle y_1^{(j)} + \alpha_1, ..., y_d^{(j)} + \alpha_d\rangle$, $\alpha_i \in \{-1, 0, 1\}$, $i = 1 \div d$. For each set Λ_{y_j} there exists no more that $3^d - 1$ neighboring sets among all $\Lambda_{y_k}, k \neq j, k = 1 \div N^d$. In this case, if the vertex x is contained in anything set Λ_{y_j}, then there are $3^d - 1$ sets Λ_{y_k} such that $(\Delta + x) \cap \Lambda_{y_k} \neq \emptyset$. Consequently, the number of vertexes $\left|\Sigma_*(\Upsilon_{j-1}, \Lambda_{y_j}; \Delta)\right|$ does not exceed $(3^d - 1) \max_{k\in\Xi_j} \left|\Sigma_*(\Lambda_{y_k}, \Lambda_{y_j}; \Delta)\right|$ for any $j = 1 \div N^d$ where Ξ_j is the set which consists of those $3^d - 1$ numbers $k \in \{1, ..., N^d\}$ for which Λ_{y_k} is a neighbor with Λ_{y_j}.

Let us estimate the number $\left|\Sigma_*(\Lambda_{y_k}, \Lambda_{y_j}; \Delta)\right|$ for two neighboring sets Λ_{y_j} and Λ_{y_k}. It is obvious that it is maximal in the case when there is the face of Λ_{y_j} with the dimension $d-1$ which divides them. It contains a^{d-1} vertexes. Let x_0 is the fixed vertex in this face. Then, we find the number of vertexes x for which simultaneous feasibility of relationships $(\Delta + x) \cap \Lambda_{y_j} \neq \emptyset$ and $(\Delta + x) \cap \Lambda_{y_k} \neq \emptyset$ is possible. In this case $x_0 \in \Delta + x$ does not exceed $|\Delta|$. Then, it is valid $\big|\Sigma_*(\Lambda_{y_k}, \Lambda_{y_j}; \Delta)\big| \leq a^{d-1}|\Delta|$. On the basis of this estimate, we obtain the following inequality

$$\big|\Sigma_*(\Upsilon_{j-1}, \Lambda_{y_j}; \Delta)\big| < (3^d - 1)a^{d-1}|\Delta| .$$

Using it and also (28) and (29), we conclude that the inequality

$$\left|\frac{\ln Z_{\Lambda(aN-1)}}{|\Lambda(aN-1)|} - \frac{\ln Z_{\Lambda(a-1)}}{|\Lambda(a-1)|}\right| < (3^d - 1)\Big(\mathsf{E}_{\Lambda(L)}\, \nu(|\tilde{\mathbf{u}}|)\Big) \cdot \|\mathsf{H}_{\Lambda(L)}\| \cdot \frac{|\Delta|}{a} . \quad (30)$$

takes place at $|\Lambda(aN - 1)| = (aN)^d$, $|\Lambda(a - 1)| = a^d$. Since the right-hand side of the inequality (30) tends to zero at $a \to \infty$, then, to complete the proof of the theorem, we show that the sequence $\langle |\Lambda(L)|^{-1} \ln Z_{\Lambda(L)}; L \in \mathcal{N}\rangle$ is the fundamental one. For this, we will prove that, for each $\varepsilon > 0$, there is such a sufficiently large number L, for which there are values a and N when for any $L' > L$ we may find $a' > a$, $N' > N$ when the following inequality

$$\left|\frac{\ln Z_{\Lambda(L')}}{|\Lambda(L')|} - \frac{\ln Z_{\Lambda(a'-1)}}{|\Lambda(a'-1)|}\right| < \varepsilon$$

takes place. It is obvious that the sequence under consideration is fundamental in this case since

$$\left|\frac{\ln Z_{\Lambda(L)}}{|\Lambda(L)|} - \frac{\ln Z_{\Lambda(L')}}{|\Lambda(L')|}\right| < 2\varepsilon , \quad L' > L . \quad (31)$$

We introduce the sets $\Lambda(aN - 1)$ and $\partial\Lambda(aN - 1) = \Lambda(L) \setminus \Lambda(aN - 1)$. Let us estimate the expression in left-hand side of (31) at $L' = L$, $a' = a$ on the basis of

$$\left|\frac{\ln Z_{\Lambda(L)}}{|\Lambda(L)|} - \frac{\ln Z_{\Lambda(a-1)}}{|\Lambda(a-1)|}\right| \leq \frac{1}{|\Lambda(L)|}\Big|\ln Z_{\Lambda(L)} - \ln Z_{\Lambda(aN-1)} - \ln Z_{\partial\Lambda(aN-1)}\Big|$$

$$+ \frac{\ln Z_{\partial\Lambda(aN-1)}}{\ln Z_{\Lambda(L)}} + \left|\frac{\ln Z_{\Lambda(aN-1)}}{|\Lambda(aN-1)|} - \frac{\ln Z_{\Lambda(a-1)}}{|\Lambda(a-1)|}\right| . \quad (32)$$

To estimate first summand, we apply (25) with $\Delta_1 = \Lambda(aN - 1)$ and $\Delta_2 = \partial\Lambda(aN - 1)$, taking into account that $\Lambda(L) = \Lambda(aN - 1) \cup \partial\Lambda(aN - 1)$,

$$\Big|\ln Z_{\Lambda(L)} - \ln Z_{\Lambda(aN-1)} - \ln Z_{\partial\Lambda(aN-1)}\Big| \leq$$

$$\leq \Big(\mathsf{E}_{\Lambda}\nu(|\tilde{\mathbf{u}}|)\Big) \cdot \|\mathsf{H}_{\Lambda}\| \cdot |\Sigma_*(\Lambda(aN - 1), \partial\Lambda(aN - 1); \Delta)| .$$

Here, $\Sigma_*(\Lambda(aN-1),\ \partial\Lambda(aN-1);\ \Delta)$ is the set of such vertexes x for which the set $\Delta + x$ contains the vertex in $\partial\Lambda(aN-1)$. Then, it follows that $\left|\Sigma_*(\Lambda(aN-1),\ \partial\Lambda(aN-1);\ \Delta)\right| \le |\partial\Lambda(aN-1)| \cdot |\Delta|$. Consequently, the inequality

$$\frac{1}{|\Lambda(L)|}\left|\ln Z_{\Lambda(L)} - \ln Z_{\Lambda(aN-1)} - \ln Z_{\partial\Lambda(aN-1)}\right| \le \delta \tag{33}$$

takes place at sufficiently large number N.

The estimate of second summand is given by the inequality

$$\frac{\ln Z_{\partial\Lambda(aN-1)}}{|\Lambda(L)|} \le \frac{|\partial\Lambda(aN-1)|}{|\Lambda(L)|} \int_{\Omega} \exp\Big(\|\mathsf{H}_{\Lambda(L)}\| \cdot \nu(|\mathbf{u}|)\Big) d\mathsf{M}(\mathbf{u}) < \delta\,, \tag{34}$$

which should be valid at sufficiently large L at fixed number N.

Finally, last summand at right-hand side of (32) is estimated by choice a sufficiently large value a in the inequality (30) for any $N \in \mathcal{N}$ so that its right-hand side may be done less that δ. Thus, by selecting $\delta < \varepsilon/3$ and, at first, choosing a suitable value a, and then choosing a sufficiently large number N so that the inequalities (33) and (34) are satisfied, we will ensure the satisfiability of the inequality (31).

Theorem 6 *If* $\mathsf{H}_\Lambda \in \mathrm{H}_\nu$, *then there exists the finite limit*

$$f(\mathsf{M}, \mathsf{H}_\Lambda) = \lim_{L\to\infty} \frac{\ln Z_{\Lambda(L)}}{|\Lambda(L)|}\,. \tag{35}$$

The limit function $f(\mathsf{M}, \mathsf{H}_\Lambda)$ *is the continuous functional in the space of* H_ν.

Proof The inequality (23) points out that the estimate

$$\frac{1}{|\Lambda|}\left|\ln Z_\Lambda[\mathsf{H}^{(1)}_{\Lambda(L)}] - \ln Z_\Lambda[\mathsf{H}^{(2)}_{\Lambda(L)}]\right| \le \Big(\mathsf{E}_\Lambda \nu(|\tilde{\mathbf{u}}|)\Big) \cdot \|\mathsf{H}^{(1)}_{\Lambda(L)} - \mathsf{H}^{(2)}_{\Lambda(L)}\|\,, \tag{36}$$

takes place for any pair of classes $\{\mathsf{H}^{(1)}_{\Lambda(L)};\ L \in \mathcal{N}\}$, $\{\mathsf{H}^{(2)}_{\Lambda(L)};\ L \in \mathcal{N}\}$ of Hamiltonians in the space H_ν.

Since the manifold $\mathrm{H}^{(0)}$ is dense in H_ν, then, for a given class of Hamiltonians $\mathsf{H}_{\Lambda(L)} \equiv \mathsf{H}^{(1)}_{\Lambda(L)} \in \mathrm{H}_\nu$, $L \in \mathcal{N}$ and for the value $\varepsilon > 0$, choosing such a class $\mathsf{H}^{(0)} \equiv \mathsf{H}^{(2)}_{\Lambda(L)}$, $L \in \mathcal{N}$ in $\mathrm{H}^{(0)}$ for which $\|\mathsf{H}_{\Lambda(L)} - \mathsf{H}^{(0)}_{\Lambda(L)}\| < \varepsilon$, we get

$$\varepsilon \mathsf{E}_\Lambda \nu(|\tilde{\mathbf{u}}|) + \frac{1}{|\Lambda|} \ln Z_\Lambda[\mathsf{H}^{(0)}_{\Lambda(L)}] > \frac{1}{|\Lambda|} \ln Z_\Lambda[\mathsf{H}_{\Lambda(L)}] > \frac{1}{|\Lambda|} \ln Z_\Lambda[\mathsf{H}^{(0)}_{\Lambda(L)}] - \varepsilon \mathsf{E}_\Lambda \nu(|\tilde{\mathbf{u}}|)\,.$$

Since, according to Theorem 5, the sequence of functions $|\Lambda(L)|^{-1} \ln Z_{\Lambda(L)}[\mathsf{H}^{(0)}_{\Lambda(L)}]$, $L \in \mathcal{N}$ converges to a fixed limit, then, going to the limit $L \to \infty$, we get an estimate for the difference between the upper and lower limits of the sequences of functions

$$\limsup_{L\to\infty} \frac{\ln Z_\Lambda[\mathsf{H}_{\Lambda(L)}]}{|\Lambda|} - \liminf_{L\to\infty} \frac{\ln Z_\Lambda[\mathsf{H}_{\Lambda(L)}]}{|\Lambda|} < \varepsilon \mathsf{E}_\Lambda \nu(|\tilde{\mathbf{u}}|)\,.$$

Taking into account the arbitrariness of the value $\varepsilon > 0$, we find that the first statement of the theorem is true.

The limit function $f(\mathsf{M}, \mathsf{H}_\Lambda)$ (35) depends functionally on the set of potentials $V_\Gamma(\mathbf{u}(\Gamma))$, $\Gamma \subset \mathcal{Z}^d$, $1 < |\Gamma| < \infty$, that is, on the class of Hamiltonians $\{\mathsf{H}_{\Lambda(L)}; L \in \mathcal{N}\}$. Since such limiting at $L \to \infty$ functions $f(\mathsf{M}, \mathsf{H}^{(m)}_{\Lambda(L)})$ exist for every pair $\{\mathsf{H}^{(m)}_{\Lambda(L)}; L \in \mathcal{N}\}$, $m \in \{1, 2\}$ of Hamiltonians classes in H_ν, then, going to the limit when $L \to \infty$ in (36), for of these limit functions, we obtain

$$\left| f(\mathsf{M}, \mathsf{H}^{(2)}_{\Lambda(L)}) - f(\mathsf{M}, \mathsf{H}^{(1)}_{\Lambda(L)}) \right| \le \left(\mathsf{E}_\Lambda \nu(|\tilde{\mathbf{u}}|) \right) \cdot \| \mathsf{H}^{(1)}_{\Lambda(L)} - \mathsf{H}^{(2)}_{\Lambda(L)} \|\,.$$

From here, it follows that the limit functional $f(\mathsf{M}, \mathsf{H}_\Lambda)$ is continuous on the space of Hamiltonians H_ν that proves the second part of the statement.

5 Conclusion

In the paper it is proved the extensiveness of the free energy $F_\Lambda[\mathsf{H}_\Lambda]$ of classical vector lattice models in statistical mechanics, that is, the presence of asymptotic behavior (9) at $\Lambda \to \mathcal{Z}^d$ for this thermodynamic function. The proved statement is valid for any classes of translationally invariant Hamiltonians of the space H_ν and for any dimension d of the immersion space of the specified type models.

It is necessary to note that investigated models are used in statistical physics only at $d = 3$ for bulk physical samples of a solid and at $d = 2$ in the study of thermodynamic phenomena on the boundaries of macroscopic physical bodies (in particular, the surface tension). Besides, in practical calculations within the framework of statistical mechanics, as a rule, Hamiltonians of *pair interaction* are used that is $V_\Gamma(\mathbf{u}(\Gamma)) \neq 0$ only when $|\Gamma| = 2$ with a summable potential.

At the same time, it should be noted that we have proved the presence of extensive asymptotic only in the special case, which is used when applying models of statistical mechanics in problems of theoretical statistical physics. Namely, the sets Λ which serve as geometric models of crystals, have the form $\Lambda = \Lambda(L)$. So, it would be desirable to extend the constructions proposed in this paper to the case when Λ sets have a more general form. It may be done if it is permissible to determine the so-called thermodynamic Van Hove limit transition (see [2]). Such a generalization is important as from the viewpoint of development of the general theory of the Gibbs random fields and as from the physical viewpoint because of the development of theoretical physics. The latter is connected with the fact that different constructions of thermodynamic limit transition may describe different physical reality. For example, if it is violated the so-called Fisher condition (see [2, Sect. 2]) when the thermodynamic limit transition is fulfilled, in particular, there are violated those conditions that are inherent in the definition of the Van Hove limit transition, then it seems that one may describe fractal solid-state structures within the framework of statistical mechanics.

In conclusion, we note that, from our opinion, the development of an alternative approach in the theory of Gibbs random fields proposed by Dobrushin [7], despite its undoubted general theoretical importance, will not lead to the elimination of the concept of thermodynamic limit transition in the traditional sense in statistical mechanics.

References

1. Minlos RA (1968) Lectures on statistical physics. Successes of mathematical sciences, vol 23, pp 133–190
2. Ruelle D (1969) Statistical mechanics. Rigorous results. Benjamin, Inc, Ney York. W.A
3. Malyshev VA, Minlos RA (1991) Gibbs random fields. Springer Science+Business Media, B.V
4. Arovas D (2013) Lecture notes on thermodynamics and statistical mechanics. University of California
5. Gallavotti G (1998) Statistical mechanics. Dipartimento di Fisica Universsitaá di Roma, Short treatise. Roma
6. Minlos RA (2002) Introduction in statistical mathematical physics. University lecture series, vol 19. American Mathematical Society
7. Dobrushin RL (1968) Gibbs random fields for lattice systems with pair interaction. Funct Anal Appl 2:31–43
8. Virchenko YP (1991) About the theory of the ground state of the Geisenberg exchange model. In: Akhiezer AI (ed) Problems of theoretical physics. Naukova Dumka, Kiev, pp 80–96
9. Berlin T, Kac M (1952) The spherical model of ferromagnet. Phys Rev 86:821–835
10. Kac M, Thompson CJ (1971) The spherical model and the infinite spin dimensionality limit. Phys Novergica. 5:163–168
11. Wilson K, Kogut J (1974) The renomalization group and the epsilon expansion. Phy Rep 2:75–199

Family of Smooth Solutions of Hyperbolic Differential-Difference Equation

Natalya V. Zaitseva

Abstract Three-parameter familie of solutions is constructed for hyperbolic differential-difference equation with shift operators of the general-type acting with respect to all spatial variables. We prove theorem showing that the solutions obtained are classical provided that the real part of the symbol of the corresponding differential-difference operator is positive. Classes of equations for which these conditions are satisfied is given.

Keywords Hyperbolic equation · Differential-difference equation · Classical solution · Shift operator · Operational scheme · Fourier transform

1 Introduction

Problems for elliptic differential-difference equations in bounded domains have been studied quite comprehensively by now; the theory for such equations was created and developed by Skubachevskii [1, 2]. Problems for elliptic differential-difference equations in unbounded domains have been studied to a much lesser extent. An extensive study of such problems is presented in Muravnik's papers [3–5]. In particular, boundary value problems for multidimensional elliptic differential-difference equations are considered in [3–5].

Problems for parabolic differential-difference equations were studied in Muravnik's monograph [6]. Vlasov and Medvedev [7] studied hyperbolic differential-difference equations for the case where the shift operators act on the time variable.

As far as the present author is aware, at present, there are few papers dealing with hyperbolic differential-difference equations containing shifts with respect to the spatial variable. In [8–10], families of classical solutions are constructed for hyperbolic equations with shifts in the space variable x; the shifts occur in the potentials.

N. V. Zaitseva (✉)
Lomonosov Moscow State University, Moscow 119991, Russia
e-mail: zaitseva@cs.msu.ru

V. Vasilyev (ed.), *Differential Equations, Mathematical Modeling and Computational Algorithms*, Springer Proceedings in Mathematics & Statistics 423,
https://doi.org/10.1007/978-3-031-28505-9_20

In the present paper, we study the existence of smooth solutions of hyperbolic differential-difference equation in the half-space $\{(x,t)|\, x \in \mathbb{R}^n,\ t > 0\}$. The equation contains a sum of differential operators and shift operators with respect to each of the spatial variables,

$$u_{tt}(x,t) = a^2 \sum_{j=1}^{n} u_{x_j x_j}(x,t) - \sum_{j=1}^{n} b_j u(x_1, \ldots, x_{j-1}, x_j - h_j, x_{j+1}, \ldots, x_n, t), \tag{1}$$

where $a, b_1, \ldots, b_n$ and $h_1, \ldots, h_n$ are given real numbers.

Definition 1 A function $u(x,t)$ is called a classical solution of Eq. (1) if the derivatives u_{tt} and $u_{x_j x_j}$ $(j = 1, \ldots, n)$ exist in the classical sense (i.e., as limits of finited-ifference ratios) at each point of the half-space $\{(x,t)|\, x \in \mathbb{R}^n,\ t > 0\}$ and if Eq. (1) holds at each point of the half-space.

2 Construction of Solutions of Equation (1)

To find solutions of the equation, we use the classical operational scheme [11, Sect.10], whereby one formally applies the Fourier transform with respect to the n-dimensional variable x to Eq. (1),

$$\widehat{f}(\xi) = \int_{\mathbb{R}^n} f(x) e^{i\xi \cdot x} dx,$$

and passes to the dual variable ξ.

In view of the formulas [12, Sect. 9]

$$F_x[\partial_x^\alpha \partial_t^\beta f] = (-i\xi)^\alpha \partial_t^\beta F_x[f], \quad F_x[f(x - x_0)] = e^{ix_0 \cdot \xi} F_x[f],$$

for the function $\widehat{u}(\xi, t) := F_x[u](\xi, t)$ we obtain the initial value problem

$$\frac{d^2 \widehat{u}}{dt^2} = -\left(a^2 |\xi|^2 + \sum_{j=1}^{n} b_j \cos(h_j \xi_j) + i \sum_{j=1}^{n} b_j \sin(h_j \xi_j) \right) \widehat{u}, \quad \xi \in \mathbb{R}^n, \tag{2}$$

$$\widehat{u}(0) = 0, \quad \widehat{u}_t(0) = 1. \tag{3}$$

For convenience, in the subsequent calculations we use the notation

$$\alpha(\xi) := \sum_{j=1}^{n} b_j \cos(h_j \xi_j), \quad \beta(\xi) := \sum_{j=1}^{n} b_j \sin(h_j \xi_j).$$

Then Eq. (2) becomes

$$\frac{d^2\widehat{u}}{dt^2} = -\left(a^2|\xi|^2 + \alpha(\xi) + i\,\beta(\xi)\right)\widehat{u}, \quad \xi \in \mathrm{R}^n,$$

and the roots of the corresponding characteristic equation are determined by the formula

$$k_{1,2} = \pm i\sqrt{a^2|\xi|^2 + \alpha(\xi) + i\,\beta(\xi)} = \pm i\rho(\xi)e^{i\,\varphi(\xi)},$$

where

$$\rho(\xi) := \left[\left(a^2|\xi|^2 + \alpha(\xi)\right)^2 + \beta^2(\xi)\right]^{1/4}, \tag{4}$$

$$\varphi(\xi) := \frac{1}{2}\mathrm{arctg}\frac{\beta(\xi)}{a^2|\xi|^2 + \alpha(\xi)}. \tag{5}$$

Thus, the general solution of Eq. (2) has the form

$$\widehat{u}(\xi, t) = C_1(\xi)e^{i\,t\,\rho(\xi)[\cos\varphi(\xi)+i\,\sin\varphi(\xi)]} + C_2(\xi)e^{-i\,t\,\rho(\xi)[\cos\varphi(\xi)+i\,\sin\varphi(\xi)]},$$

where $C_1(\xi)$ and $C_2(\xi)$ are arbitrary constants depending on the parameter ξ; to determine these constants, we substitute the function $\widehat{u}(\xi, t)$ into the initial conditions (3). From the system

$$\begin{cases} C_1(\xi) + C_2(\xi) = 0, \\ C_1(\xi) - C_2(\xi) = (i\,\rho(\xi)[\cos\varphi(\xi) + i\,\sin\varphi(\xi)])^{-1}, \end{cases}$$

we find the values of these constants,

$$C_1(\xi) = \frac{e^{-i\,\varphi(\xi)}}{2i\,\rho(\xi)}, \quad C_2(\xi) = -\frac{e^{-i\,\varphi(\xi)}}{2i\,\rho(\xi)}.$$

As a result, the solution of problem (2), (3) is given by the formula

$$\begin{aligned} \widehat{u}(\xi, t) &= \frac{e^{-i\,\varphi(\xi)}}{2i\,\rho(\xi)}\left[e^{i\,t\,\rho(\xi)[\cos\varphi(\xi)+i\,\sin\varphi(\xi)]} - e^{-i\,t\,\rho(\xi)[\cos\varphi(\xi)+i\,\sin\varphi(\xi)]}\right] = \\ &= \frac{e^{-i\,\varphi(\xi)}}{2i\,\rho(\xi)}\left[e^{-t\,\rho(\xi)\sin\varphi(\xi)}e^{i\,t\,\rho(\xi)\cos\varphi(\xi)} - e^{t\,\rho(\xi)\sin\varphi(\xi)}e^{-i\,t\,\rho(\xi)\cos\varphi(\xi)}\right] = \\ &= \frac{1}{2i\,\rho(\xi)}\left[e^{-t\,\rho(\xi)\sin\varphi(\xi)}e^{i(t\,\rho(\xi)\cos\varphi(\xi)-\varphi(\xi))} - e^{t\,\rho(\xi)\sin\varphi(\xi)}e^{-i(t\,\rho(\xi)\cos\varphi(\xi)+\varphi(\xi))}\right] = \\ &= \frac{1}{2i\,\rho(\xi)}\left[e^{-t\,G_1(\xi)}e^{i(t\,G_2(\xi)-\varphi(\xi))} - e^{t\,G_1(\xi)}e^{-i(t\,G_2(\xi)+\varphi(\xi))}\right], \end{aligned} \tag{6}$$

where we use the notation

$$G_1(\xi) := \rho(\xi)\sin\varphi(\xi), \quad G_2(\xi) := \rho(\xi)\cos\varphi(\xi). \tag{7}$$

Now we formally apply the inverse Fourier transform F_ξ^{-1} to relation (6) and obtain

$$u(x,t) = \frac{1}{(2\pi)^n}\int\limits_{\mathbf{R}^n} \frac{1}{2i\,\rho(\xi)}\left[e^{-t\,G_1(\xi)}e^{i(t\,G_2(\xi)-\varphi(\xi))} - e^{t\,G_1(\xi)}e^{-i(t\,G_2(\xi)+\varphi(\xi))}\right]e^{-ix\cdot\xi}d\xi =$$

$$= \frac{1}{2i\,(2\pi)^n}\int\limits_{\mathbf{R}^n} \frac{1}{\rho(\xi)}\left[e^{-t\,G_1(\xi)}e^{i(t\,G_2(\xi)-\varphi(\xi)-x\cdot\xi)} - e^{t\,G_1(\xi)}e^{-i(t\,G_2(\xi)+\varphi(\xi)+x\cdot\xi)}\right]d\xi.$$

Since the functions $\alpha(\xi)$, $\rho(\xi)$, and $G_2(\xi)$ are even and the functions $\beta(\xi)$, $\varphi(\xi)$, and $G_1(\xi)$ are odd in each of the variables ξ_j, we transform the last expression as follows:

$$\frac{1}{2i\,(2\pi)^n}\int\limits_{\mathbf{R}^n} \frac{1}{\rho(\xi)}\left[e^{-t\,G_1(\xi)}e^{i(t\,G_2(\xi)-\varphi(\xi)-x\cdot\xi)} - e^{t\,G_1(\xi)}e^{-i(t\,G_2(\xi)+\varphi(\xi)+x\cdot\xi)}\right]d\xi =$$

$$= \frac{1}{2i\,(2\pi)^n}\left[\int\limits_{\mathbf{R}^n_-} \frac{1}{\rho(\xi)}\left[e^{-t\,G_1(\xi)}e^{i(t\,G_2(\xi)-\varphi(\xi)-x\cdot\xi)} - e^{t\,G_1(\xi)}e^{-i(t\,G_2(\xi)+\varphi(\xi)+x\cdot\xi)}\right]d\xi + \right.$$

$$\left.\int\limits_{\mathbf{R}^n_+} \frac{1}{\rho(\xi)}\left[e^{-t\,G_1(\xi)}e^{i(t\,G_2(\xi)-\varphi(\xi)-x\cdot\xi)} - e^{t\,G_1(\xi)}e^{-i(t\,G_2(\xi)+\varphi(\xi)+x\cdot\xi)}\right]d\xi\right] =$$

$$= \frac{1}{2i\,(2\pi)^n}\left[\int\limits_{\mathbf{R}^n_+} \frac{1}{\rho(\xi)}\left[e^{t\,G_1(\xi)}e^{i(t\,G_2(\xi)+\varphi(\xi)+x\cdot\xi)} - e^{-t\,G_1(\xi)}e^{-i(t\,G_2(\xi)-\varphi(\xi)-x\cdot\xi)}\right]d\xi + \right.$$

$$\left.\int\limits_{\mathbf{R}^n_+} \frac{1}{\rho(\xi)}\left[e^{-t\,G_1(\xi)}e^{i(t\,G_2(\xi)-\varphi(\xi)-x\cdot\xi)} - e^{t\,G_1(\xi)}e^{-i(t\,G_2(\xi)+\varphi(\xi)+x\cdot\xi)}\right]d\xi\right] =$$

$$= \frac{1}{2i\,(2\pi)^n}\int\limits_{\mathbf{R}^n_+} \frac{1}{\rho(\xi)}\left[2i\,e^{t\,G_1(\xi)}\sin\left(t\,G_2(\xi)+\varphi(\xi)+x\cdot\xi\right) + \right.$$

$$\left. 2i\,e^{-t\,G_1(\xi)}\sin\left(t\,G_2(\xi)-\varphi(\xi)-x\cdot\xi\right)\right]d\xi =$$

$$= \frac{1}{(2\pi)^n}\int\limits_{\mathbf{R}^n_+} \frac{1}{\rho(\xi)}\left[e^{t\,G_1(\xi)}\sin\left(t\,G_2(\xi)+\varphi(\xi)+x\cdot\xi\right) + e^{-t\,G_1(\xi)}\sin\left(t\,G_2(\xi)-\varphi(\xi)-x\cdot\xi\right)\right]d\xi.$$

We use the resulting representation to prove the following assertion.

3 Existence of Smooth Solutions of the Equation (1)

Theorem 1 *Under condition*

$$a^2|\xi|^2 + \sum_{j=1}^{n} b_j \cos(h_j \xi_j) > 0, \tag{8}$$

for all $\xi \in \mathrm{R}^n$, *the functions*

$$F(x, t; \xi) := e^{t\, G_1(\xi)} \sin(t\, G_2(\xi) + \varphi(\xi) + x \cdot \xi), \tag{9}$$

$$H(x, t; \xi) := e^{-t\, G_1(\xi)} \sin(t\, G_2(\xi) - \varphi(\xi) - x \cdot \xi), \tag{10}$$

where $\varphi(\xi)$ *is determined by formula (5) and* $G_1(\xi)$ *and* $G_2(\xi)$ *are determined by relations (7), satisfy Eq. (1) in the classical sense.*

Proof First, let us substitute the function (9) directly into Eq. (1). To this end, we find the derivatives

$$F_{x_j}(x, t; \xi) = \xi_j e^{t\, G_1(\xi)} \cos(t\, G_2(\xi) + \varphi(\xi) + x \cdot \xi),$$

$$F_{x_j x_j}(x, t; \xi) = -\xi_j^2 e^{t\, G_1(\xi)} \sin(t\, G_2(\xi) + \varphi(\xi) + x \cdot \xi),$$

$$F_t(x, t; \xi) = G_1(\xi) e^{t\, G_1(\xi)} \sin(t\, G_2(\xi) + \varphi(\xi) + x \cdot \xi) + \\ + G_2(\xi) e^{t\, G_1(\xi)} \cos(t\, G_2(\xi) + \varphi(\xi) + x \cdot \xi),$$

$$F_{tt}(x, t; \xi) = \left[G_1^2(\xi) - G_2^2(\xi)\right] e^{t\, G_1(\xi)} \sin(t\, G_2(\xi) + \varphi(\xi) + x \cdot \xi) + \\ + 2G_1(\xi) G_2(\xi) e^{t\, G_1(\xi)} \cos(t\, G_2(\xi) + \varphi(\xi) + x \cdot \xi).$$

Now let us evaluate the expressions $2G_1(\xi)G_2(\xi)$ and $G_1^2(\xi) - G_2^2(\xi)$. Since $G_1(\xi)$ and $G_2(\xi)$ are defined in (7), we conclude that

$$2G_1(\xi)G_2(\xi) = \rho^2(\xi) \sin 2\varphi(\xi).$$

It follows from formula (5) that $|2\varphi(\xi)| < \pi/2$ and hence $\cos 2\varphi(\xi) > 0$. Then we have

$$\sin 2\varphi(\xi) = \frac{\operatorname{tg} 2\varphi(\xi)}{\sqrt{1+\operatorname{tg}^2 2\varphi(\xi)}} =$$

$$= \operatorname{tg}\left(\operatorname{arctg}\frac{\beta(\xi)}{a^2|\xi|^2+\alpha(\xi)}\right)\left[1+\operatorname{tg}^2\left(\operatorname{arctg}\frac{\beta(\xi)}{a^2|\xi|^2+\alpha(\xi)}\right)\right]^{-1/2} =$$

$$= \frac{\beta(\xi)}{a^2|\xi|^2+\alpha(\xi)}\left[1+\frac{\beta^2(\xi)}{\left(a^2|\xi|^2+\alpha(\xi)\right)^2}\right]^{-1/2} =$$

$$= \frac{\beta(\xi)}{a^2|\xi|^2+\alpha(\xi)}\left[\frac{\left(a^2|\xi|^2+\alpha(\xi)\right)^2}{\left(a^2|\xi|^2+\alpha(\xi)\right)^2+\beta^2(\xi)}\right]^{1/2} =$$

$$= \frac{\beta(\xi)}{a^2|\xi|^2+\alpha(\xi)}\,\frac{|a^2|\xi|^2+\alpha(\xi)|}{\rho^2(\xi)}.$$

By virtue of condition (8), from the last relation we obtain

$$\sin 2\varphi(\xi) = \frac{\beta(\xi)}{a^2|\xi|^2+\alpha(\xi)}\,\frac{a^2|\xi|^2+\alpha(\xi)}{\rho^2(\xi)} = \frac{\beta(\xi)}{\rho^2(\xi)},$$

and hence

$$2G_1(\xi)G_2(\xi) = \beta(\xi). \tag{11}$$

With the inequality $\cos 2\varphi(\xi) > 0$ established above and under condition (8), now we find

$$\begin{aligned} G_1^2(\xi) - G_2^2(\xi) &= \rho^2(\xi)\left[\sin^2\varphi(\xi) - \cos^2\varphi(\xi)\right] = \\ &= -\rho^2(\xi)\cos 2\varphi(\xi) = -\frac{\rho^2(\xi)}{\sqrt{1+\operatorname{tg}^2 2\varphi(\xi)}} = \end{aligned} \tag{12}$$

$$= -\rho^2(\xi)\left[\frac{\left(a^2|\xi|^2+\alpha(\xi)\right)^2}{\left(a^2|\xi|^2+\alpha(\xi)\right)^2+\beta^2(\xi)}\right]^{1/2} = -a^2|\xi|^2 - \alpha(\xi).$$

In view of the expressions (11) and (12), the function F_{tt} becomes

$$\begin{aligned} F_{tt}(x,t;\xi) = \big[&-(a^2|\xi|^2+\alpha(\xi))\sin\left(t\,G_2(\xi)+\varphi(\xi)+x\cdot\xi\right)+ \\ &+\beta(\xi)\cos\left(t\,G_2(\xi)+\varphi(\xi)+x\cdot\xi\right)\big]\,e^{t\,G_1(\xi)}. \end{aligned}$$

Now let us substitute the derivatives $\widetilde{F}_{tt}$ and $\widetilde{F}_{x_jx_j}$ into Eq. (1),

$$F_{tt}(x,t;\xi) - a^2\sum_{j=1}^{n} F_{x_jx_j}(x,t;\xi) =$$

$$= [-(a^2|\xi|^2 + \alpha(\xi))\sin(t\,G_2(\xi) + \varphi(\xi) + x\cdot\xi) +$$
$$+\beta\cos(t\,G_2(\xi) + \varphi(\xi) + x\cdot\xi) +$$
$$+a^2\sum_{j=1}^{n}\xi_j^2\sin(t\,G_2(\xi) + \varphi(\xi) + x\cdot\xi)]e^{t\,G_1(\xi)} =$$
$$= -[\alpha(\xi)\sin(t\,G_2(\xi) + \varphi(\xi) + x\cdot\xi) -$$
$$-\beta(\xi)\cos(t\,G_2(\xi) + \varphi(\xi) + x\cdot\xi)]e^{t\,G_1(\xi)} =$$
$$= -\left[\sum_{j=1}^{n} b_j\cos(h_j\xi_j)\sin(t\,G_2(\xi) + \varphi(\xi) + x\cdot\xi) -\right.$$
$$\left.-\sum_{j=1}^{n} b_j\sin(h_j\xi_j)\cos(t\,G_2(\xi) + \varphi(\xi) + x\cdot\xi)\right]e^{t\,G_1(\xi)} =$$
$$= -\sum_{j=1}^{n} b_j\sin(t\,G_2(\xi) + \varphi(\xi) + x\cdot\xi - h_j\xi_j)e^{t\,G_1(\xi)} =$$
$$= -\sum_{j=1}^{n} b_j\sin(t\,G_2(\xi) + \varphi(\xi) + x_1\xi_1 + \cdots + x_n\xi_n - h_j\xi_j)e^{t\,G_1(\xi)} =$$
$$= -\sum_{j=1}^{n} b_j\sin(t\,G_2(\xi) + \varphi(\xi) + x_1\xi_1 + \cdots + x_{j-1}\xi_{j-1} +$$
$$+(x_j - h_j)\xi_j + x_{j+1}\xi_{j+1} + \cdots + x_n\xi_n)e^{t\,G_1(\xi)} =$$
$$= -\sum_{j=1}^{n} b_j\sin(t\,G_2(\xi) + \varphi(\xi) +$$
$$+(x_1,\ldots,x_{j-1},x_j - h_j,x_{j+1},\ldots,x_n)\cdot\xi)e^{t\,G_1(\xi)} =$$
$$= -\sum_{j=1}^{n} b_jF(x_1,\ldots,x_{j-1},x_j - h_j,x_{j+1},\ldots,x_n,t;\xi).$$

Next, let us substitute the function (10) into Eq. (1). To this end, we find the derivatives

$$H_{x_j}(x,t;\xi) = -\xi_je^{-t\,G_1(\xi)}\cos(t\,G_2(\xi) - \varphi(\xi) - x\cdot\xi),$$

$$H_{x_jx_j}(x,t;\xi) = -\xi_j^2e^{-t\,G_1(\xi)}\sin(t\,G_2(\xi) - \varphi(\xi) - x\cdot\xi),$$

$$H_t(x,t;\xi) = -G_1(\xi)e^{-t\,G_1(\xi)}\sin(t\,G_2(\xi) - \varphi(\xi) - x\cdot\xi) +$$
$$+G_2(\xi)e^{-t\,G_1(\xi)}\cos(t\,G_2(\xi) - \varphi(\xi) - x\cdot\xi),$$

$$H_{tt}(x,t;\xi) = \left[G_1^2(\xi) - G_2^2(\xi)\right] e^{-t\,G_1(\xi)} \sin\left(t\,G_2(\xi) - \varphi(\xi) - x\cdot\xi\right) -$$
$$-2G_1(\xi)G_2(\xi)e^{-t\,G_1(\xi)} \cos\left(t\,G_2(\xi) - \varphi(\xi) - x\cdot\xi\right) =$$
$$= \left[-(a^2|\xi|^2 + \alpha(\xi)) \sin\left(t\,G_2(\xi) - \varphi(\xi) - x\cdot\xi\right) -\right.$$
$$\left.-\beta(\xi)\cos\left(t\,G_2(\xi) - \varphi(\xi) - x\cdot\xi\right)\right] e^{-t\,G_1(\xi)}.$$

Now let us substitute the derivatives H_{tt} and $H_{x_jx_j}$ into Eq. (1),

$$H_{tt}(x,t;\xi) - a^2\sum_{j=1}^{n} H_{x_jx_j}(x,t;\xi) =$$
$$= [-(a^2|\xi|^2 + \alpha(\xi)) \sin\left(t\,G_2(\xi) - \varphi(\xi) - x\cdot\xi\right) -$$
$$-\beta(\xi)\cos\left(t\,G_2(\xi) - \varphi(\xi) - x\cdot\xi\right) +$$
$$+a^2\sum_{j=1}^{n}\xi_j^2 \sin\left(t\,G_2(\xi) - \varphi(\xi) - x\cdot\xi\right)]e^{-t\,G_1(\xi)} =$$
$$= -[\alpha(\xi)\sin\left(t\,G_2(\xi) - \varphi(\xi) - x\cdot\xi\right) +$$
$$+\beta(\xi)\cos\left(t\,G_2(\xi) - \varphi(\xi) - x\cdot\xi\right)]e^{-t\,G_1(\xi)} =$$
$$= -\left[\sum_{j=1}^{n} b_j\cos(h_j\xi_j)\sin\left(t\,G_2(\xi) - \varphi(\xi) - x\cdot\xi\right) +\right.$$
$$\left.+\sum_{j=1}^{n} b_j\sin(h_j\xi_j)\cos\left(t\,G_2(\xi) - \varphi(\xi) - x\cdot\xi\right)\right] e^{-t\,G_1(\xi)} =$$
$$= -\sum_{j=1}^{n} b_j\sin\left(t\,G_2(\xi) - \varphi(\xi) - x\cdot\xi + h_j\xi_j\right)e^{-t\,G_1(\xi)} =$$
$$= -\sum_{j=1}^{n} b_j\sin\left(t\,G_2(\xi) - \varphi(\xi) - x_1\xi_1 - \cdots - x_n\xi_n + h_j\xi_j\right)e^{-t\,G_1(\xi)} =$$
$$= -\sum_{j=1}^{n} b_j\sin\left(t\,G_2(\xi) - \varphi(\xi) - x_1\xi_1 - \cdots - x_{j-1}\xi_{j-1} -\right.$$
$$\left.-(x_j - h_j)\xi_j - x_{j+1}\xi_{j+1} - \cdots - x_n\xi_n\right)e^{-t\,G_1(\xi)} =$$
$$= -\sum_{j=1}^{n} b_j\sin\left(t\,G_2(\xi) - \varphi(\xi) -\right.$$
$$\left.+(x_1,\ldots,x_{j-1},x_j - h_j,x_{j+1},\ldots,x_n)\cdot\xi\right)e^{-t\,G_1(\xi)} =$$
$$= -\sum_{j=1}^{n} b_j H(x_1,\ldots,x_{j-1},x_j - h_j,x_{j+1},\ldots,x_n,t;\xi).$$

A straightforward substitution into Eq. (1) shows that the function $H(x, t; \xi)$ satisfies this equation in the classical sense.

Note that the functions (4) and (5) are well defined for any $\xi \in \mathrm{R}^n$ under condition (8), because the radicand in formula (4) is always positive, and the denominator in the argument of the arctangent in (5) does not vanish. This means that the functions (9) and (10) are smooth solutions of the Eq. (1).

The proof of the theorem is complete.

Corollary 1 *Under condition (8), the family of functions*

$$G(x, t; A, B, \xi) := A\, e^{t\, G_1(\xi)} \sin\left(t\, G_2(\xi) + \varphi(\xi) + x \cdot \xi\right) + \\ + B\, e^{-t\, G_1(\xi)} \sin\left(t\, G_2(\xi) - \varphi(\xi) - x \cdot \xi\right), \qquad (13)$$

where $\varphi(\xi)$ is given by (5) and $G_1(\xi)$ and $G_2(\xi)$ are given by (7), satisfies Eq. (1) in the classical sense for any real values of the parameters A, B, and ξ.

We represent the condition (8) in the form

$$\left(a^2\xi_1^2 + b_1 \cos\left(h_1\xi_1\right)\right) + \ldots + \left(a^2\xi_n^2 + b_n \cos\left(h_n\xi_n\right)\right) > 0.$$

Each of the n terms on the left side of this inequality will be positive if the conditions

$$0 < b_j h_j^2 \le 2a^2, \quad j = \overline{1, n}.$$

For $\xi = \overrightarrow{0}$ the condition (8) will be satisfied if the coefficients at the nonlocal potentials satisfy the inequality

$$\sum_{j=1}^{n} b_j > 0.$$

Condition (8), holds for any shifts $h_1, \ldots, h_n$ and any values $\xi_1, \ldots, \xi_n$ if the coefficients and the shifts of the equation satisfy the conditions

$$\sum_{j=1}^{n} b_j > 0, \quad 0 < b_j h_j^2 \le 2a^2, \quad j = \overline{1, n}.$$

These conditions are sufficient conditions that ensure the existence of a family of smooth solutions (13) to Eq. (1).

References

1. Skubachevskii AL (1997) Elliptic functional differential equations and applications. Birkhäuser, Basel-Boston-Berlin
2. Skubachevskii AL (2016) Boundary-value problems for elliptic functional-differential equations and their applications. Russ Math Surv 71(5):801–906. https://doi.org/10.1070/RM9739
3. Muravnik ÀB (2020) Elliptic differential-difference equations in the half-space. Math Notes 108(5):727–732. https://doi.org/10.4213/mzm12737
4. Muravnik ÀB (2021) Elliptic differential-difference equations of general form in a half-space. Math Notes 110(1):92–99. https://doi.org/10.1134/S0001434622030270
5. Muravnik ÀB (2021) Elliptic differential-difference equations with differently directed translations in half-spaces. Ufa Math J 13(3):104–112. https://doi.org/10.13108/2021-13-3-104
6. Muravnik ÀB (2016) Functional differential parabolic equations: integral transformations and qualitative properties of solutions of the Cauchy problem. J Math Sci 216(3):345–496. https://doi.org/10.1007/s10958-016-2904-0
7. Vlasov VV, Medvedev DA (2010) Functional-differential equations in Sobolev spaces and related problems of spectral theory. J Math Sci 164(5):659–841. https://doi.org/10.1007/s10958-010-9768-5
8. Zaitseva NV (2021) Classical solutions of hyperbolic equations with nonlocal potentials. Dokl Math 103(3):127–129. https://doi.org/10.1134/S1064562421030157
9. Zaitseva NV (2021) Hyperbolic differential-difference equations with nonlocal potentials. Ufa Math. J. 13(3):36–43. https://doi.org/10.13108/2021-13-3-36
10. Zaitseva NV (2021) Classical solutions of hyperbolic differential-difference equations in a half-space. Differ Equ 57(12):1629–1639. https://doi.org/10.1134/S0012266121120090
11. Gel'fand IM, Shilov GE (1953) Fourier transforms of rapidly growing functions and questions of uniqueness of the solution of the Cauchy problem. Usp Mat Nauk 8(6)(58):3–54
12. Vladimirov VS (1981) Uravneniya matematicheskoi fiziki (Equations of mathematical physics). Nauka, Moscow